Optical Computing

WILEY SERIES IN MICROWAVE AND OPTICAL ENGINEERING

KAI CHANG, Editor
Texas A&M University

INTRODUCTION TO ELECTROMAGNETIC COMPATIBILITY
Clayton R. Paul

OPTICAL COMPUTING: AN INTRODUCTION
Mohammad A. Karim and Abdul A. S. Awwal

COMPUTATIONAL METHODS FOR ELECTROMAGNETICS AND MICROWAVES
Richard C. Booton, Jr.

FIBER-OPTIC COMMUNICATION SYSTEMS
Govind P. Agrawal

Optical Computing: An Introduction

MOHAMMAD A. KARIM
Center for Electro-Optics
University of Dayton

ABDUL A. S. AWWAL
Computer Science & Engineering Department
Wright State University

A WILEY-INTERSCIENCE PUBLICATION
JOHN WILEY & SONS, INC.
NEW YORK / CHICHESTER / BRISBANE / TORONTO / SINGAPORE

Library of Congress Cataloging in Publication Data:

Karim, Mohammad A.

Optical computing: an introduction/Mohammad A. Karim and Abdul A. S. Awwal.

p. cm.—(Wiley series in microwave optical engineering)

Includes bibliographical references and index.

ISBN 0-471-52886-2

1. Optical data processing. I. Awwal, Abdul A. S. II. Title. III. Series.

TA1630.K36 1992 91-30533

621.36′7—dc20 CIP

Printed in the United States of America

10 9 8 7 6 5 4 3 2 1

Printed and bound by Courier Companies, Inc.

This book is dedicated to the memory of those brilliant men and women whose eternal gift of "zero" and many other related mathematical discoveries keep the present-day computers humming, while the gift of "first-real" optics is at last taking the computers through a new frontier.

Preface

Optics is perhaps the most sought-after multidisciplinary area in both sciences and engineering. In particular, the area of optical computing provides a serious challenge to the educators since it encompasses not one simple field but many diverse areas, such as geometrical optics, digital systems design, computer architecture, digital image processing, Fourier optics, and integrated optics. The confusion is evident from the very fact that many different and unrelated schemes have been dubbed "optical computing" or "optical data processing." However, these friendly disagreements reflect the nature and rate with which optical computing is undergoing changes. It is obvious that both students and researchers are in dire need of a single comprehensive book to direct them to fully comprehend the scope of this subject.

The goal of this book is to give the readers an introduction to the principles of optical computing and its main emphasis is on applying the characteristics of both all-optical and electro-optic devices/systems to the solution of optical computing. The book consists of an accumulation of lecture notes prepared for a first course on this subject and is intended for graduate students, with an undergraduate preparation in either electrical engineering, mathematics, computer science/engineering, or physics, who are somewhat familiar with geometrical optics and differential equations. A minimum list of prerequisites may be needed by the readers with varying background preparation. Familiarity with basic optics is helpful but not essential for this fairly self-contained book.

To facilitate the understanding of this very diverse interdisciplinary subject, we take the readers through the pertinent concepts of classical optics, Fourier optics, digital logic, and digital image processing. The 13 chapters of this book have been organized to cover four particular aspects of optical computing: preliminaries (Chapters 1 and 2), analog optical computing (Chapters 3–5), digital logic (Chapters 6–9), and digital optical computing (Chapters 10–13). Except for Chapter 10, each of the chapters is equipped with a collection of exercise problems, while all the chapters list up-to-date references and materials for further reading. It is our hope that the readers of this book will make every effort to read related journal papers, only a handful of which have been listed here in the book.

Chapters 1 and 2 review the essentials of mathematical preliminaries and digital image processing, which are used extensively in the subsequent chapters. The next three chapters are devoted exclusively to topics that form the area of analog optical computing. Chapter 3 introduces Fourier-transform-based linear optical processing systems and techniques. The ability of analog optical systems to perform nonlinear optical data processing and arithmetic operations is developed in Chapter 4. Finally, Chapter 5 is devoted exclusively to optical pattern recognition schemes and systems. The next four chapters cover all the pertinent digital systems design concepts that are applicable to optical computing systems. Chapter 6 introduces number systems, while Chapter 7 covers the various logic gates as well as logic function minimization techniques. Chapters 8 and 9, respectively, cover the combinational and sequential digital systems. The last four chapters form the crux of this book and can be categorized as "digital optical computing," which serves as the source of inspiration for most of the current optical computing research. Chapter 10 introduces and discusses the physics and engineering of various optical and optoelectronic devices that are used in optical computing applications. Chapter 11 develops the technique of shadow-casting and symbolic substitution for realizing both arithmetic and image processing. The theory, application, and devices pertaining to optical matrix processing are covered in Chapter 12. Finally, Chapter 13 discusses the role that optics can play in neural networks and artificial intelligence.

We would like to thank all those optical scientists and engineers whose works had stimulated us to keep on believing in the immense potential of optics. They are just too many to mention here. The material of this book, as usual, had been changed and updated because of our many interactions with the numerous graduate students who volunteered to try optical computing. We are particularly indebted to Farid Ahmed and Jamal Ahmed of Wright State University, and Ming Gao, Khan Iftekharuddin, and Song Zheng of the University of Dayton for producing many of the illustrations that we used. And, of course, we would not have been able to do much without the understanding of the members of our families: Aliya, Lamya, Lutfi, and Setara—the Karims; and Ibrahim and Rownak—the Awwals. Both of us are ever indebted to them all and thank God Almightly for the endless bounties given us.

MOHAMMAD A. KARIM
ABDUL A. S. AWWAL

Dayton, Ohio
January 1992

Contents

DIGITAL LOGIC

Optical Computing

MATHEMATICAL AND DIGITAL IMAGE FUNDAMENTALS

CHAPTER ONE

Mathematical Preliminaries

1.1 INTRODUCTION

In this chapter, we develop the basic tools that are necessary for analyzing optical processes in the analog and digital domains. These mathematical tools enable one to estimate the output of a linear optical system given its input and system characteristics. The tools are mainly of three types: continuous domain transform, discrete domain transform, and wave domain integral equation. The Fourier transform operation, in particular, is described in terms of both continuous and discrete domains. The discrete domain representation is important since it enables us to simulate or verify results obtained in analytic (continuous or wave) domain using a digital computer. In wave domain, the output of an optical system is seen as a phenomenon of wave propagation with some given boundary condition. Under certain special conditions, however, we shall see that wave domain input–output relationships for some of the selected optical systems are linked to the Fourier transform operation. Optical and digital techniques of peforming transforms add more flexibility in the understanding of the optical computing systems that we consider in this book. Computer programs for obtaining discrete transforms will enable the reader to test or implement the novel optical processing techniques rather easily.

Basic diffraction theory is discussed next to provide analytical tools for understanding optical systems as well as to validate the Fourier transform approach. For readers with diverse backgrounds in engineering, physics, mathematics, or computer science, reconciliation of all the three different tools is the main purpose that this chapter serves. This is not the last word since, as will be shown later, a lens is capable of performing a Fourier transform on the input. As light propagates it performs a Fourier transform on itself (aperture function). No wonder that an optical scientist/engineer is so interested in Fourier transforms.

1.2 FOURIER TRANSFORM

1.2.1 Definition

Transforms are orthogonal expansions of functions such that a given function can be represented uniquely by a combination of some elementary basis functions. The linear independence of orthogonal functions provides the advantage of identifying or recognizing a particular function uniquely by means of its transform. Sometimes the properties of such functions give rise to some desirable characteristics such as position invariance. Furthermore, novel transforms may be sought that have some desirable attributes to facilitate certain applications. Very often, transforms leads to an easy way of performing certain operations that are otherwise difficult to perform. For example, to avoid the mathematical difficulty of performing a convolution operation, one might prefer to do a simple multiplication in the Fourier[1] transform domain. Another strong motivation for dealing with transforms, in particular the Fourier transform, is the treatment of an optical system as a linear system. This simplicity provides more flexibility, insight (an engineer can easily visualize the effect of a system transform on the output), and faster analyzing capability. One can certainly treat optical signals as electromagnetic waves and analyze an optical problem but Fourier transform techniques described by linear systetm theory can probably solve the same problem with much less mathematical ambiguity and elegant physical envisionment. For example, the input–output relationship of a system may be described by a simple product of input Fourier transform and the system transform.

All the functions referred to above (input, output, and system function) are mathematically realizable and physically comprehensible. For example, consider a system function to describe an infinitesimally small pinhole in an opaque medium. Mathematically it is a close approximation of the Dirac delta function, while physically it is a pinhole that can be used to perform imaging. To verify that a pinhole is nothing but a linear system represented by a delta function, one can take a picture using a pinhole camera. The output picture is a reproduction of the input, which is only possible for a system with a delta-like impulse response.

The one-dimensional (1-D) Fourier transform of a spatial function $f(x)$ is described mathematically by the following continuous summation or integration.

$$G(u) = \int_{-\infty}^{\infty} f(x)e^{-j2\pi ux}\,dx \tag{1.1}$$

For this relation to be true, the function $f(x)$ must be sectionally continuous in the finite region and absolutely integrable. This condition, referred to as the Dirichlet condition, is usually satisfied by most physically existing functions. Most optical

[1]Jean Baptiste Joseph Fourier (1768–1830), a French mathematician, was famous for his contributions to the theory of heat conduction by what came to be known as the Fourier expansion.

systems are appropriate for handling information in two-dimensional (2-D) form. The two-dimensional Fourier transform is given by

$$G(u, v) = \int_{-\infty}^{\infty} \int_{-\infty}^{\infty} g(x, y) e^{-j2\pi(ux + yv)} \, dx \, dy \tag{1.2}$$

The domain described by x and y is traditionally referred to as the space domain while the domain described by u and v is referred to as the frequency domain. Mathematical simplification of the above formulation occurs when the function $g(x, y)$ is separable in the orthogonal directions; the two-dimensional transform can be expressed as the product of two one-dimensional transforms. Providing that function $g(x, y)$ can be expressed as a product of two one-dimensional functions $g_x(x)$ and $g_y(y)$, the two-dimensional Fourier transform is given by

$$\begin{aligned} G(u, v) &= \int_{-\infty}^{\infty} \int_{-\infty}^{\infty} g_x(x) g_y(y) e^{-j2\pi(ux + yv)} \, dx \, dy \\ &= \int_{-\infty}^{\infty} g_x(x) e^{-j2\pi ux} \, dx \int_{-\infty}^{\infty} g_y(y) e^{-j2\pi vy} \, dy \\ &= G(u) G(v) \end{aligned} \tag{1.3}$$

Since one-dimensional transforms are easy to evaluate, the transforms of two-dimensional separable functions are easier to calculate using Eq. (1.3). For continuous domain Fourier transforms, the above definition suffices. However, the utility of the above discussion will become more apparent when one studies the different properties of the Fourier transform.

1.2.2 Properties

The Fourier transform operation involves a complex integral. Properties of the integrals give rise to behaviors such as linearity, scaling, translation (in space or spatial frequency), convolution, and correlation. Each one of these properties has some optical implications.

(*a*) *Linearity* Let $F_i(u, v)$ be the two-dimensional Fourier transform of a two-dimensional image $f_i(x, y)$. In short notation, the relationship is represented as $f_i(x, y) \Leftrightarrow F_i(u, v)$. According to the property of linearity, the Fourier transform of the linear combination of several such functions is the sum of the Fourier transforms of the individual functions. Mathematically,

$$\mathfrak{F}\{c_1 f_1(x, y) + c_2 f_2(x, y) + \cdots\} = c_1 F_1(u, v) + c_2 F_2(u, v) + \cdots \tag{1.4}$$

where c_1 c_2, · · · are arbitrary constants and $\mathfrak{F}$ is the Fourier transform operation. This property allows one to estimate the spectrum of a sum of two or more signals from a knowledge of the spectrum of individual signals.

(*b*) *Scaling* Magnification in one domain is accompanied by a reduction in the corresponding transform domain. If $f(x, y) \Leftrightarrow F(u, v)$,

$$\mathfrak{F}\{f(ax, by)\} = \frac{1}{ab} G\left(\frac{u}{a}, \frac{v}{b}\right) \tag{1.5}$$

Accordingly, if the size of an optical image is increased, the spectrum becomes smaller. This scaling property also explains why certain optical pattern recognition operations, which are based on Fourier transform, are not scale invariant.

(*c*) *Translation* If an object is translated in the spatial domain, there is a phase shift in the corresponding frequency domain. However, the amplitude and location of the Fourier transform remain unchanged. If, for example, $f(x) \Leftrightarrow F(u)$, then

$$\mathfrak{F}\{f(x - a)\} = e^{-j2a\pi u}F(u) \tag{1.6}$$

This property enables one to perform shift invariant optical pattern recognition since the Fourier spectrum is always formed at the same location although the image may have shifted.

(*d*) *Frequency Domain Translation* A shift in the frequency domain is obtained when the input object is multiplied by a space-dependent phase factor. This is mathematically expressed as

$$\mathfrak{F}\{f(x, y)e^{-j2\pi(xu_0 + yv_0)}\} = F(u + u_0, v + v_0) \tag{1.7}$$

(*e*) *Transform of a Conjugate* If $f(x, y) \Leftrightarrow F(u, v)$, then

$$\mathfrak{F}\{f^*(x)\} = F^*(-u) \tag{1.8}$$

(*f*) *Transform of Convolution* The transform of a convolution of two signals is the product of the transforms. For example, if $f_i(x, y) \Leftrightarrow F_i(u, v)$, then provided the convolution of the two function $f_1(x, y)$ and $f_2(x, y)$ is defined as

$$f_1(x, y) * f_2(x, y) = \iint f_1(x', y')f_2(x - x', y - y')\, dx'\, dy' \tag{1.9}$$

the transform of the convolution is given by

$$\mathfrak{F}[f_1(x, y) * f_2(x, y)] = F_1(u, v)F_2(u, v) \tag{1.10}$$

This is the oft-quoted simplification formula, which avoids the convolution route to find the output by performing a simple multiplication to generate the output from the input and system transform. In general, the output of a linear optical system is given by the convolution of the input with the system impulse response. In transform domain the output transform is given by the product of input transform and the system transform.

(*g*) *Transform of a Transform* Transform of a transform reproduces the original function with its axes reversed. If $f(x, y) \Leftrightarrow F(u, v)$, then

$$\mathfrak{F}\{F(u, v)\} = f(-x, -y) \tag{1.11}$$

If a function is transformed twice, we get the original function back, only the two axes are flipped. In other words, the inverse transform can be found by taking a forward transform on the Fourier transform. This theorem is applicable for optical Fourier and inverse transforms since it enables one to use the Fourier transform lens for obtaining the inverse transform.

(*h*) *Transform of a Product* If $f(x) \Leftrightarrow F(u)$ and $g(x) \Leftrightarrow G(u)$, then

$$\mathfrak{F}\{f(x)g(x)\} = F(u) * G(u) \tag{1.12}$$

(*i*) *Correlation* The Fourier transform of correlation between two inputs f and g is given by the product of the corresponding transform and their complex conjugate; that is,

$$\mathfrak{F}\{f(x) \circledast g(x)\} = F(u)G^*(u) \tag{1.13}$$

where the correlation is defined as

$$f_1(x, y) \circledast f_2(x, y) = \iint f_1(x', y')f_2^*(x' - x, y' - y)\, dx'\, dy'$$

Sometimes when the two functions are dissimilar, the correlation operation is referred to as the cross-correlation. When the two functions are similar, it is termed instead the autocorrelation.

The correlation theorem enables one to perform the correlation operation in the frequency domain optically by multiplying the conjugate of the to-be-correlated function. When a Fourier transform of an autocorrelation function is performed, intensity distribution of the Fourier transform results: that is,

$$\mathfrak{F}\{f(x) \circledast f(x)\} = |F(u)|^2 \tag{1.14}$$

The inverse of this intensity distribution generates an ouput pattern that is called autocorrelation output. This process of multiplying the Fourier transform by F^* and then taking the inverse transform is termed matched filtering.

1.2.3 Applications

1. *Image Processing* Fourier transform based image processing can be performed by modifying the frequency spectrum. For example, a portion of the

frequency spectrum may be blocked to remove information contained in that portion of the spectrum. This may perform noise removal if the noise happens to be in that portion of the spectrum.

2. *Image Enhancement* Once the Fourier transform of an image is known, the image may be enhanced by providing gain to the high-frequency region or by reducing the dc components.
3. *Pattern Recognition* The autocorrelation theorem justifies that a correlation operation may be performed in the Fourier domain by a simple multiplication of the input spectrum with the complex conjugate of the spectrum of the to-be-recognized image.
4. *Holography* Holograms for storing three-dimensional or phase information may be produced by recording the interference of an image transform with another light beam.

1.3 DISCRETE FOURIER TRANSFORM

Integration is nothing but a summation operation with infinite precision. If one sacrifices the precision of samples, one may get a new definition of the Fourier transform, which may be used for its evaluation numerically. This alternate discrete approximation of the continuous domain Fourier transform is referred to as discrete Fourier transform (DFT). The one-dimensional DFT is expressed as the following summation:

$$F(n) = \sum_{k=0}^{N-1} f_k e^{-(j2\pi/N)nk}, \qquad n = 0, 1, \cdots, N-1 \tag{1.15}$$

The inverse DFT can be found from the DFT as follows:

$$f(k) = \frac{1}{N} \sum_{n=0}^{N-1} F_n e^{(j2\pi/N)kn}, \qquad k = 0, 1, \cdots, N-1 \tag{1.16}$$

The two-dimensional extension of the DFT is given by

$$\begin{aligned} G(n, m) &= \sum_{k=0}^{N-1} \sum_{j=0}^{N-1} g_{m,n} e^{-(j2\pi/N)(nk+mj)} \\ &= \sum_{k=0}^{N-1} \left\{ \sum_{j=0}^{N-1} g_{m,n} e^{-(j2\pi/N)(nk)} \right\} e^{-(j2\pi/N)mj}, \\ & m, n = 0, 1, \cdots, N-1 \end{aligned} \tag{1.17}$$

The linearity of the summation operator yields a rather efficient method of evaluating the two-dimensional DFT. From Eq. (1.17) one notices that an efficient way of performing a two-dimensional DFT consists of the following two steps.

1. Perform a one-dimensional column transformation of the two-dimensional image; and then
2. Perform a one-dimensional row transform of the above transformed image.

For faster and reduced computation, a technique known as fast Fourier transform (FFT) developed by Cooley and Tukey[2] may be used. Subroutines based on FFT can be used to compute the two-dimensional FFT. By using the FFT subroutine repeatedly on each column of an image, one can develop a one-dimensional FFT subroutine. Then the one-dimensional transformed image can be transposed and again operated on by the one-dimensional routine and retransposed to get a two-dimensional FFT of an image.

```
C       A FORTRAN PROGRAM FOR PERFORMING TWO-
C       DIMENSIONAL FFT.
C ************** THE MAIN PROGRAM **************
C THE MAIN PROGRAM IS WRITTEN TO PERFORM FILTERING
C OF AN INPUT IMAGE OF 64x64 SIZE, WITH A FILTER
C GIVEN BY AMTF. INPUT IS STORED IN FILE
C FOROO(INFL).DAT DATA GRAY LEVELS ARE ASSUMED TO
C BE BETWEEN 0 TO 255, THE AMTF MUST BE EITHER
C CALCULATED OR SUPPLIED BY THE USER. (NO INPUT
C STATEMENT IS SHOWN FOR AMTF)
C
C SUBROUTINES USED:
C
C        TWOFFT: 2-D FFT IS USED TO MAKE SUCCESSIVE
C        CALLS TO 1-D
C FFT AND   OTHER SUBROUTINES
C
COMPLEX F(64,64),FFF
INTEGER NIM(64,64)
REAL AMTF(64,64)
N=64
M=6
PRINT*, 'INPUT INFL NUMBER'
READ*,INFL
DO 5 I=1,N
DO 5 J=1,N
READ(INFL,*) B
AI(I,J)=B
5 F(I,J)=CMPLX(B*(-1)**(I+J),0.)/N**2
CALL TWOFFT(F,M,N,-1)
```

[2]In 1965, Cooley and Tukey published a famous paper showing that the DFT can be efficiently computed by $N \log_2 N$ complex multiplications and additions where N is a power of 2.

```
C
C PERFORMING A FILTERING OPERATION
C FILTER OR MTF IS GIVEN BY AMTF
C
DO 6 I=1,N
DO 6 J=1,N
6 F(I,J)=F(I,J)*AMTF(I,J)
C
C INVERSE TRANSFORM
C
CALL TWOFFT(F,M,N,1)
C SCALING
DO 30 I=1,64
DO 30 J=1,64
NIM(I,J)=ABS((F(I,J))*4096)
30 WRITE(12,31) NIM(I,J)
31 format(i4)
STOP
END
C
C
*****************************************************
**************SUBROUTINE TRANSPOSE **************
C THIS IS USED TO INTERCHANGE BETWEEN ROWS AND
C COLUMNS OF A MATRIX SO THAT THE SAME 1-D FFT CAN
C BE USED TO FIND THE 2-D FFT.
C
C  PARAMETERS:  FX(N,N)---- INPUT MATRIX
C               FY(N,N)---- OUTPUT MATRIX
C               (TRANSPOSED)
C N      ---- THE SIZE
C
C
SUBROUTINE TRANSP(FX,FY,N)
COMPLEX FX(64,64),FY(64,64)
DO 10 I=1,N
DO 10 J=1,N
  10 FY(I,J)=FX(J,I)
RETURN
END
C****************************************************
C**************  SUBROUTINE TWO FFT  **************
C
C IT IS USED TO CALL VARIOUS SUBROUTINES WHICH
C COMPOSED THE 2-D FFT
```

```
C PARAMETERS:    M,N,AND,ISIGN ARE THE SAME AS USED
C                FOR 1-D FFT
C                F IS THE INPUT/OUTPUT 2-D FFT
C THIS DOES THE FOLLOWING:
C     1--  IT CALLS 1-D FFT (TO FIND FFT OF ROWS)
C     2--  TRANPOSES THE OBTAINED RESULT TO BE
C          READY TO FIND 1-D FFT OF COLUMNS
C     3--  CALLS  1-D FFT (TO FIND FFT OF COLUMNS)
C     4--  TRANSPOSES THE RESULT BACK TO ITS
C          ORIGINAL FORM
C
C
C
SUBROUTINE TWOFFT(F,M,N,ISIGN)
COMPLEX F(64,64),XX(64,64)
CALL ONEFFT(F,M,N,ISIGN)
CALL TRANSP(F,XX,N)
CALL ONEFFT(XX,M,N,ISIGN)
CALL TRANSP(XX,F,N)
RETURN
END
C*****************************************************
C*************     SUBROUTINE  ONE FFT  *************
C  PARAMETERS:   SAME AS FFT SUBROUTINE
C                TEMP IS A WORKING ARRAY TO STORE
C                EACH ROW/COLUMN
C                 AND THEN SEND IT TO FFT ROUTINE
C
C
SUBROUTINE ONEFFT (Y,M,N,ISIGN)
COMPLEX Y(64,64),TEMP(64)
DO 20 I=1,N
C    DO 10 IS STORING EVERY ROW/COLUMN IN TEMP
C
DO 10 J=1,N
 10 TEMP(J)= Y(I,J)
CALL FFT(TEMP,M,N,ISIGN)
C
C      DO 15 IS RESTORING EACH ROW/COLUMN
C
C
DO 15 J=1,N
 15 Y(I,J)=TEMP(J)
 20 CONTINUE
RETURN
```

```
END
C **************    SUBROUTINE FFT   ***************
C THIS ROUTINE PERFORMS FFT ON AN ARRAY OF DATA
C SUPPLIED BY A(I) OF DIMENSION N, THE RESULT IS
C RETURNED TO THE SAME ARRAY.
C IF = -1 FOR FORWARD FFT, IF=1, FOR INVERSE FFT.
C USUALLY N = 2**M.
C THE FFT ALGORITHM IS BASED ON DECIMATION IN
C FREQUENCY
C (OPPENHEIM AND SCHAFER, 1975)
C
SUBROUTINE FFT(A,M,N,IF)
COMPLEX A(N),U,W,P,T
PI=4.*ATAN(1.0)
SIGN=IF
J=1
NR=N-1
C
DO 17 I=1,NR
IF (I .GT. J) GO TO 25
P=A(J)
A(J)=A(I)
A(I)=P
  25 K=N/2
  30   IF(K .GE. J) GO TO 35
J=J-K
K=K/2
GO TO 30
  35 J=J+K
DO 20 L=1,M
LT=2**L
U=(1.0,0.0)
W=CMPLX(COS(2*pi/LT),SIN(2*PI*SIGN/LT))
LEHALF=LT/2
DO 45 J=1, LEHALF
DO 40 I=J,N,LT
IP=I+LEHALF
T=A(IP)*U
A(IP)=A(I)-T
  40 A(I)=A(I)+T
  45 U=U*W
IF(SIGN .LT. 0.0) RETURN
DO 18 I=1,N
  A(I)=A(I)/FLOAT(N)
  18 CONTINUE
```

```
RETURN
END
```

1.4 BASIC DIFFRACTION THEORY

Light, being an electromagnetic wave, must satisfy Maxwell's equation. Maxwell's equation describes the propagation of light when both space and time vary. Although according to the description of light by Maxwell's equation the electric and magnetic fields are coupled, they are treated independently, for simplicity, in the scalar diffraction theory of light. The vector theory, or more rigorous diffraction theory, illustrates the interdependence between the two fields.

It can be shown that the temporal variation of the electric or magnetic field of a monochromatic wave at a point P may be described by

$$U(P, t) = U_0 e^{-j\omega t} \tag{1.18}$$

where ω is the frequency of oscillation of the electromagnetic wave. However, this form of the field does not describe the spatial variation of light at any particular instant. Substituting the temporal variation in the time-dependent Maxwell equation,

$$\nabla^2 U - \frac{1}{c^2}\frac{\partial^2 U}{\partial t^2} = 0 \tag{1.19}$$

one finds that the time-independent amplitude variation must satisfy a stationary wave equation, also known as Helmholtz equation:

$$(\nabla^2 + k^2)U_0 = 0 \tag{1.20}$$

where $k = 2\pi/\lambda$ is referred to as the wave number and c is the velocity of light in free space. To perform any kind of processing using optics, light must be propagated through space or some kind of modulator or through optical elements such as lenses. Light travels in a straight-line path; however, when it strikes an obstacle the optical field is disturbed and it deviates from the straight line. Such an optical arrangement is analogous to a boundary problem. Since the wave equation is valid for optical amplitude variation, it is used to find the field immediately behind the boundary. Thus the boundary effects can be determined using the stationary wave equation and the corresponding boundary conditions. To solve the second-degree differential equation, however, one must know the field at the boundary with the first-order derivative. For example, it is necessary to have a knowledge of the value of the function at the boundary and the first derivative for solving the initial value problem.

According to Huygen's principle, a wave is composed of a summation of spherical waves. In other words, each point on a wavefront acts as a point source of light, and these point sources add up to make new wave surfaces. Later on, Fresnel showed that the field at any plane may be expressed as a superposition summation of spherical waves. A spherical wave acts as a Greens function (same as impulse response or point spread function) for the wave equation. In other words, if the wave equation describes a system function that produces the optical wave propagation phenomenon and if an impulse input (an infinitesimal width point source of infinitesimal duration) is then applied to such a system, the result is a spherical wave. Accordingly, if any other input is subjected to the field equation, the output can be calculated by a superposition sum of inputs with the impulse response. Consequently, the resulting spherical wave acts as a set of functions that can be used to construct any wave shape, which is the basic notion of Huygens' formulation. Thus the description of the wave expressed in terms of a spherical wave is given by

$$U(x, y) = \int_{\text{aperture}} U(x_0, y_0)\gamma \frac{e^{jkr}}{r}\, dx_0\, dy_0 \tag{1.21}$$

where γ is called the obliquity factor, indicating that light has a preferred direction of travel, and r is the distance of the observation point (x_i, y_i) at the x-y plane from the input at x_0y_0 plane as shown in Fig. 1.1. The theory (obliquity factor takes care of it) prohibits light from traveling backward, restricting it to a forward propagation of Huygens wavelets. Thus the secondary point sources are not isotropic but rather directive.

The basic diffraction theory allows one to evaluate the integral given by Eq. (1.21) at different distances from a given aperture, where the size of the aperture plays a role in defining the limits of the integral. Depending on the distance from the aperture, the diffraction is classified as Fresnel or Fraunhofer diffraction. At a

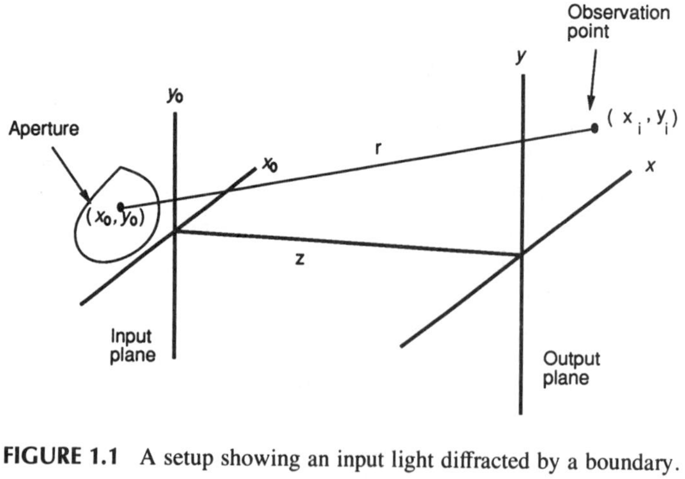

FIGURE 1.1 A setup showing an input light diffracted by a boundary.

distance of a few wavelengths from the aperture in the direction of light propagation starts the Fresnel region. The Fraunhofer region is also contained within the Fresnel region; however, it does not start until after a larger distance, when the radiation pattern approaches that of the Fourier transform of the aperture. The elegant nature of the integral allows us to concentrate only on the distance r; we do not change U because our final expression has U in its unchanged form. Also note that r appears in two places of the integral, contributing to both phase and amplitude. For the amplitude part, one may approximate r as z, in particular, when the point of observation in the output plane is very close to the optic axis (z-axis). This approximation is termed the paraxial approximation. The r contribution to the phase cannot be ignored so easily since it is multiplied by a very large number, k. Accordingly, to mathematically describe the diffraction region, the distance r (of the phase part) can be expressed as follows:

$$\begin{aligned} r &= \sqrt{z^2 + (x_0 - x_i)^2 + (y_0 - y_i)^2} \\ &= z\sqrt{1 + \frac{(x_0 - x_i)^2 + (y_0 - y_i)^2}{z^2}} \end{aligned} \tag{1.22}$$

When considering a small region around the optical axis such that $[(x_0 - x_i)^2 + (y_0 - y_i)^2]^{1/2}$ is less than the distance z, a binomial expansion can be used in Eq. (1.22). The distance z satisfying this condition describes the beginning of the Fresnel region when

$$\begin{aligned} r &= z + \frac{(x_0 - x_i)^2 + (y_0 - y_i)^2}{2z} \\ &\quad - \frac{[(x_0 - x_i)^2 + (y_0 - y_i)^2]^2}{8z^3} + \cdots \\ &\simeq z + \frac{x_i^2 + y_i^2}{2z} + \frac{x_0^2 + y_0^2}{2z} - \frac{x_i x_0 + y_i y_0}{z} \end{aligned} \tag{1.23}$$

Usually the last term may be neglected on the basis that the total phase change produced by this term is less than 1 radian. Accordingly, when

$$\frac{\pi}{\lambda} \frac{[(x_0 - x_i)^2 + (y_0 - y_i)^2]^2}{4z^3} \ll 1 \tag{1.24}$$

Eq. (1.21) reduces to what is known as the Fresnel integral, given by

$$U(x, y) = \frac{e^{jkz}}{j\lambda z} \iint_{\text{aperture}} U(x_0, y_0) e^{(jk/2z)[(x_0 - x_i)^2 + (y_0 - y_i)^2]} \, dx_0 \, dy_0 \tag{1.25}$$

If in addition $z \geq \frac{1}{2}k(x_0^2 + y_0^2)_{\max}$, the second quadratic term of Eq. (1.23) may be neglected to give

$$r = z + \frac{x_i^2 + y_i^2}{2z} + \frac{x_0^2 + y_0^2}{2z} - \frac{x_i x_0 + y_i y_0}{z}$$

$$\simeq z + \frac{x_i^2 + y_i^2}{2z} - \frac{x_i x_0 + y_i y_0}{z} \tag{1.26}$$

Then Eq. (1.25) becomes

$$U(x, y) = \frac{e^{jkz}}{j\lambda z} \iint_A U(x_0, y_0) e^{(jk/2z)[x_i^2 + y_i^2 - 2y_0 y_i - 2x_0 x_i]} \, dx_0 \, dy_0$$

$$= \frac{e^{jkz}}{j\lambda z} e^{(jk/2z)[x_i^2 + y_i^2]} \iint_A U(x_0, y_0) e^{(-jk/z)[y_0 y_i + x_0 x_i]} \, dx_0 \, dy_0 \tag{1.27}$$

This equation is in the form of a Fourier transform of the field at the aperture (A). Thus when the distance z between the aperture and the observation plane satisfies the criterion given by Eq. (1.26), the field distribution is given by the Fourier transform of the aperture function. This distance must be very large in order to satisfy the so-called Fraunhofer condition. However, this distance may be reduced even further either by using a lens after the aperture or by illuminating the aperture with a converging beam of light.

1.5 FOURIER TRANSFORM PROPERTY OF LENS

When light passes through an ordinary convex lens, it integrates the light and converges it to the focus. The lens functions as a phase element and causes a phase delay such that each segement of the light experiences a thickness-dependent phase delay. This effect is equivalent to having a linear phase factor present in the integral form of the corresponding Fourier transform. Accordingly, the corresponding integral combined with a phase factor produces an effect similar to that of Fourier transform on the input.

Consider a thin lens with focal length f as shown in Fig. 1.2. Rays parallel to the optical axis will eventually converge to the focus on the right side of the lens, at a distance f from the lens. At a distance f from the focus, the radius of curvature of the converging ray will be f. Suppose this ray passes through the lens at a height x from the optic axis, when the thickness of the lens is Δ. From the diagram it is obvious, that

$$f^2 = x^2 + (f - \Delta)^2 \tag{1.28}$$

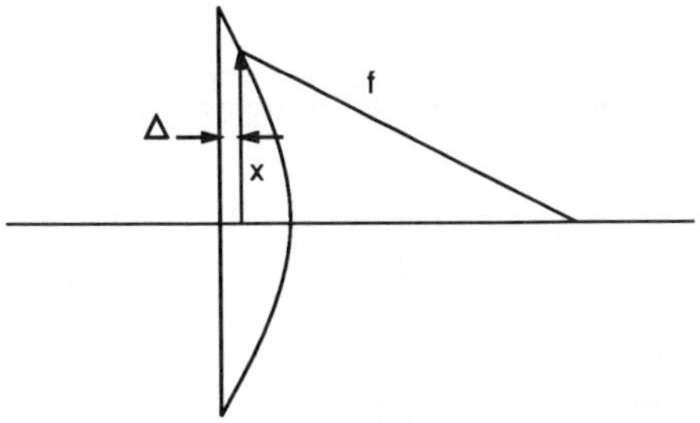

FIGURE 1.2 Fourier transform property of a lens.

Using a thin lens approximation, Δ^2 may be neglected when Eq. (1.28) reduces to

$$\Delta = \frac{x^2 + \Delta^2}{2f}$$

$$= \frac{x^2}{2f} \tag{1.29}$$

The above approximation is also known as a sagittal approximation. The phase change caused by the lens is $k\Delta = k\,[(x_i^2 + y_i^2)/2f]$. If a lens is placed at the aperture of Fig. 1.1, the Fresnel integral takes the form

$$U(x, y) = \frac{e^{jkz}}{j\lambda z} \iint_A U(x_0, y_0) e^{(jk/2z)[(x_0 - x_i)^2 + (y_0 - y_i)^2]} e^{-jk(x_i^2 + y_i^2)/2f}\, dx_0\, dy_0 \tag{1.30}$$

When $z = f$, the quadratic factor cancels out, reducing Eq. (1.30) to

$$U(x, y) = e^{j\lambda f} e^{(jk/2f)(x_0^2 + y_0^2)} \iint_A U(x_0, y_0) e^{(-j2\pi/\lambda f)[y_0 y_i + x_0 x_i]}\, dx_0\, dy_0 \tag{1.31}$$

Thus the lens performs a Fourier transform of the aperture field at one focal length away from the aperture. This plane is often referred to as the Fourier plane. This Fourier transform property of a lens is the basis for most analog optical processing. A 64 × 64 FFT that takes 1 ms on a fast FFT processor can be performed in just 0.33 ns using a 10-cm focal length lens.

1.6 PROBLEMS

1. Calculate the Fourier transform of the following functions:

$$\textbf{(a)}\ \text{rect}\,(x/a) = \begin{cases} 1, & \text{when } |x| \le a \\ 0, & \text{elsewhere} \end{cases}$$

$$\textbf{(b)}\ \text{tri}\,(y/b) = \begin{cases} 1 - \dfrac{y}{b}, & \text{when } |y| \le b \\ 0, & \text{elsewhere} \end{cases}$$

2. Write a two-dimensional FFT subroutine and plot the transform of apertures in the shape of (a) a square, (b) a circle, (c) the letter E, and (d) the letter F.

3. For each part of Problem 2, set the magnitude of the Fourier transform to unity and then take an inverse transform of the complex spectrum. Plot the function. Discuss and justify the nature of the output.

4. Repeat Problem 3 after having corrupted the amplitude with some random noise.

5. Suppose a lens is placed at different locations between the input and the output plane. Find the analytical expression for the output at one focal length f_1 away from the lens using the diffraction formula, when (a) the lens is against the input object, (b) the lens is at a distance f_1 from the input, and (c) the lens is behind the object at a distance f_1. Discuss the results. How do you eliminate the phase factor that appears in front of one or more of the expressions.

1.7 SUGGESTED READINGS

Banerjee, P. P., and T. C. Poon, *Principles of Applied Optics*, Aksen Associates, Homewood, 1991.

Born, M., and E. Wolf, *Principles of Optics*, Pergamon Press, New York, 1964.

Bracewell, R., *The Fourier Transform and Its Application*, McGraw-Hill, New York, 1956.

Gaskill, J. D., *Linear Systems, Fourier Transforms, & Optics*, Wiley, New York, 1978.

Goodman, J. W., *Introduction to Fourier Optics*, McGraw-Hill, New York, 1968.

Jong, M. T., *Methods of Descrete Signal and Systems Analysis*, McGraw-Hill, New York, 1982.

Karim, M. A., *Electro-Optical Devices and Systems*, PWS–Kent Publishing, Boston, 1990.

Lizuka, K., *Engineering Optics*, 2nd ed., Springer-Verlag, New York, 1983.

Lohmann, A. W., *Optical Information Processing*, 2nd ed., Lohmann, Erlangen, Germany, 1979.

O'Neil, E. L., *Introduction to Statistical Optics*, Addison-Wesley, Reading, MA, 1963.

Oppenheim, A. V., and R. W. Schafer, *Digital Signal Processing*, Prentice-Hall, Englewood Cliffs, NJ, 1975.

Yu, F. T. S., *Optical Information Processing*, 2nd ed., Kreiger Publishing Melbourne, FL, 1990.

CHAPTER TWO

Digital Image Fundamentals

2.1 INTRODUCTION

The aim of this chapter is to develop background for digital image processing techniques. Review of optical computing related publications will clearly point out the significance of digital image processing. Computer-aided digital image processing techniques provide excellent, simple, and highly flexible tools for exploring and testing novel ideas in optical computing and processing. The concept of phase-only filters (covered in Chapter 5), for example, known for its use in pattern recognition, was first realized in the digital domain. Based on its initial success, it was implemented later in optics. Furthermore, the economic aspect of digital processing cannot be underestimated. To avoid procurement of expensive optical equipment and at the same time explore the enormous possibilities of optical processing, one may instead test the schemes using digital simulations. Homomorphic image processing (covered in Chapter 4) is another example that was first implemented in the digital domain; then the idea was carried into the optics arena. By using the powerful optical computing algorithm, purely digital techniques, such as histogram equalization, are now being proposed for implementation in the realm of optics. On the other hand, holography (covered in Chapter 3), which started out in purely classical optics, is rather frequently used now in the form of digitized holograms. The marriage of such techniques involving optics and digital systems is continually resulting in many stimulating signal processing experiences.

2.2 SAMPLING AND QUANTIZATION

Both for optical and digital processing, the input to a digital/optical processing system must be sampled. For digital representation of real-world, continuous, two-dimensional information, both the space and amplitude of signals are discretized. Another incentive of sampling involves suppressing enormous redundancy present in any two-dimensional pictorial information. For example, in an image, a certain background region may contain only a constant gray level. Therefore it is only sufficient to get one pixel of information representing that region and be aware of the boundary of that region. Considerations similar to these make it possible to realize tremendous compression of two-dimensional information. The routinely used FAX machine, for example, makes use of such compression schemes. Again, examining it from the frequency domain point of view, one notes that all information may be represented by a limited number of frequencies. Since information does not vary more rapidly (lines per inch) than a certain maximum rate, it is possible to skip some information (i.e., take samples) without sacrificing the information content or quality. Moreover, since only a finite number of samples may be processed in a reasonable amount of time, it restricts the total number of samples that can be retained. Again, for a gray level image, the number of gray levels that should be retained also affects the total number of samples. If each sample is represented by m bits, it results in 2^m numbers of gray levels. Thus an $n \times n$ image with m bit data requires $n \times n \times m$ bits of memory. For example, a 64×64 image with 256 gray levels requires a total of $64 \times 64 \times 8$ (i.e., 32,768) bits of storage.

The exact number of samples as well as the exact number of corresponding gray levels necessary for proper retention of information are determined by the signal bandwidth, the memory limitations, the computation/processing time, and the task to be performed. Increasing both n and m will make the picture look like the original; however, the storage requirement and the processing time will increase dramatically. Depending on the size and frequency content of the image, a minimum number of samples needs to be retained to faithfully reconstruct the image. Usually, it requires a minimum of two samples per period for the highest possible frequency. This minimum requirement is referred to as the Nyquist rate. For example, if a periodic signal is being displayed, the sampling frequency must exceed the highest frequency by at least a multiple of 2. Increasing the gray level reduces the quantization error; sometimes it may also reduce contrast.

The relative importance of the number of samples and the number of gray levels can be quantified using psychological image quality measurements in terms of isopreference curves. Such experiments support the conclusion that increasing both n and m increases the quality. Sometimes decreasing m may improve quality, however, by increasing the contrast. As the number of samples increases, fewer gray levels are required to keep the quality the same. Figure 2.1 demonstrates the effect of gray level and sampling size.

If a continuous image $f(x, y)$ is sampled, its two-dimensional discrete representation is referred to as $f(i, j)$, where each $f(i, j)$ value is an element of an $n \times n$ matrix. Each of these elementary picture elements is known as a pixel. Most images do not have a uniform distribution of all frequencies throughout the image. Some portions have relatively low frequency while some other segments contain very highly varying gray level information. It is therefore necessary to retain more samples where oscillations of gray levels are observed and low samples near a background. Thus one might adaptively choose the samples according to the rate of variation. Similarly, nonrectangular grids may also be used to sample an image. The advantage of such grids may be a smaller number of samples to be retained.

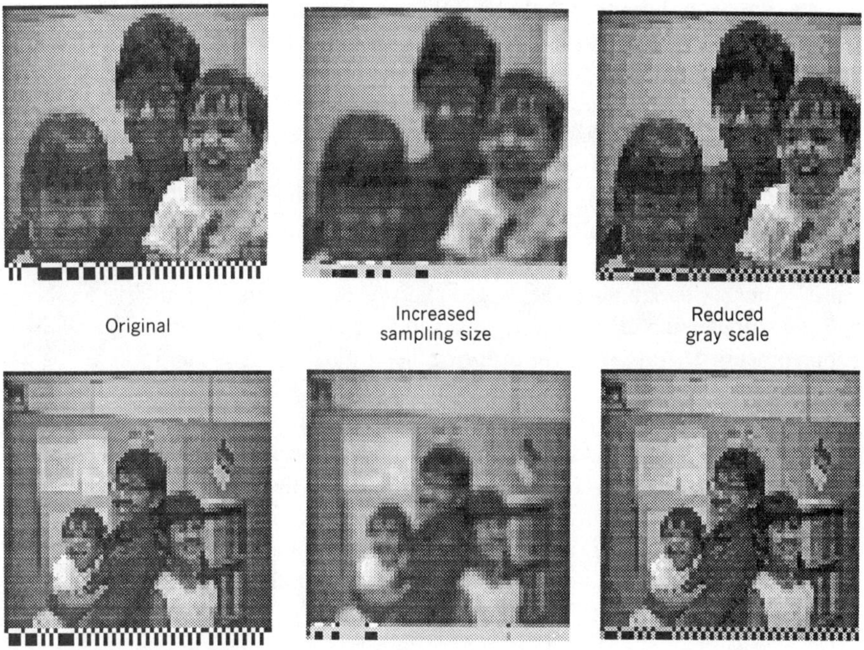

FIGURE 2.1 Effect of gray levels and sampling size.

2.3 IMAGE ENHANCEMENT

2.3.1 Spatial Domain Versus Frequency Domain

Spatial domain refers to the gray level image itself. Spatial domain filtering, in particular, refers to the operation on the gray level image itself without performing any Fourier transform. One of the attractive features of the spatial domain technique is that, in spatial domain processing, one can avoid the transformation to and from transform domain. Most of the linear filtering operations that are done in frequency domain have their equivalents in the spatial domain. Spatial domain filtering even enables one to perform nonlinear, highly complex, operations that are impossible to realize in the frequency domain. For example, by varying the threshold in the edge detection by a Laplacian operator scheme, one can manipulate the size and shape of the edges; however, it has no equivalent operation in the transform domain. The very popular and powerful noise removal technique of median filtering, for example, cannot be duplicated or even explained in terms of frequency domain filtering operations.

2.3.2 Histogram Equalization

Statistically, in an image no specific remarks may be applied as to how the gray levels in an image should be distributed. Estimating the histograms of different images, one would observe that they follow a wide variety of probability density functions: from uniform to highly nonuniform or even binary. Accordingly, it is extremely difficult to recommend, for any specific image, how its histogram should be shaped. However, when an image is being degraded as a result of some recording process, the mechanism of the particular process may suggest the specific way in which degradation may have taken place. For example, in an overexposed negative, the higher levels of intensity are easily saturated to give almost equal density for wide variations of high intensities. We also note that the recording process directly affects the gray levels of the image, which maps most white intensities in very dark areas. The way to undo such a process is to involve a scheme that redistributes the existing gray levels to all possible gray levels. This redistribution process of gray levels for restoring the original image is known as histogram modification. If based on a simple assumption that all gray levels of the input scene were equally distributed, the gray level may be equalized, which is a specific case of histogram modification known as histogram equalization. Histogram equalization is a specific kind of transformation that converts the probability density function of the gray level of an image into a uniform one.

Let g be the gray level of an image that is to be transformed into gray level h, such that the transformed gray levels become uniform. This transformation can be described as

$$h = T(g) \tag{2.1}$$

where T is the transformation operator. If the density of g is $p_g(g)$ and that of h is $p_h(h)$, then

$$p_h(h) = \left[p_g(g) \frac{dg}{dh} \right]_{g = T^{-1}(h)} \tag{2.2}$$

Accordingly, one can verify that a transformation of the form

$$h = \int_0^g p_g(\alpha) d\alpha \tag{2.3}$$

can transform the distribution of g to a uniform one:

$$\frac{dh}{dg} = p_g(g) \tag{2.4}$$

Then, from the previous equations,

$$\begin{aligned} p_h(h) &= \left[p_g(g) \frac{1}{p_g(g)} \right]_{g = T^{-1}(h)} \\ &= [1]_{g = T^{-1}(h)} \\ &= 1 \end{aligned} \tag{2.5}$$

where $0 \leq g \leq 1$. Thus $p_h(h)$ is a uniform density. Therefore making a transformation that is the same as calculating cumulative frequency distribution results in a uniform density function. For discrete images, if the kth gray level g_k appears q_k times out of a total of mn pixels, then the probability of gray level g_k is

$$p_g(g_k) = \frac{q_k}{mn} \tag{2.6}$$

where k can take values from 0, 1, . . . , L, where L is the number of gray levels. Then the discrete representation of Eq. (2.3) is given by

$$\begin{aligned} h_k &= T(g_k) \\ &= \sum_{i=0}^{k} \frac{q_i}{mn} \\ &= \sum_{i=0}^{k} p_g(g_i) \end{aligned} \tag{2.7}$$

The inverse of the above transformation,

$$g_k = T^{-1}(h_k) \tag{2.8}$$

is useful for arbitrary histogram modification. In an arbitrary histogram modification, the histogram of the given image is first equalized. Then the arbitrary histogram is equalized. Now the inverse relationship relating the to-be-obtained gray level to the equalized one is applied to convert the equalized histogram of the given image to the desired histogram.

2.3.3 Image Smoothing

Smoothing of an image is necessary to reduce the effects of "salt-and-pepper" noise, transmission noise, random noise, and so on. Three different techniques are discussed: namely, median filtering, lowpass filtering, and averaging. All these are based on the assumption that the average gray level characteristics of an image remain invariant even after addition of noise.

2.3.3.1 Median Filtering

One of the most attractive, nonlinear, space-variant techniques is known as median filtering. Median is defined as the central value of a distribution, when the values are arranged in ascending order of magnitudes. It is most effective in removing large noise spikes. Since no averaging is applied, it only eradicates the point without averaging or diluting the large spike among its neighboring pixels. One of its nice features is its capability of removing salt-and-pepper noise without blurring or disturbing the edges. A salt-and-pepper noise is characterized by the fact that it always stands apart from the gray levels of its neighborhood. Thus the noise can never be the same as the value of the median of its neighboring pixels.

For applying the median operation, a 5 × 5, 3 × 3, or cross-shaped window may be chosen. The gray levels of the windows are first arranged in any order of magnitudes, and then the gray value of the central pixel of the window is replaced by the median value of the window. Due to the operation involved in arranging the pixel values, optical implementation of median filtering is difficult. However, using a scheme referred to as the symbolic substitution, which will be described in Chapter 11, median filtering can be defined as a process of selecting an output corresponding to a specific input combination. Thus, using an optical look-up truth-table, median filtering can be optically implemented.

Figure 2.2a shows an original image, and Fig. 2.2b shows the same image when it has been corrupted by salt-and-pepper noise. When the image of Fig. 2.2b is processed with a 3-pixel median window, the median filtered image is obtained as shown in Fig. 2.2c. Note that as a result, most of the salt-and-pepper noise has been eliminated.

2.3.3.2 Lowpass Filtering

Sometimes the noise in an image resides in the high-frequency region, whereas Fourier transform of an image reveals that most of the information is contained in the low-frequency region. Therefore by filtering out the high-frequency region, one can get rid of such noise in an image.

Lowpass filtering can easily be applied in the frequency domain. By multiplying the Fourier transform of the image by the transfer function of the filter and then inverse transforming the product, one obtains the lowpass filtered image. The resulting image is given by

$$f_l(x, y) = \mathfrak{F}^{-1}\{F(u, v)H(u, v)\} \tag{2.9}$$

where $H(u, v)$ represents the lowpass filter, and u and v represent the horizontal and vertical spatial frequencies, respectively. Since the low-frequency signal is passed without any alteration, this process is called lowpass filtering.

An ideal lowpass filter has very sharp cutoff characteristics. Realization of an ideal lowpass filter in the spatial domain, like its one-dimensional temporal domain

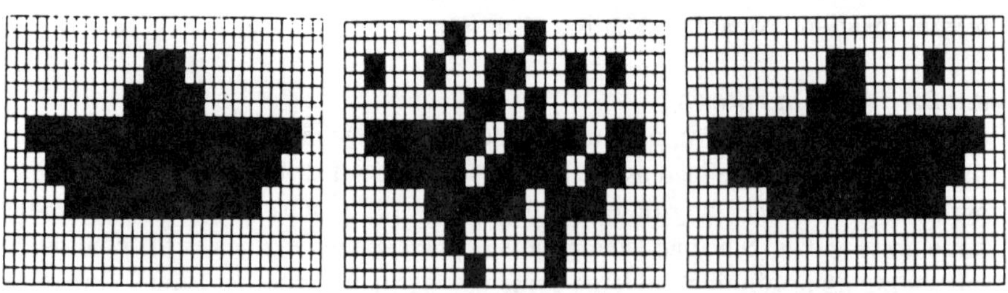

(a) (b) (c)

FIGURE 2.2 Median filtered image.

counterpart, requires a very large window or a very large number of coefficients. However, the frequency domain implementation of an ideal lowpass filter is much more simple using the 2-D FFT. Once the image is transformed into the frequency domain, lowpass filtering can be achieved by suppressing all frequency components outside a window of cutoff frequency r_0, while keeping all the components inside the region untouched, such that

$$H(u, v) = \begin{cases} 1, & u^2 + v^2 \leq r_0 \\ 0, & \text{otherwise} \end{cases} \tag{2.10}$$

where r_0 is a constant defining the radius of the all-pass region, the sharp cutoff frequency of the lowpass filter. Lowpass filtering, in addition to removal of noise, also causes blurring of the sharp edges and ringing. The blurring is due to the attenuation of the high-frequency content, which preserves the edge information. The ringing or blurring can also be explained, if one sees that multiplication by a circular window is equivalent to convolution by a Bessel-type function, which has an infinite extent of black and white ring structure. A single point will appear to have a ringlike shape. Since the image can be expressed as a two-dimensional spatial sum of individual delta functions, the overall effect is blurring of the edges. Convolution is also a process of averaging, which explains why lowpass filtering smoothes the edges.

In order to counter the ring effect, one can smooth the transition between the pass band and the stop band of the lowpass filter. Filters such as the Butterworth and Chebyshev filters may be applied to this effect.

2.3.3.3 Averaging

Statistically, the mean, median, and mode all represent measures of central tendency of a distribution. Arithmetic mean and median are the most common ones used for image smoothing. While median is evaluated by arranging the gray levels as a sequence of increasing or decreasing progression, the arithmetic mean is a more direct and faster way to perform smoothing. Median filtering helps eliminate the highly localized spike noise without disturbing the spatial features of an image; the mean tends to reduce the widely distributed global type of noise. To realize averaging a window is first chosen, as in the medium on which the operation of running average is performed. The central pixel value is replaced by the arithmetic mean of the gray levels inside the window. Obviously, other types of means such as geometric or harmonic mean may be applied for image smoothing.

Averaging can also be done on multiple images in order to reduce the effect of random noise. Usually, when the average of a sequence of noisy images is taken, the noise in each pixel tends to average out. Images of the same object taken by multiple sources can be used to perform averaging. Care must be taken to align the different images, so that the same pixels of image information are averaged from different images; otherwise, the averaged image will experience more deterioration.

2.3.4 Image Sharpening

Averaging or saturation in the image recording medium may result in blurring of the edges of an image. Differentiating techniques may be applied to regain or emphasize the image edge information. One can apply differentiation in the spatial domain using discrete approximation of continuous derivatives. Alternatively, highpass filtering may be employed in the frequency domain to enhance the edge information.

2.3.4.1 Differentiation

The derivative operator can easily be approximated by a forward, backward, or central difference. A one-dimensional difference operator may be used to sharpen the horizontal or vertical edges. For example, the x-directional operator that will detect the vertical edges in an image is given by

$$\frac{\partial f}{\partial x} = \frac{f(x + \Delta x) - f(x)}{\Delta x} \tag{2.11}$$

where f is the image gray level, and Δx denotes the distance between two neighborhood pixels in x direction. To sharpen or detect edges in another orientation, operators in both the x and y directions may be combined in some fashion.

The gradient G of an image f points in the direction of maximum change. It is given by

$$\mathbf{G}(x, y) = \left(\mathbf{i}\frac{\partial}{\partial x} + \mathbf{j}\frac{\partial}{\partial y}\right) f \tag{2.12}$$

and its magnitude is calculated as

$$\begin{aligned} G(x, y) &= \left[\left(\frac{\partial f}{\partial x}\right)^2 + \left(\frac{\partial f}{\partial y}\right)^2\right]^{1/2} \\ &= \{[f(x, y) - f(x + h, y)]^2 + [f(x, y) - f(x, y + h)]^2\}^{1/2} \end{aligned} \tag{2.13}$$

Now when the value of the gradient replaces the image gray level, the background with unchanged gray level becomes zero. To preserve the background while performing image sharpening, one may replace the image by its gradient, only if its gradient exceeds a certain threshold value, and leave the image unchanged otherwise. The other option in image sharpening could be just highlighting the edge information only, while the background may be assigned a different value. The Roberts operator, in particular, uses 4 pixels to determine the edges. It is given by

$$\begin{aligned} G[f(x, y)] = \{&[f(x, y) - f(x + h, y + h)]^2 \\ &+ [f(x + h, y) - f(x, y + h)]^2\}^{1/2} \end{aligned} \tag{2.14}$$

Sometimes a 2-Norm absolute value is used instead.

2.3.4.2 Highpass Filtering

A high-contrast image is characterized by the presence of distinct sharp edges. The sharpest transition of an image usually corresponds to the high spatial frequency of information. Consequently, by suppressing the low-frequency information by a highpass filter while allowing the high frequency to pass, one may get edge sharpening. This is analogous to replacing the image by its gradient when it exceeds a certain threshold, while all frequencies less than that value are replaced by zero. By blocking the dc, the constant background can be eliminated. A Butterworth or ideal highpass filter may be utilized to this effect.

2.4 IMAGE RESTORATION

Image restoration techniques are usually associated with an image degradation model. Image restoration techniques use the degradation model to undo the effect of degradation. Such technique differs from image enhancement, which tries to either amplify or deamplify some features of an image for detection or identification purposes or purely to make an image more acceptable or pleasing to the human eye. In it there is a sense of overdoing rather than undoing. However, it may be difficult to distinguish between restoration and enhancement operations. Image enhancement may work without any degradation model; however, some image enhancement tasks may use a priori degradation mechanisms for better enhancement. Histogram equalization may be classified as either an image enhancement or image restoration based on the nature of the input image. Sometimes, in the process of restoring an image after applying some restoration scheme, image enhancement may be applied to further restore the image. Chapter 3 shows an example of image restoration when implemented optically.

2.4.1 Inverse Filter

Degradation Model In general, in a visual information transmission system, the degradation function is modeled as a linear operator. The noise is assumed to be additive. Graphically, it is shown in Fig. 2.3. One can easily express the output $g(x, y)$ in terms of the input $f(x, y)$ and noise $n(x, y)$ by

$$g(x, y) = Df(x, y) + n(x, y) \tag{2.15}$$

where D is a degradation operator. Assuming $d(x, y)$ to be the degradation impulse response, Eq. (2.15) can be rewritten as

$$g(x, y) = \int_{-\infty}^{\infty}\int_{\infty}^{\infty} f(\alpha, \beta)d(x - \alpha, y - \beta)d\alpha\, d\beta + n(x, y) \tag{2.16}$$

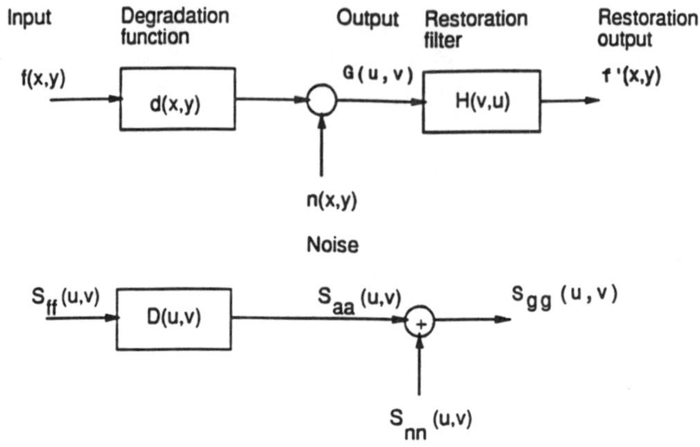

FIGURE 2.3 Wiener model.

Taking the Fourier transform, we get

$$G(u, v) = D(u, v)F(u, v) + N(u, v) \tag{2.17}$$

Assuming noise-free degradation, that is $n(u, v) = 0$, Eq. (2.16) reduces to

$$F(u, v) = \frac{G(u, v)}{D(u, v)} \tag{2.18}$$

Then the restored image is given by

$$f(x, y) = \mathfrak{F}^{-1}\left\{\frac{G(u, v)}{D(u, v)}\right\} \tag{2.19}$$

One of the difficulties of the inverse filter is that the degradation function usually has a finite extent beyond which there exists zeros in the frequency domain of the function. During the degradation process, some parts of the frequency content are transformed into zeros (as in the linear motion for photographic film). Therefore information about some of the frequency content is permanently lost. During the frequency domain restoration process, the indeterminate condition causes the inverse filters to explode. One possible way of alleviating the problem is to add a small bias to the denominator before inverting the frequency domain function. Alternatively, the frequency domain function can be restricted to only finite values in the frequency domain.

2.4.2 Weiner Filter

The Weiner filter is functionally better called a least-square filter since it minimizes the square of the difference between the original and restored image. Assume a homogeneous random field, a known power spectrum, and zero mean independent noise. However, neither of these can easily be verified. They can, however, be estimated or modeled by defining the error as

$$e = f(x, y) - f'(x, y) \tag{2.20}$$

where e is the error term, $f(x, y)$ is the desired output, and $f'(x, y)$ is the estimated output after using a restoration filter $h(x, y)$. The mean square error is given by

$$E[e^2] = E[\{f(x, y) - f'(x, y)\}^2] \tag{2.21}$$

The restoration filter is assumed to be linear where the output is given by

$$E[e^2] = E[\{f(x, y) - \iint_{-\infty}^{\infty} h(x - x', y - y')g(x', y')dx'\, dy'\}^2] \tag{2.22}$$

where $h(\cdot)$ is the restoration filter and $g(\cdot)$ is the output of the system. Equation (2.22) is minimized where

$$E[\{f(x, y) - \iint_{-\infty}^{\infty} h(x - x', y - y')g(x', y')dx'\, dy'\}g(x'', y'')] = 0 \tag{2.23}$$

Changing sides, one obtains

$$E[f(x, y)g(x'', y'')] = \iint_{-\infty}^{\infty} h(x - x', y - y')E[g(x', y')g(x'', y'')]dx'\, dy' \tag{2.24}$$

and substituting the autocorrelation and cross-correlation terms, one finds

$$R_{fg}(x - x'', y - y'') = \iint_{-\infty}^{\infty} h(x - x', y - y')R_{gg}(x' - x'', y' - y'')dx'\, dy' \tag{2.25}$$

Defining $x - x'' = \alpha$, $y - y'' = \beta$, $x' - x'' = \lambda$, and $y - y'' = \tau$, we find that

$$R_{fg}(\alpha, \beta) = \iint_{-\infty}^{\infty} h(\alpha - \lambda, \beta - \tau)R_{gg}(\lambda, \tau)d\lambda\, d\tau \tag{2.26}$$

When white noise, which has a flat spectrum in all frequencies, is present, Eq. (2.25) reduces to

$$R_{gg}(\lambda, \tau) = R_{aa}(\lambda, \tau) + \frac{N_0}{2}\,\delta(\lambda, \tau) \tag{2.27}$$

$$R_{fg}(\alpha, \beta) = \iint_{-\infty}^{\infty} h(\alpha - \lambda, \beta - \tau)R_{aa}(\lambda, \tau)d\lambda\, d\tau + \frac{N_0}{2}\,h(\alpha, \beta) \tag{2.28}$$

The above integral is known a Fredholm integral of the second kind. When no white noise component is present,

$$R_{fg}(\alpha, \beta) = \iint_{-\infty}^{\infty} h(\alpha - \lambda, \beta - \tau)R_{aa}(\lambda, \tau)d\lambda\, d\tau \tag{2.29}$$

The above is a superposition integral known as the Fredholm integral of the first kind. Taking the Fourier transform of Eq. (2.26), we obtain

$$S_{fg}(u, v) = H(u, v)S_{gg}(u, v) \tag{2.30}$$

where $S_{fg}(u, v)$ is the cross-spectral density and $S_{gg}(u, v)$ is the spectral density. Note that for our image model, the following equations hold:

$$S_{gg}(u, v) = |D(u, v)|^2 S_{ff}(u, v) + S_{nn}(u, v) \tag{2.31}$$

$$S_{fg}(u, v) = D^*(u, v)S_{ff}(u, v) \tag{2.32}$$

Therefore

$$D^*(u, v)S_{ff}(u, v) = H(u, v)[|D(u, v)|^2 S_{ff}(u, v) + S_{nn}(u, v)] \tag{2.33}$$

Changing sides, we find

$$H(u, v) = \frac{D^*(u, v)S_{ff}(u, v)}{[|D(u, v)|^2 S_{ff}(u, v) + S_{nn}(u, v)]} \tag{2.34}$$

Rearranging, we have

$$H(u, v) = \frac{1}{D(u, v)}\,\frac{|D(u, v)|^2}{|D(u, v)|^2 + \dfrac{S_{nn}(u, v)}{S_{ff}(u, v)}}. \tag{2.35}$$

When no noise is present, the Weiner filter is the inverse filter. On the other hand, with large noise, $S_{nn}(u, v) >> S_{ff}(u, v)$, when $D(u, v) \to 0$.

The gain of the filter varies according to the ratio $S_{nn}(u, v)/S_{ff}(u, v)$. Often when little a priori statistical information is present, one can approximate $D(u, v)$ as

$$H(u, v) = \frac{1}{D(u, v)} \frac{|D(u, v)|^2}{|D(u, v)|^2 + \Gamma} \tag{2.36}$$

where Γ is a constant approximating the noise-to-signal ratio. By supplying an estimate of Γ a Wiener filter may be constructed.

2.4.3 Spatial Domain Filter

A spatial domain filter for image restoration is attractive for its simplicity and speed of implementation. Optically, since the Fourier transform is done very simply, there is great preference to use Fourier domain filters rather than using spatial ones. Recently, however, spatial domain filters have also been implemented optically.

2.5 PROBLEMS

1. Use a Gaussian random number generator to generate random noise of standard deviation 2, 4, and 8 to add to an image at every fourth pixel. Then use (a) a median filter, (b) a mean filter, (c) a geometric mean filter, and (d) a harmonic mean filter to filter the random noise. Compare each of the outputs with the original image.
2. Fill up to 20% of the pixels of a 64×64 image with Gaussian random noise. Now use the 2-D FFT routine to filter this noise and only keep the noise (a) within a small frequency range in the low-frequency region and (b) within the high-frequency region with a highpass filter. (c) Use a bandpass filter to limit the noise in a certain pass band in-between. Next, use the bandpass noise to corrupt the images and then use all the filters that you know to find the best way to remove the noise. Specifically, test the performance of the Weiner filter by assuming that you have an a priori knowledge of the noise spectrum.
3. Use the degradation function (Cherri et al., 1991) to degrade your image for a velocity of (a) 2 mm/s and (b) 8 mm/s and then use an inverse filter to restore the images. What happens when the velocity of the restoration filter is (c) 4 mm/s and (d) 10 mm/s.
4. Use the sinc function to degrade the image as in a photographic film. Then restore it using (a) an inverse filter and (b) a Weiner filter.

2.6 SUGGESTED READINGS

Angel, E. S., and A. K. Jain, Restoration of Images Degraded by Spatially Varying Point Spread Functions by a Conjugate Gradient Method, *Appl. Opt.*, **17,** 2186 (1978).

Armand, A., A. A. Sawchuk, and T. C. Strand, Nonlinear Optical Processing with Halftones: Accurate Predictions for Degradation and Compensation, *Appl. Opt.*, **26,** 1007–1014 (1987).

Bryngdahl, O., Halftone Images: Spatial Resolution and Tone Reproduction, *J. Opt. Soc. Am.*, **68,** 416–422 (1978).

Cherri, A. K., A. A. S. Awwal, M. A. Karim, and D. L. Moon, Restoration of Moving Binary Images Degraded Owing to Phosphor Persistence, *Appl. Opt.*, **30,** 3734–3739 (1991).

Dashiell, S. R., and A. A. Sawchuk, Nonlinear Optical Processing: Analysis and Synthesis, *Appl. Opt.*, **16,** 1009–1015 (1977).

Gonzalez, R. C., and P. Wintz, *Digital Image Processing*, 2nd ed., Addison-Wesley, Reading, MA, 1987.

Huang, T. S. PCM Picture Transmission, *IEEE Spectrum*, **2**(12), 57–63 (1965).

Just, D., R. Hauck, and O. Bryngdahl, Rational Carrier-Period for Binarization of Sampled Images and Holograms, *Opt. Commun.*, **60,** 359–363 (1986).

Karim, M. A. Fabrication of Precompensated Contact Screens for Finite-Gamma Recording Media, *J. Appl. Photogr. Eng.*, **9,** 100–104 (1983).

Karim, M. A., Realization of Precompensated Monotonic Contact Screens, *Opt. Lett.*, **9,** 527–529 (1984).

Karim, M. A., and H. K. Liu, Linear Versus Logarithmic Spatial Filtering in the Removal of Multiplicative Noises, *Opt. Lett.*, **6,** 207–209 (1981).

Karim, M. A., and H. K. Liu, All-Optical Homomorphic Image Processing System, *Opt. Lett.*, **7,** 371–373 (1982).

Oppenheim, A. V., and R. W. Schafer, *Digital Signal Processing*, Prentice-Hall, Englewood Cliffs, NJ, 1975.

Peng, F., M. A. Karim, and A. A. S. Awwal, Halftone Image Generation Using Two Carrier Periods, *Opt. Commun.*, **84,** 242–246 (1991).

Yu, F. T. S., *Optical Information Processing*, 2nd ed., Kreiger Publishing, Melbourne, FL, 1990.

ANALOG OPTICAL COMPUTING

CHAPTER THREE

Linear Optical Processing

3.1 INTRODUCTION

Without doubt, optics and optical systems first demonstrated their data processing capability in the analog domain. In these systems, the data introduced to the systems were continuous-tone images while output data were also in the form of two-dimensional images. Most often, the images were introduced and collected in the form of photographic transparencies with transmittance values ranging from completely opaque to completely transparent. In more recent times, however, newer and more sophisticated spatial light modulators (SLMs) have been developed to replace photographic films. These SLMs can be used to store and output two-dimensional images in nearly real-time. Consequently, the scope of the resulting linear optical systems appears to be more supportive for developing optical computing schemes that could be used to process digital data as well. It must be stressed that physics of photographic film is different from that of the various SLMs but their manifestations are not all that unrelated.

In all analog optical processing systems, the Fourier transform operation plays the most significant role. The laser provides the coherent light beam for properly transporting the information through the system and the biconvex lenses perform rather easily the Fourier transform operation whenever necessary. Once the input is introduced into the system, the wavefront of light travels through the input as well as through all other processing elements in parallel.

This chapter first introduces the optical properties of a recording medium such as photographic film, for example, and then discusses the various spatial filtering operations that can be realized using a linear optical processor. While the most basic spatial filtering operations involve binary filters, the more sophisticated filtering operations require complex filters. Consequently, holography is discussed next as a means for synthesizing complex filters. In succeeding sections, two particular examples of the application of complex spatial filters are elaborated. Inverse filters are used to restore motion-degraded images obtained in phosphor-based display devices and Wiener filters are used for restoring blurred images.

3.2 PHOTOGRAPHIC FILM

The photographic recording medium consists of a base, made of either acetate film or glass plate, and a layer of photographic emulsion, where photosensitive silver halides are suspended in a gelatin support. When exposed to light, the silver halide particles undergo a physical alteration, reducing to tiny metallic silver particles. The process of reduction is concluded (or *developed*) and then *fixed* by a set of prescribed chemical processes that extract the remaining silver halide, leaving behind the silver grains. The transmittance of the developed film depends strictly on the presence of silver grains that are basically opaque at the optical frequencies. The intensity transmittance T of the developed photographic transparency is specified as

$$T(x, y) = \langle I_{\text{out}}(x, y)/I_{\text{in}}(x, y)\rangle \tag{3.1}$$

where I_{out} and I_{in}, respectively, represent the output and the input irradiance and $\langle\ \rangle$ denotes the localized ensemble average over an area that is large compared to the size of the film grain but small compared to an area within which the output intensity changes drastically. The density of the transparency D, defined as the silver mass per unit area, has been determined by Hurter–Driffield to be

$$D = -\log(T) \tag{3.2}$$

The most commonly used photographic recording characteristic is the Hurter–Driffield (H-D) curve, which relates the density with the logarithm of the energy used to expose the photographic medium. Figure 3.1 shows a typical H-D curve having three distinct regions: shoulder, linear, and toe regions. In both the toe and shoul-

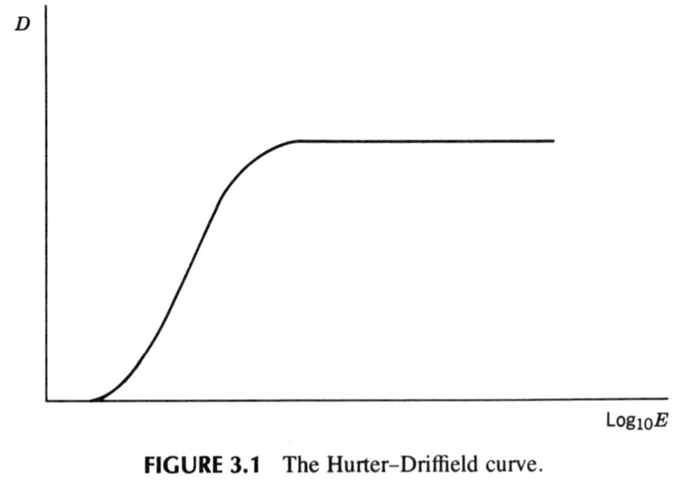

FIGURE 3.1 The Hurter–Driffield curve.

der regions, density is quite independent of the exposure. However, in the linear region, the density is proportional to the logarithm of the energy. The linear region is described by

$$D = g \log(E) - D' \tag{3.3}$$

where g, referred to as the *film gamma*, is the slope of the linear region of the H-D curve and $-D'$ is the intercept of this line with the density axis. A photographic medium with a large g value is referred to as a high-contrast recording medium, while that with a small g value is called a low-contrast recording medium. The photographic gamma is a function of not only the type of emulsion but also of the developing time and developing chemicals.

Since $E = It$, where t is the time of exposure, the intensity transmittance can be determined from Eq. (3.3) using Eq. (3.2) as

$$T = cI^{-g} \tag{3.4}$$

where $c = 10^{D'}t^{-g}$ is a positive constant. It is apparent that a one-step photographic process provides an intensity transmittance that is nonlinear with respect to the input irradiance. However, a two-step process, consisting of taking an intermediate transparency at first and then passing the light through the negative to obtain the final transparency (also known as the positive transparency), yields a different option. Using Eq. (3.4) repeatedly, we find the intensity transmittance of the positive transparency to be

$$T_p = c_p(I)^{g'} \tag{3.5}$$

where c_p is a positive constant, $g' = g_1 g_2$, and g_i is the film gamma corresponding to the ith step involved. By choosing $g_1 g_2 = 1$, it is possible to obtain a linear relationship between the intensity transmittance and the input irradiance.

Whenever the photographic medium is used as an element in the coherent optical processors, it is more reasonable to use amplitude transmittance rather than intensity transmittance. It is important to realize that neither the base thickness nor the emulsion thickness is uniform. In fact, emulsion thickness varies in accordance with the spatial density, which in turn depends on the exposure variations. The amplitude transmittance T_a is given by

$$T_a(x, y) = [T_i(x, y)]^{1/2} \exp[j\phi(x, y)] \tag{3.6}$$

where T_i is the intensity transmittance and $\phi(x, y)$ accounts for the random phase shifts. However, film thickness variations that are mostly uncontrollable are usually not desirable in most practical applications. Such variations are often avoided by means of an index-matching liquid gate, as shown in Fig. 3.2, where the film emulsion is immersed in index-matching fluid (having a refractive index very sim-

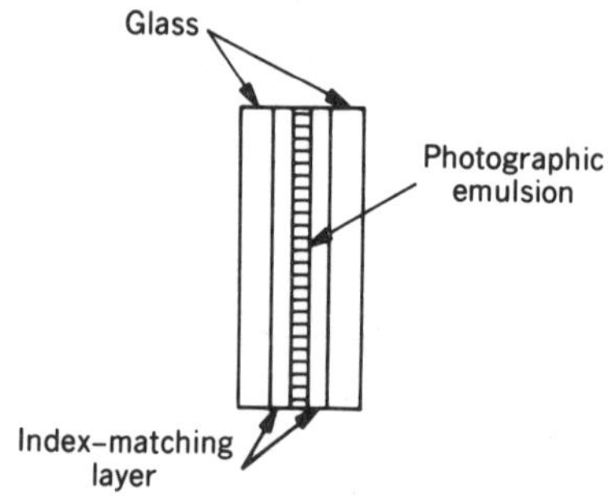

FIGURE 3.2 A liquid gate.

ilar to that of the emulsion) placed between two parallel glass plates. The amplitude transmittance of the index-matching liquid gate can thus be written

$$T_a(x, y) = [T_i(x, y)]^{1/2} \tag{3.7}$$

A two-step contact printing process would reveal as before that a linear relationship exists between the amplitude transmittance and the amplitude of the recording optical field when $g_1 g_2 = 1$.

While the photographic film is perhaps the best recording medium, the photographic process is not instantaneous. It takes very little time to expose the film but it takes a significant time to complete the chemical processing. The chemical processing, on the other hand, is neither clean nor uniform. To overcome this limitation of the photographic film, one can make use of the spatial light modulators (SLMs) to imprint a pattern on a beam of light in nearly real time. Examples of such SLMs are the liquid crystal light valve (LCLV), magneto-optic SLM, and liquid crystal display. All the variations of SLM follow only variations of the same physical principle.

The term *liquid crystal* (LC) refers to a particular class of materials whose rheological behavior is similar to that of the liquids, but whose optical behavior is similar to that of the crystalline solids over a given range of temperature. A type of particular LC, referred to as nematic LC, is commonly used in LCLV as well as in most other LC devices. Currently, ferroelectric LCs are also being used in real-time image displays. In comparison, ferroelectric LCs are faster in response than the nematic LCs. With an electric field applied to a LC layer, the LC molecules are reoriented in space due to both field and ionic conduction effects.

A typical SLM is shown in Fig. 3.3, where the cell is organized in the form of a quarter-wave plate and is placed between a mirror and a polarizer (with its polarization direction, shown by the arrows, oriented along the vertical axis). Consider an incoming light that is linearly polarized at an angle of 45° to the direction of alignment. The transmitted light is found to be circularly polarized, but the reflected light that passes back through the quarter-wave cell becomes polarized in a direction perpendicular to that of the incident light. The first polarizer acts as an analyzer and thus blocks the returning light. By changing the external voltage, however, the amount of light blocked can be varied. In particular, transmission is

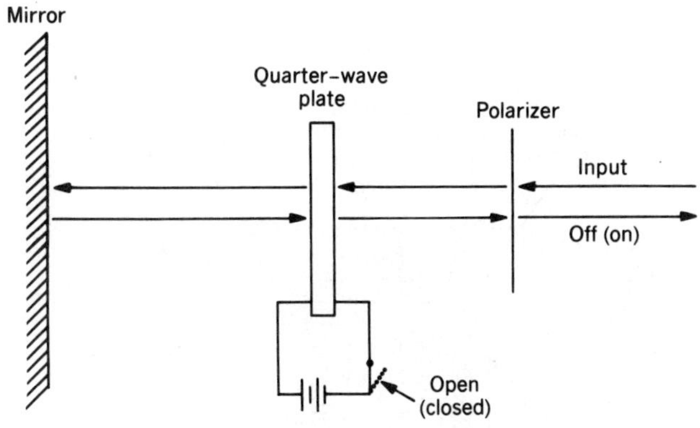

FIGURE 3.3 Schematic of a spatial light modulator.

zero when voltage is zero, but transmission reaches a maximum at a high enough voltage, as shown in Fig. 3.4. The transmission is nonbinary when voltage is set between the two extreme voltage limits. The LCLV requires only a little energy to produce an output, but it requires a large energy to yield its maximum output. Devices made using this SLM configuration are generally very sensitive to variations in cell thickness and light wavelength.

LCLVs produce a positive image when addressed. Optical addressing of a LCLV allows the image to be fed into the SLM in parallel. Thus, in the case of a LCLV, the frame time is the same as the response time of 1 pixel. A typical optically addressed LCLV is shown in Fig. 3.5. A photoconductive material is used to transfer the optical input to an equivalent voltage on the LC layer. The photocon-

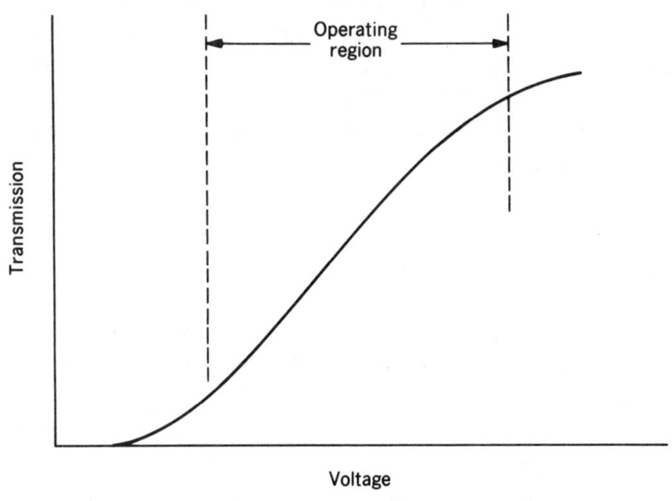

FIGURE 3.4 LCLV transmission characteristics.

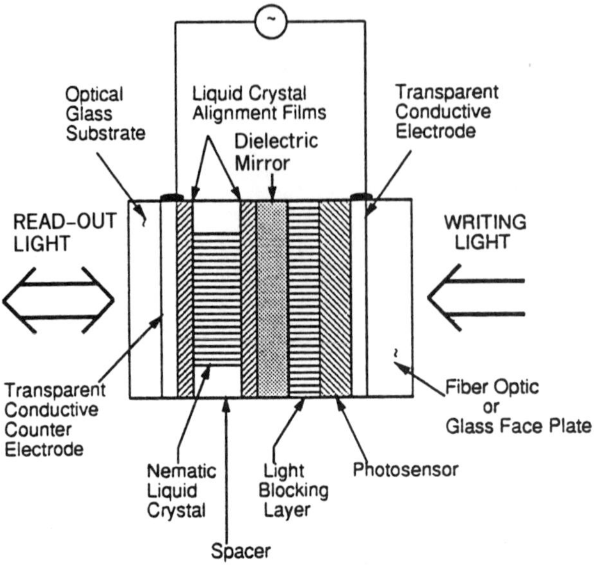

FIGURE 3.5 LCLV schematic.

ductor is highly resistive, and it thus uses most of the voltage when there is no incident light. In the presence of incident light, the resistivity of the photoconductor decreases, as a result of which more voltage appears across the LC layer. Again, the coherent read-out beam illuminating the back of LCLV is reflected back but modulated by the birefringent LC layer. Thus the input optical image is transferred as a spatial modulation of the read-out beam. In practice, however, the LCLV is driven by an ac voltage. For frequencies with periods less than the molecular response time, the LCLV responds to the rms value of the voltage. It is possible to feed the optical input data by means of a CRT, fixed masks, or even an actual real-time imagery. A typical LCLV has a 25-mm diameter aperture, a 3-μm LC layer, a 15-ms response time, 60-lines/mm resolution at 50% modulation transfer function, a 100 : 1 average contrast ratio, and a lifetime of several thousand operating hours.

3.3 SPATIAL FILTERING USING BINARY FILTERS

Spatial frequency filtering is an information processing operation in which certain desired spatial frequencies of an input image are either removed or altered by filters placed at the Fourier domain plane. Consider the coherent optical information processing system of Fig. 3.6. A coherent beam diverging from a pinhole spatial filter placed at the focal plane of lens L_1 results in a plane wavefront. This wavefront is used next to illuminate the input image transparency. Lens L_2 (with focal length f) transforms the input image (which is located at its front focal plane) and, as a

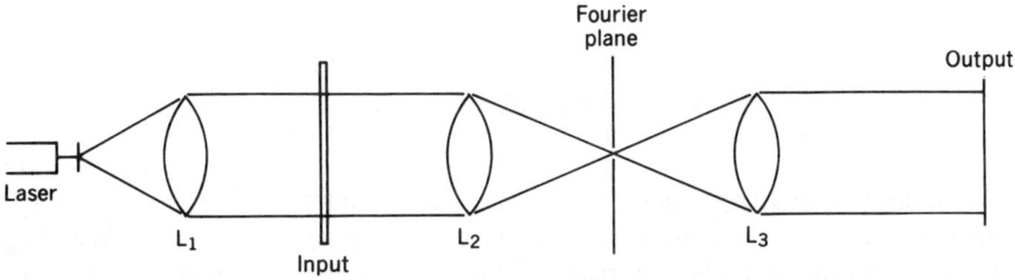

FIGURE 3.6 An optical information processing system.

result, the frequency spectrum of the input is displayed in the back focal plane of L_2. The back focal plane of L_2, referred to as the Fourier plane, also happens to be the front focal plane of lens L_3. Therefore L_3 retransforms (i.e., inverse Fourier transforms) the Fourier plane spectrum and forms an image at the output plane of the system. The filters necessary for the particular spatial filtering operation are placed in the Fourier plane and thus a filtered image appears at the output plane.

Several kinds of spatial filters are used in spatial filtering operations: binary, amplitude, phase, and complex filters. *Binary* filters constitute the simplest type of filter. Stops and slits are typical examples of such a filter. The transmittances of stops and slits are either unity or zero. Those among the binary filters that pass only lower spatial frequency components are referred to as *lowpass filters*. Similarly, those that pass higher spatial frequency components are referred to as *highpass filters*. Finally, certain binary filters, know as *bandpass filters*, are designed to pass only selected spatial frequency components.

Consider a line diffraction grating of finite extension b, which consists of slits of width a and cell period X. The amplitude transmittance of such a 50–50 duty cycle line grating is given by

$$T(x) = \sum_{n=-\infty}^{\infty} \text{rect}[(x - nX)/a] \, \text{rect}\,(x/b) \tag{3.8}$$

where

$$\text{rect}(z) = \begin{cases} 1, & |z| \leq \frac{1}{2} \\ 0, & \text{otherwise} \end{cases} \tag{3.9}$$

But

$$\begin{aligned} \text{rect}(x/a) * \text{comb}(x/X) &= X \int \text{rect}(x'/a) \sum_{n=-\infty}^{\infty} (x - nX - x')dx' \\ &= X \sum_{n=-\infty}^{\infty} \text{rect}[(x - nX)/a] \end{aligned} \tag{3.10}$$

since

$$\text{comb}(x/X) = \sum_{n=-\infty}^{\infty} \delta((x/X) - n) = X \sum_{n=-\infty}^{\infty} \delta(x - nX) \tag{3.11}$$

When this line grating, as described by Eq. (3.8), is introduced in the input object plane of the coherent optical information processing system, the corresponding field distribution at the Fourier plane (obtained by taking the Fourier transform of the input) is found to be proportional to

$$\begin{aligned} F\{T(x)\} &= (1/X)F\{[\text{rect}(x/a) * \text{comb}(x/X)]\text{rect}(x/b)\} \\ &= (1/X)F\{\text{rect}(x/a) * \text{comb}(x/X)\} * F\{\text{rect}(x/b)\} \\ &= (1/X)[F\{\text{rect}(x/a)F\{\text{comb}(x/X)\}] * F\{\text{rect}(x/b)\} \end{aligned} \tag{3.12}$$

using the theorem

$$F\{g_1(x, y)\} * F\{g_2(x, y)\} = F\{g_1(x, y)\}F\{g_2(x, y)\} \tag{3.13}$$

and the fact that

$$F\{g_1(x, y)g_2(x, y)\} = F\{g_1(x, y)\}F\{g_2(x, y)\} \tag{3.14}$$

where $F\{g(x, y)\}$ represents the Fourier transform of function $g(x, y)$. However,

$$F\{\text{rect}(x/a)\} = \{\sin(\pi u_x a)\}/(\pi u_x) = a \text{ sinc}(au_x) \tag{3.15}$$

and

$$F\{\text{comb}(x/X)\} = X \sum_{n=-\infty}^{\infty} \delta(u_x X - n) = X \text{ comb}(u_x X) \tag{3.16}$$

where $u_x = x/\Gamma f$, f is the focal length of the lens, and Γ is the wavelength of light. Accordingly, Eq. (3.12) can now be evaluated as follows:

$$\begin{aligned} F\{T(x, y)\} &= (ab/X) \{\text{sinc}(au_x) \sum_{n=-\infty}^{\infty} \delta(u_x - (n/X))\} * \text{sinc}(bu_x) \\ &= (ab/X) \sum_{n=-\infty}^{\infty} \text{sinc}(an/X)\, \delta(u_x - (n/X)) * \text{sinc}(bu_x) \\ &= (ab/X) \sum_{n=-\infty}^{\infty} \text{sinc}(an/X) \text{ sinc}(b\{u_x - (n/X)\}) \\ &= (ab/X) \text{ sinc}(bu_x) + \sum_{n \neq 0}^{\infty} \text{sinc}(an/X) \text{ sinc}(b\{u_x - (n/X)\}) \end{aligned} \tag{3.17}$$

The frequency spectra of the input image, as will be seen in the Fourier plane, thus consist of a bright central spot and many higher-order diffraction spots whose intensity diminishes more or less continually at higher diffraction orders. The third lens L_3 performs a Fourier transform of this diffraction pattern and thus reproduces an output image that is proportional to the input image with its coordinate system reflected. The reflected system of the coordinate justifies the fact that an inversion has been caused by double Fourier transformations. The finite size of the aperture at the Fourier plane may limit the transmitted spatial frequencies and, in addition, may cause the system to modify the transmitted image.

Now consider introducing a particular binary filter at the Fourier plane to cause what has been referred to as spatial filtering. Assume further that $b >> X$ such that there is no significant overlapping of diffraction orders. First, let us introduce a rather restrictive lowpass filter that passes only the zeroth diffraction order. According to Eq. (3.17) therefore, only the Fourier plane information corresponding to $\{ab \operatorname{sinc}(bu_x)\}/X$ centered at $u_x = 0$ is transmitted by this binary filter. The amplitude field distribution at the output plane is thus given by

$$f(x, y)|_{n=0} = F\{(ab/X)\operatorname{sinc}(bu_x)\} = (a/X)\operatorname{rect}(x/b) \tag{3.18}$$

The output consists of a broad luminating line of width b and contains no information at all about the input periodicity. Next, consider a lowpass filter that allows the two first-order diffraction spots as well as the central spot to pass through. The amplitude transmittance of the resulting output is accordingly given by

$$\begin{aligned} f(x, y)|_{n=0,1-1} &= F\{(ab/X)\,[\operatorname{sinc}(bu_x) + \operatorname{sinc}(a/X) \\ &\quad \{\operatorname{sinc}[b(u_x + X^{-1})] + \operatorname{sinc}[b(u_x - X^{-1})]\}]\} \\ &= (a/X)\operatorname{rect}(x/b)\,[1 + 2\cos(2\pi x/X)\operatorname{sinc}(a/X)] \end{aligned} \tag{3.19}$$

This output amplitude distribution, unlike that described by Eq. (3.18), has the same periodicity as that of the input. A pinhole spatial filter like the one used in the system of Fig. 3.6 is an example of this latter binary filter. Accordingly, the pinhole spatial filter removes spurious fringe patterns that are otherwise present in the laser output. With a lowpass filter that allows more diffraction orders to pass through the filter, the output becomes identical to the input. Interestingly, if a bandpass filter capable of passing only the first orders is introduced at the Fourier plane, the resulting output image appears to have a periodicity that is half of what it is in the input grating.

Finally, consider the effect of having a highpass binary filter that blocks only the central diffraction spot. The output amplitude distribution of this spatially filtered image is found to be given by

$$\begin{aligned} f(x, y)|_{n\neq 0} &= F\{(ab/X) \sum_{n\neq 0}^{\infty} \operatorname{sinc}(an/X)\operatorname{sinc}[b(u_x - (n/X))]\} \\ &= X^{-1}\{\operatorname{rect}(x/b)\,[\operatorname{rect}(x/a) * \operatorname{comb}(x/X)] - a\operatorname{rect}(x/b)\} \end{aligned} \tag{3.20}$$

The first term of Eq. (3.20) has been normalized to unity. Consequently, the coefficient of the second term determines the actual characteristics of the output amplitude. When $a/X = \frac{1}{2}$, that is, when the line diffraction has a 50–50 duty cycle, the field amplitude at the output plane is either $\frac{1}{2}$ or $-\frac{1}{2}$ at all x such that $f(x, y)$ is in phase with $T(x, y)$. The resulting output is characterized by a uniform intensity and not by periodicity.

Another interesting case is realized when $a/X > \frac{1}{2}$. The output field amplitude is the either $1 - (a/X)$ or $-(a/X)$, such that $f(x, y)$ and $T(x, y)$ are in phase. The corresponding output is characterized by an intensity distribution such that the darker portions appear brighter at the output and the brighter portions of the input appear darker at the output. Thus a contrast reversal is obtained via such selective highpass spatial filtering. However, when $a/X < \frac{1}{2}$, the output intensity distribution and the input intensity distribution are both found to be in phase. Highpass spatial filters block the lower diffraction orders. Consequently, the output image obtained using a highpass spatial filter consists of only the edges of the input. As more central diffraction orders are blocked, the detail of the edge of the object is continually enhanced.

3.4 HOLOGRAPHY

While a photograph records the real image of an object as formed by a lens, a hologram records the field distribution of the light scattered by an object. But, more importantly, a hologram provides a means whereby phase information is also frozen into the holographic plate by superimposing a reference beam coming directly from the source and the scattered field coming from the object.

The interferometric experimental setup of Fig. 3.7 can be used to describe the recording of a hologram. The point source, placed at the focal point S of lens L_1, produces a parallel beam of light of wavelength Γ, which is then split into two portions. A portion passing through the prism P contributes to a parallel beam of light that strikes the photographic film placed in the Fourier filter plane at an angle θ. The other portion of light passes through a transparency of transmittance $h(x, y)$, which is positioned at the focal plane in front of the lens L_2. Consequently, the information recorded at the filter plane is obtained from the interference be-

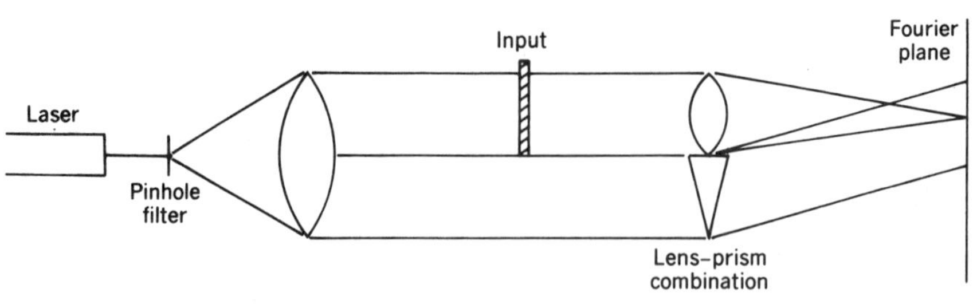

FIGURE 3.7 System for synthesizing a complex filter.

tween the prism-deflected plane beam of light and the Fourier transform of $h(x, y)$, $H(u_x, u_y)$. The amplitude distribution at the Fourier filter plane is given by

$$S(u_x, u_y) = H_0 e^{-j\phi} + A e^{-j2\pi\beta x} \tag{3.21}$$

where $\beta = \sin\theta/\Gamma$, and H_0 and ϕ represent the amplitude and phase of $H(u_x, u_y)$, respectively. The developed film negative will have a transmittance given by

$$\begin{aligned} T(x, y) &= 1 - S(u_x, u_y) S^*(u_x, u_y) \\ &= 1 - H_0^2 - A_0^2 - 2AH_0 \cos(2\pi\beta x - \phi) \end{aligned} \tag{3.22}$$

Thus the recorded transparency (prepared using either a positive or a negative film) is associated with the phase information. The technique for recording such phase information is known as *holography*. The fine-grain photographic film used to record the object information (for both amplitude and phase) acts basically as a diffraction screen, also referred to as a hologram.

To make a hologram, the coherent laser beam is split into two separate beams, one of which has a complex amplitude given by $A_0 e^{j\phi}$; it is the reflected wavefront referred to as the object field. The other beam, referred to as the reference beam, arrives at the holographic recording medium without striking the object. A hologram recording system is shown in Fig. 3.8a. The recording medium is exposed simultaneously to both the reference beam, $Ae^{j2\pi(\mu x + \epsilon y)}$, and the reflected beam. The resulting interference pattern recorded by the film constitutes a hologram. In fact, a hologram contains all the information needed to reproduce the wave field of the object. It can be used subsequently for reconstructing the object (in its three-dimensional form) by means of a coherent optical setup that is the same as the one shown in Fig. 3.8b.

When the developed hologram is illuminated with a beam of light $Ae^{j2\pi(\mu x + \epsilon y)}$ proportional to the reference beam, the transmitted wave field becomes

$$\begin{aligned} t(x, y) &= [A_0^2 + A^2 + 2AA_0 \cos(\phi - 2\pi\{\mu x + \epsilon y\})] \\ &\quad Ae^{j2\pi(\mu x + \epsilon y)} \\ &= (A_0^2 + A^2) A e^{j2\pi(\mu x + \epsilon y)} + A^2 A_0 e^{j\phi} \\ &\quad + A^2 A_0 e^{j4\pi(\mu x + \epsilon y)} e^{-j\phi} \end{aligned} \tag{3.23}$$

Hologram reconstruction thus involves three components: a direct beam and two first-order diffracted beams on either side of the direct beam. The first term of Eq. (3.23) corresponds to the direct beam, whereas the second term, which is proportional to the complex amplitude of the object beam, represents one of the diffracted beams responsible for producing the virtual image. The last term of Eq. (3.23) is responsible for yielding a real image of the object. There are a multitude of ap-

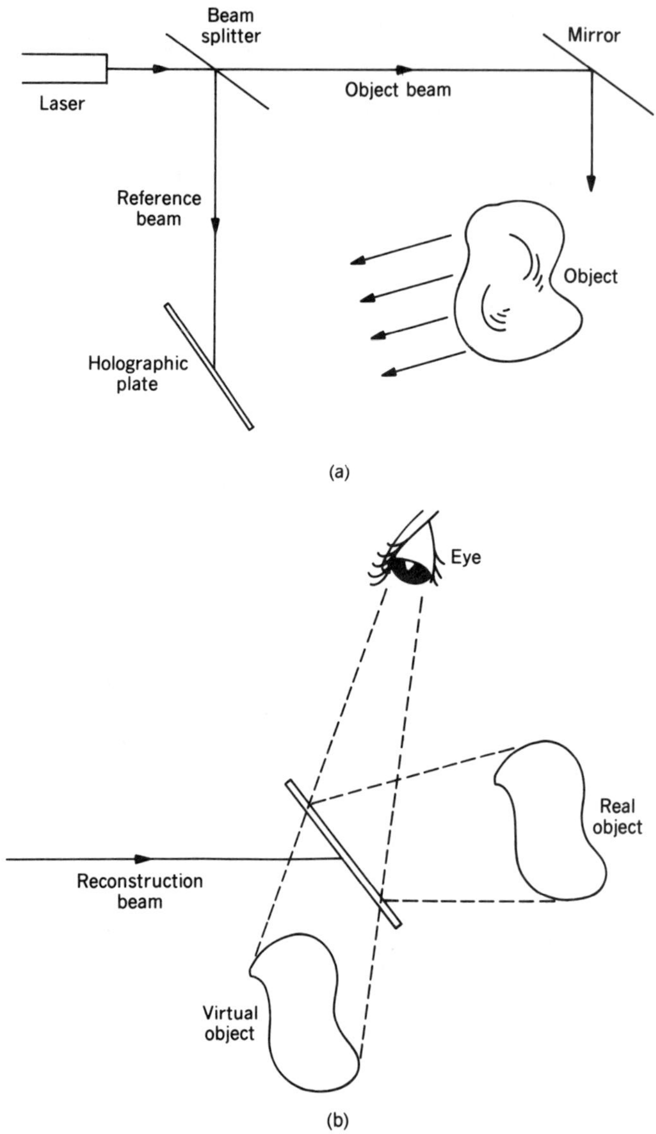

FIGURE 3.8 Holography: (a) recording and (b) reconstruction.

plications of holograms; but most important of all is the fabrication of holographic complex spatial filters, which are used in optical information processing systems.

3.5 INVERSE FILTERING

Holograms are used to fabricate many different kinds of complex filters. One of these is the inverse filter, which can be used, for example, in pattern recognition (discussed in Chapter 5). The Fourier transform of an image is given by

$$O(u_x, u_y) = I(u_x, u_y)H(u_x, u_y) \tag{3.24}$$

where $O(u_x, u_y)$ and $I(u_x, u_y)$ are the Fourier transforms, respectively, of output and input images and $H(u_x, u_y)$ is the coherent transfer function. For an ideal system, the value of the coherent transfer function is constant and unity for all spatial frequencies. A real system, however, has too many built-in characteristics, due to which the output turns out to be different from the input. However, if a filter with transmittance characteristics $T(u_x, u_y) = [H(u_x, u_y)]^{-1}$ were to be introduced and multiplied to the output spectrum, the output can be made exactly equal to the input. This particular filter is referred to as an inverse filter and the scheme of restoring the input is referred to as inverse filtering.

In this section, we illustrate the use of an inverse filter in the restoration of motion-degraded images. When the input scene, consisting of dynamic images, is displayed using phosphor-based display devices, the output images appear to be degraded. But to determine the coherent transfer function of the system, it is important to understand the characteristics of the system involved.

In general, imaging systems and, in particular, displays are evaluated in a static mode. For many of the imaging applications, the temporal characteristics of the imaging system may not be very critical. In dynamic environments, however, the temporal response of an imaging system plays a more important role than the static response in determining the quality of the display. The absence of temporal response considerations in the analysis of a traditional imaging system, in situations such as driving and piloting tasks, fails to explain the cause of degradation of the modulation contrast. This degradation becomes more significant at higher spatial frequencies. In such situations, for instance, targets begins to blend with background and the user loses the ability to discriminate between the targets and their background. To fully characterize an electro-optical imaging system therefore, both static and dynamic aspects need to be considered.

In a dynamic environment where there is relative motion between the target and the sensor, a longer phosphor persistence value is able to reduce the modulation contrast and eventually causes the loss of gray levels. Often the loss of one gray level at a high spatial frequency may be very significant, for instance, in cases where there is only enough modulation contrast to provide one or two gray levels in a static environment. The smearing of the displayed image may be properly quantified by having considered the temporal response in the imaging system analysis. Also, the phosphor type and the value of the relative motion between the target and the sensor play significant roles in image degradation.

Figure 3.9 shows a block diagram of an imaging system where an input object $i(x, y, t)$ is operated on by the point-impulse function $h(x, y, t)$ to produce an output image $g(x, y, t)$. Provided the point-impulse response function $h(x, y, t)$ can be expressed as a product of a purely spatial function $s(x, y)$ and a purely temporal function $\Delta(t)$, such as

$$h(x, y, t) = s(x, y)\Delta t \tag{3.25}$$

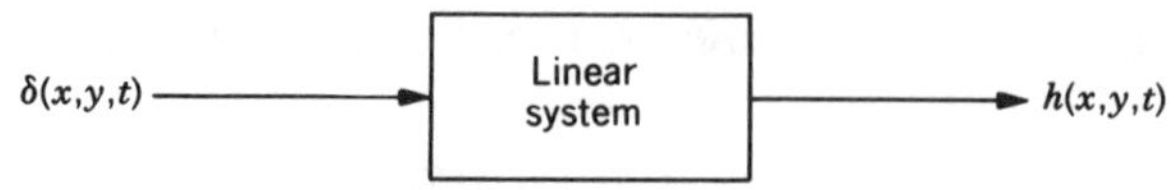

FIGURE 3.9 Block diagram of an imaging system.

one may evaluate the Fourier transform of $g(x, y, t)$ as

$$G(u, v, f) = D(f)S(u, v)I(u, v)\delta(f - um - vn) \quad (3.26)$$

where m and n denote velocity components, respectively, along the x and y directions, u and v, respectively, represent the spatial frequencies along the x and y directions, f is the temporal frequency, and $D(f)$, $S(u, v)$, and $I(u, v)$ are, respectively, the dynamic transfer function, the static transfer function, and the spectrum of the static image. The presence of the factor $\delta(f - um - vn)$ implies that, during evaluation of $G(u, v, f)$, the variable f in $D(f)$ needs to be replaced with $um + vn$. The overall system block diagram of the electro-optical display is shown in Fig. 3.10, which includes both static and dynamic transfer functions. The product $D(f)S(u, v)\delta(f - um - vn)$ is referred to as the DMTF. This new form for the spatiotemporal transfer function of the imaging system can be used to characterize the motion-degraded images. It is important to understand that Eq.(3.26) is correct as long as the separability condition of Eq. (3.25) is valid. Fortunately, this separability condition remains valid as long as phosphor-based electro-optic display systems are considered.

A well-known phosphor screen characteristic is provided by its exponential decay behavior, which can be approximated by

$$\Delta(t) = A_1 e^{-t/\tau_1} + A_2 e^{-t/\tau_2} + A_3 e^{-t/\tau_3} + \cdots \quad (3.27)$$

where A_i is a factor representing the proportion of the phosphor in the ith state and τ_i is the corresponding time constant defined as the duration in which the phosphorescence decays to $(1/e)$th of its maximum value A_i. In general, the last two terms of Eq. (3.27) represent fluorescence modes of a phosphor light output. When compared to the first term representing the phosphorescence mode of the light output, the contribution of the last two terms is insignificant. Accordingly, Eq. (3.26) may be approximated by

$$\Delta(t) = \tau^{-1} \exp(-t/\tau) \quad (3.28)$$

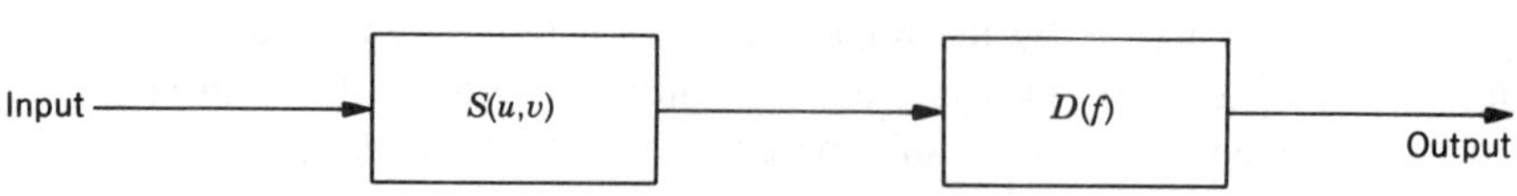

FIGURE 3.10 Overall transfer function of a display.

where, for simplicity, A_1 is set equal to $1/\tau$. By taking the Fourier transform of Eq. (3.28), the dynamic component of the DMTF for the phosphor screen can be obtained:

$$D(f) = (1 + j2\pi f\tau)^{-1} \tag{3.29}$$

Equation (3.29), representing the temporal transfer function of a phosphor screen, can be introduced into Eq. (3.26) to predict the overall degradation of motion-involved displayed images. However, in order to use $D(f)$, one needs to estimate the specific value of the time constant of the phosphor used in the E-O display under consideration. One may make use of the fact that the Type P-1052 phosphor, for instance, takes 20 ms to decay to 10% of its maximum value. Accordingly, we may set $\tau = 8.68$ ms for the case of Type P-1052 phosphor.

To comprehend the extent of image degradation in an electro-optical display, a particular image intensifier tube, such as that used in night-vision goggles, may be considered. The calculated value of m and the experimentally determined value for the time constant of Type P-1052 phosphor, for example, is used in Eq. (3.29) to calculate the corresponding MTF of the dynamic response. A computer program may be used to simulate the effect of the DMTF on images. Figure 3.11 shows a 64×64 pixel-sized binary input image that is subjected to different head motion degradations. Figure 3.12 shows the corresponding degraded images when the head movement speeds along the x-direction approach 4, 8, and 15 cm/s, respectively. It is evident that with increasing speed the image degrades more and more.

FIGURE 3.11 A 64×64 input binary image.

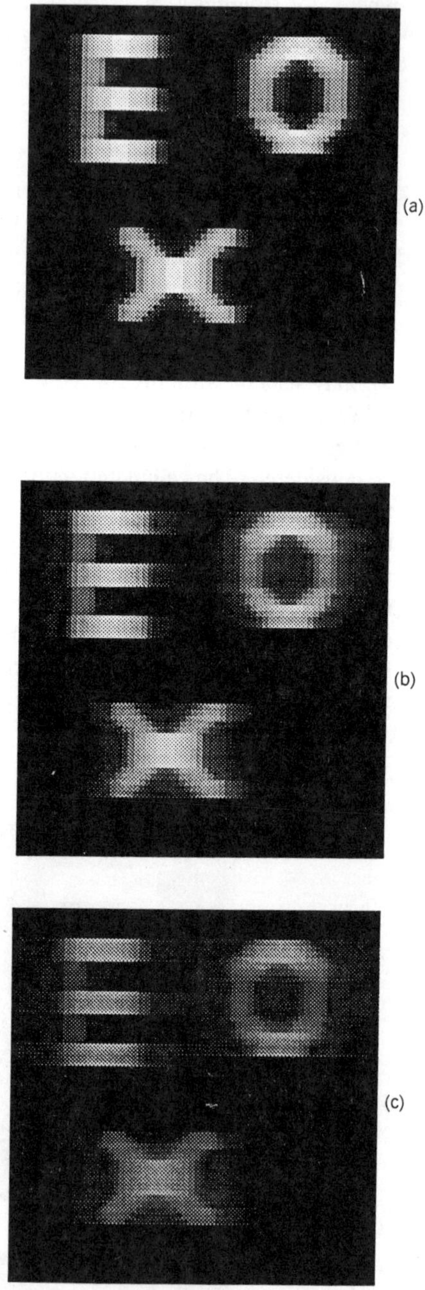

FIGURE 3.12 Images degraded due to horizontal speeds of (a) 4 cm/s, (b) 8 cm/s, and (c) 15 cm/s.

Since motion is directed only along the x axis, the degradation of vertical edges becomes more significant. Similarly, corresponding motion along the vertical axis would have deteriorated only the horizontal edges.

Figure 3.12 shows that the DMTF can degrade input images even in the absence of noise to a degree that the object is hardly recognizable. To restore the degraded images, consider the degradation model shown in Fig. 3.10. In this model, the object image $i(x, y, t)$ is operated on by a degradation function $H(u, v, f)$, which produces a degraded output image $g(x, y, t)$. The main objective is to approximate $i(x, y, t)$ given $g(x, y, t)$ and some knowledge of the degradation function $H(u, v, f)$. However, the degradation function as given by Eq. (3.26) consists of two separable functions $S(u, v)$ and $D(f)$, where $S(u, v)$ is a purely static (spatial) function and $D(f)$ is a purely dynamic (temporal) function. Therefore, by eliminating the effects of $D(f)$, a reasonably good displayed image can be obtained. This is accomplished simply by dividing $H(u, v, f)$ by $D(f)$. Such a restoration technique, based on the concept of inverse filtering, can be applied in a straightforward fashion since the modulus of $D(f)$ is nonzero except at infinity. If this were not true, such a condition would have posed a major restriction to inverse filtering. The simulation of the restoration process is summarized as follows:

1. A FFT of the degraded image $g(x, y)$ is obtained. This gives $S(u, v)D(f)$.
2. Given the velocity of the object and its direction, the corresponding inverse filter, $[D(f)]^{-1}$, is calculated.
3. The results obtained in steps 1 and 2 are multiplied (point by point) in the frequency domain.
4. An inverse FFT of the result of step 3 yields the restored image.

To restore the degraded image, one needs to know (1) the type of phosphor and hence the corresponding time constant(s) and (2) the value and the direction of the relative motion between the sensor and the target. This information is necessary to realize $[D(f)]^{-1}$, as discussed earlier. While there is no problem in meeting the first requirement, often it may be difficult to exactly estimate the value of the relative motion.

The real-time restoration model is shown in Fig. 3.13a. In this model, both magnitude and direction of the relative motion are determined first. This information is used in the design of the restoration filter. A simple hybrid setup to perform the restoration process is illustrated in Fig. 3.13b, where a spatial light modulator (SLM) can be used to imprint the inverse filter. Note that this real-time restoration model is suitable in applications where both speed and direction of the relative motion can be reasonably estimated.

For applications where the relative motion between the target and the sensor is not known exactly, a simpler restoration model like the one shown in Fig. 3.14 can be used. In this simple system, a fixed Fourier hologram may be used to store the inverse filter (for a single specific speed or a range of speeds). The image of Fig. 3.12c, for example, may be restored to some degree, as in Fig. 3.15a, using

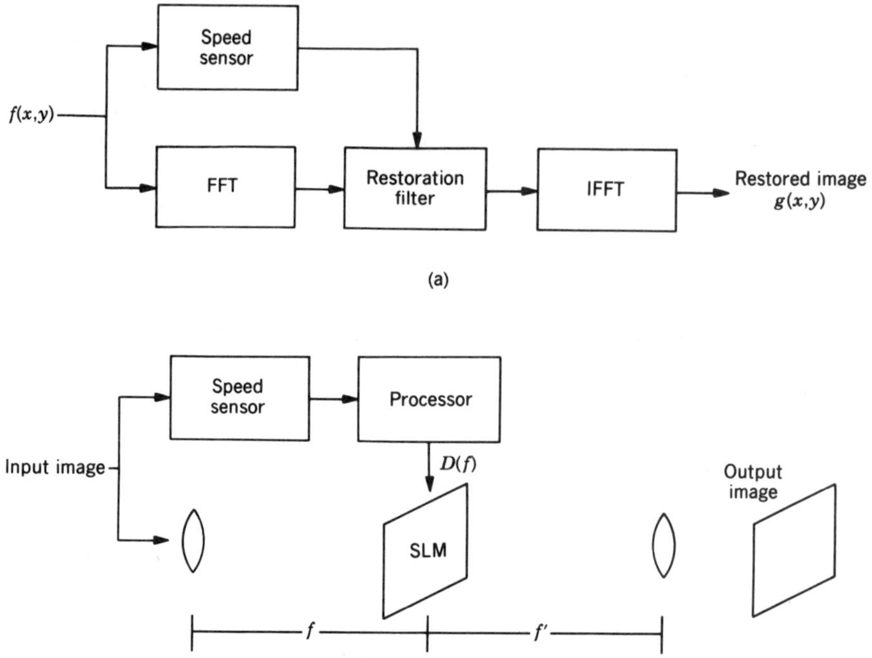

FIGURE 3.13 Adaptive restoration: (a) model and (b) system.

an inverse filter designed for a speed value of 12 cm/s, for example, or as in Fig. 3.15b using an inverse filter designed for a speed of 7 cm/s. Observations suggest that it is not always necessary to know the exact value of the relative velocity. A value close to the relative speed of the target is more likely to give a better output image of an electro-optical display than that without using an inverse filter.

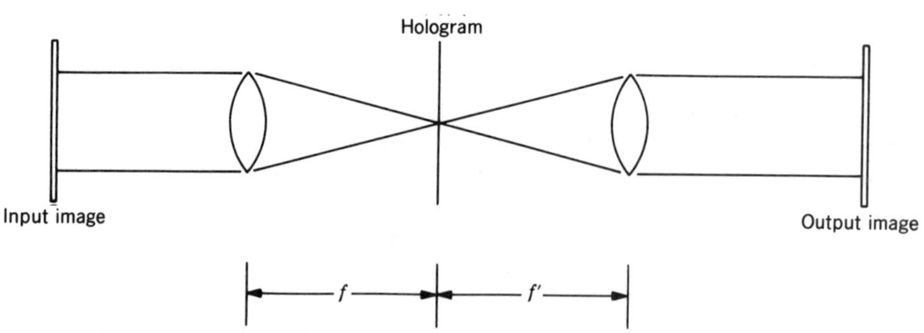

FIGURE 3.14 Nonadaptive restoration with Fourier hologram used as an inverse filter.

(a)

(b)

FIGURE 3.15 Restoration of image of Fig. 3.12c using inverse filters corresponding to (a) 12 cm/s and (b) 7 cm/s.

3.6 DEBLURRING

As pointed out in earlier sections, a complex spatial filter may be realized by the combination of an amplitude filter and a phase filter. This is relatively easily obtained using holographic techniques. A complex spatial filter can also be synthesized for correcting blurred images. Consider that an object intensity distribution is blurred by some degradation associated with the system that forms its image. Provided the optical system involved is linear and space invariant, the Fourier transform $O(u_x, u_y)$ of the blurred image $o(x, y)$ can be expressed by

$$O(u_x, u_y) = I(u_x, u_y)B(u_x, u_y) \tag{3.30}$$

where $I(u_x, u_y)$ is the Fourier transform of the undistorted input image $i(x, y)$ and $B(u_x, u_y)$ is the Fourier transform of the distorting function $b(x, y)$.

The deblurring operation involves estimating $i'(x, y)$ of the true input image $i(x, y)$ from the measured output image $o(x, y)$ or, equivalently, to performing a deconvolution operation. One general approach to such a problem is the use of inverse filtering, where the inverse filter transfer function for deblurring is expected to be

$$H(u_x, u_y) = [B(u_x, u_y)]^{-1} \tag{3.31}$$

The estimate of the true object may be obtained by passing the output $o(x, y)$ through a linear space-invariant filter. While, for the particular case of motion-degraded images obtained using a phosphor-based display device, inverse filters are easily synthesized, it is not that easily realized in the cases of deblurring.

Consider, for example, that the transmission function of a linear smeared point image is approximated by

$$f(x) = \begin{cases} 1, & -[2L]^{-1} \leq x \leq [2L]^{-1} \\ 0, & \text{otherwise} \end{cases} \tag{3.32}$$

where L is the smear length. If the transparency described by Eq. (3.32) is introduced at the input plane of a coherent optical information processing system, the corresponding Fourier plane field is given by its Fourier transform,

$$F(u_x) = (2/u_x) \sin (u_x L/2) \tag{3.33}$$

It can readily be seen from Eq. (3.33) that the Fourier spectrum is not only bipolar but has infinite isolated zeros, thereby requiring the corresponding inverse filter to have poles. Furthermore, the inverse filter takes no account of the noise inherent in the measurement of the output image $o(x, y)$. Consequently, the resulting deblurred image may have its noisy components enhanced as to be entirely useless. However, if we are to be partially successful in restoration, an approximate restoration filter may be realized by using a least-mean-square error filter, referred to as a Wiener filter. A Wiener filter produces an estimate $i'(x, y)$ that minimizes the quantity

$$E = \iint |i'(x, y) - i(x, y)|^2 \, dx \, dy. \tag{3.34}$$

In the most general case, the output image $O(x, y)$ is related to the true input image $i(x, y)$ by

$$O(x, y) = i(x, y) * h(x, y) + n(x, y) \tag{3.35}$$

where $*$ is the convolution operation and $n(x, y)$ is a statistically stationary noise function with power spectral density $P_n(u_x, u_y)$. It must be noted that the power spectral density of any function is equivalent to the Fourier transform of its autocorrelation. The transfer function $H(u_x u_y)$ of the Wiener filter that makes the value of E a minimum is

$$H(u_x, u_y) = B^*(u_x, u_y)/[|B(u_x, u_y)|^2 + \{P_n(u_x, u_y)/P_i(u_x, u_y)\}] \quad (3.36)$$

where $P_i(u_x, u_y)$ is the power spectral density of the input image $i(x, y)$. For those frequencies at which the signal-to-noise ratio P_i/P_n is large compared with $|B(u_x, u_y)|^{-2}$, $H(u_x, u_y)$ reduces to $|B(u_x, u_y)|^{-1}$, that is an inverse filter. However, at frequencies where the signal-to-noise ratio is very small, the filter transmission is minimal and, consequently, the noise present in the deblurred image is drastically reduced.

3.7 PROBLEMS

1. Identify the coherent optical processor and the particular spatial filter for which the output, corresponding to the input image given by $T(x, y) = [\sin\{(x^2 + y^2)^{1/2}\} + 2]/4$, will have its contrast maximized.
2. Determine the intensity distribution of the output when the input line diffraction grating, of 50–50 duty cycle, is subjected to a bandpass binary filter capable of passing only the first orders. Comment on the periodicity of the output.
3. Determine and describe what will be recorded by a photographic film of gamma 2, placed at the Fourier plane, if the input image of the coherent optical processor consists of two images $T_1(x, y)$ and $T_2(x, y)$ placed side by side at a distance of X.
4. What will be the output characteristics when a sinusoidal grating, having amplitude transmittance given by

$$T(x', y') = [1 + \cos(\sigma x')]/2$$

for all y', is introduced at the Fourier plane of the coherent optical processor described in Problem 3.
5. Repeat Problem 4 when the sinusoidal grating has an amplitude transmittance given by

$$T(x', y') = [1 + \sin(\sigma x')]/2$$

for all y'.
6. Consider that a transparency consisting of two slits described by

$$T(x, y) = \text{rect}[(x - (a/2))/x_0] + \text{rect}[(x + (a/2))/x_0]$$

is introduced at the input plane of a coherent optical processor. Determine the corresponding Fourier plane characteristics. Assume further that the input is illuminated by a light beam containing two frequencies u_1 and u_2. Determine the ratio of u_1/u_2 for which the second off-axis diffraction order corresponding to u_1 coincides with the third off-axis diffraction order corresponding to u_2.

3.8 SUGGESTED READINGS

Awwal, A. A. S., A. K. Cherri, M. A. Karim, and D. L. Moon, Dynamic Modulation Transfer Function of a Display System, *Appl. Opt.*, **30,** 201–205 (1991).

Banarjee, P. P., and T.-C. Poon, *Principles of Applied Optics*, Aksen Associated, Homewood, 1991.

Celaya, L., and S. Mallick, Incoherent Processor for Restoring Images Degraded by a Linear Smear, *Appl. Opt.*, **17,** 2191–2194 (1978).

Cherri, A. K., A. A. S. Awwal, M. A. Karim, and D. L. Moon, Restoration of Moving Binary Images Degraded Due to Phosphor Persistence, *Appl. Opt.*, **30,** 3734–3739 (1991).

Goodman, J. W., *Introduction to Fourier Optics*, McGraw-Hill, New York, 1968.

Karim, M. A., *Electro-Optical Devices and Systems*, PWS–Kent Publishing, Boston, 1990.

Karim, M. A., A. M. Hanafi, F. Hussain, S. Mustafa, Z. Samberid, and N. M. Zain, Realization of a Uniform Circular Source Using a Two-Dimensional Binary Filter, *Opt. Lett.*, **10,** 470–472 (1985).

Leith, E. N., Image Deblurring Using Diffraction Gratings, *Opt. Lett.*, **5,** 70–72 (1980).

Rhodes, W. T., Acousto-Optic Signal Processing: Convolution and Correlation, *Proc. IEEE*, **69,** 65–79 (1981).

Vachss, F., and L. Hesselink, Synthesis of a Holographic Image Velocity Filter Using the Nonlinear Photorefractive Effect, *Appl. Opt.*, **27,** 2887–2894 (1988).

Yu, F. T. S., *Optical Information Processing*, Wiley, New York, 1983.

CHAPTER FOUR

Analog Optical Arithmetic

4.1 INTRODUCTION

In the previous chapter, we introduced coherent optical processors. The key element of such a processor is the Fourier lens, which, under proper conditions, is capable of performing a two-dimensional Fourier transform operation of an input image. Such systems are immensely powerful since they are able to perform the Fourier transform operation in parallel. In this chapter, we extend the application of such optical systems for realizing arithmetic operations.

First, the halftone screening technique is introduced for preprocessing input images. As an example, halftone screens are shown realizing a particular nonlinear information processing. This particular nonlinear processing involves a series of three steps: logarithmic, linear, and exponential operations. Theta modulation is also discussed for realizing nonlinear operations. Finally, specific Fourier-based optical systems are described for realizing arithmetic operations involving input images that are not preprocessed. Instead, Fourier domain filters are designed for realizing the operations. In particular, addition, subtraction, multiplication, division, averaging, differentiation, and integration operations are addressed in this chapter.

4.2 HALFTONE PROCESSING

A halftone scheme is capable of converting a continuous-tone two-dimensional image into a binarized one for either analog or digital implementations. Such binarized images, referred to as halftone images, have already found use in many nonlinear image processing applications. In the graphic arts industry, for example, this is accomplished by photographing the given image through a halftone screen of a single period onto a high-contrast recording medium, which in turn thresholds the processed information. The resulting halftone image is thus a function of three parameters: the transmittance profile of the screen, period of the screen, and threshold value. Basically, the halftone positive (or negative) contains an array of opaque dots with sizes that vary according to the optical density of the original. Printing plates can be then photoreproduced from the halftone photograph.

In principle, the halftone scheme does not have to use photographic films but it may instead use real-time spatial light modulators. However, since most of the real-time devices do not have distinct threshold characteristics, precompensated halftone screens may have to be used instead of the regular halftone screens. Halftone screening can be used rather easily to implement several nonlinear optical image processing operations. In this context, it may be noted that linear operations such as those treated in the last chapter can be realized relatively easily with simple optical processors. It should be noted that digital processors can readily achieve nonlinear transformations. Accordingly, any extension of the capabilities of optical systems to include nonlinear operations enhances their suitability as viable alternatives to purely digital systems.

Nonlinear optical image processing using contact screens consists of either one or both of the following steps. The first step involves a mapping of the input intensity I_{in} through a mapping function f such that the output intensity $I_{out} = f(I_{in})$. It is important to note that both the input and output intensities are functions of spatial coordinates. The mapping function f can be made either linear or nonlinear through the use of contact screens. The output is generally in the form of a pulse-width or pulse-area modulated nonlinear image of the original. In the second step, when the modulated binary output in transparency form is placed at the input plane of a coherent optical image processing system as shown in Fig. 4.1, many orders of diffraction patterns appear at the Fourier plane. A spatial filter can then be placed at the Fourier plane to select or to modify any one or several of the diffraction orders of the spectrum. The corresponding output image can then be found at the focal plane of the second lens. The intensity at any point in this plane can be made to vary as a nonlinear function of the intensity at the corresponding point in the input image. Thus the overall nonlinear relationship between the input and the output depends on spatial filtering as well as on the contact screen processing used in the first step.

The most important element used in the contact screen process is the contact screen itself. It normally consists of a periodic array of continuous-tone cell patterns on a glass or plastic substrate. The contact screen processing involves contact-printing the original continuous-tone input image through the screen onto a

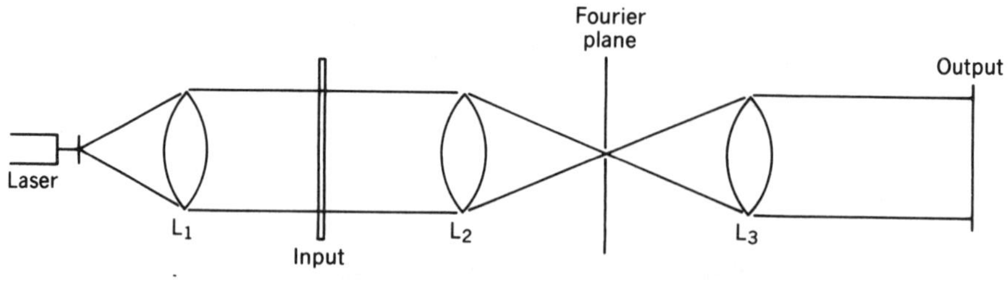

FIGURE 4.1 A coherent optical information processing system.

high-contrast recording medium (ideally having $g = \infty$). When the exposed high-contrast film is developed, the original gray tones can be represented by the average transmittance of either area-modulated opaque dots, if a two-dimensional screen is used, or pulse-width modulated opaque bars, if a one-dimensional screen is used.

The contact screen process can be illustrated using a specific one-dimensional screen with the help of Fig. 4.2. In Fig. 4.2a, the transmittance $T^s(x)$ of several unit cells of the screen are shown plotted against x. Assume that the continuous-tone original has a maximum spatial frequency of less than the frequency of the screen. This assures that the input intensity transmittance T^p remains constant over

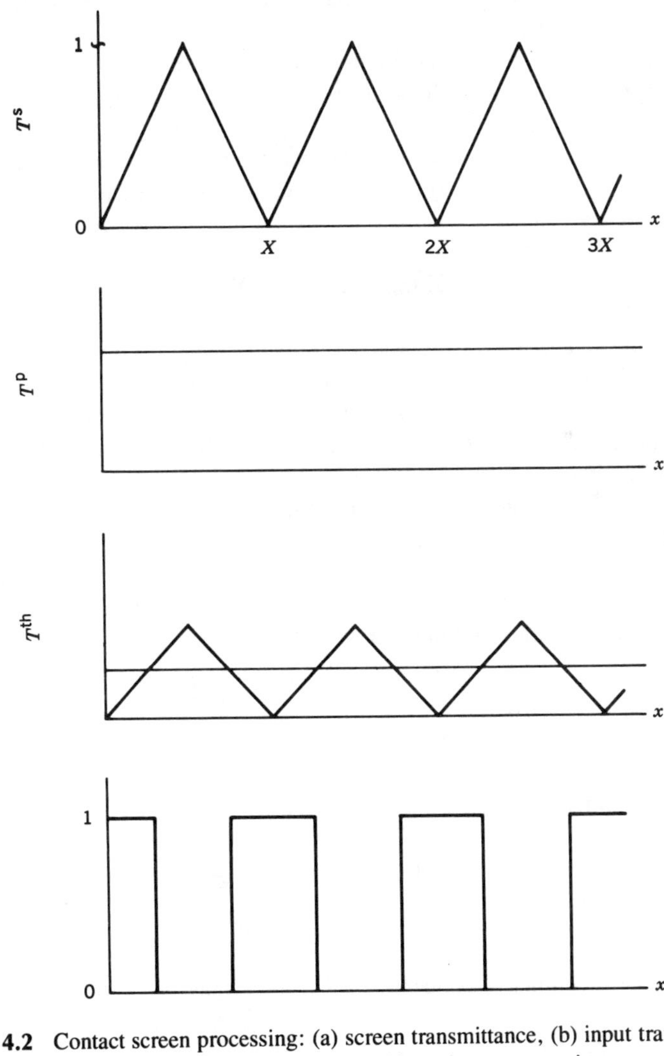

FIGURE 4.2 Contact screen processing: (a) screen transmittance, (b) input transmittance, (c) transmittance prior to thresholding, and (d) halftone image transmittance.

the region of at least one unit cell, provided the edge effects are not considered. During the contact-printing process, the screen is sandwiched between the continuous-tone original and the recording film. When the screen is in contact with the original, and when the effective transmittance $T^{s}(x)T^{p}$ is equal to or greater than the threshold transmittance T^{th}, the developed film has a transmittance of nearly zero; otherwise, its transmittance is equal to unity, as shown in Fig. 4.2d. For convenience, the threshold transmittance may be defined as

$$T^{th} = I_{th}/I_{in} \tag{4.1}$$

where I_{th} and I_{in} are the threshold energy and the energy of the input illumination, respectively. The resulting binarized image is called a halftone image, which is a spatially sampled version of the original.

For simplicity, the derivations of screen transmittance corresponding to a particular nonlinear mapping of the input are limited to screens of one-dimensional cell patterns only. As shown in Fig. 4.2d, each group of equal-width halftone transmittance pulses corresponds to a constant value of input photographic transmittance T^{p}. The screen transmittance is generally assumed to be periodic, such that

$$T^{s}(x) = T^{s}(x + nX) \tag{4.2}$$

where X is the screen period and n is any integer.

The overall average transmittance T^{H} of a unit cell can be defined as the ratio of the width of the transparent region to the total width of the unit cell. When the original input is sufficiently sampled, T^{H} represents the local average transmittance of the halftone image.

For $T^{p} \geq T^{th}/T^{s}$, the width of the transparent bar within each unit cell of the halftone image is equal to $X - \delta$, where δ is the particular value of x for which $T^{s}(x)T^{p} = T^{th}$. When $T^{p} < T^{th}/T^{s}$, the film is not sufficiently exposed, and therefore the corresponding halftone image region has a transmittance of nearly unity. Based on these rationales, T^{H} may be written

$$T^{H} = \begin{cases} 1 - (x/X), & T^{s}T^{p} \geq T^{th} \\ 1, & \text{otherwise} \end{cases} \tag{4.3}$$

while the screen transmittance is given by

$$T^{s}(x) = \begin{cases} T^{th}/T^{p}, & T^{p} \geq T^{th} \\ 1, & \text{otherwise} \end{cases} \tag{4.4}$$

In general, the functional dependence of T^{H} on T^{p} may be expressed as

$$T^{H} = f(T^{p}) \tag{4.5}$$

where the function f determines the actual nonlinear relationship. Accordingly, one can determine x in terms of T^{p} as

$$x = [1 - f(T^{\mathrm{p}})]X \tag{4.6}$$

when $T^{\mathrm{p}} \geq T^{\mathrm{th}}$.

Consider two specific nonlinearity cases for the purpose of illustration: logarithmic and exponential. When the logarithmic functional relationship $f(T^{\mathrm{p}}) = \alpha - \beta \ln(T^{\mathrm{p}})$ is substituted into Eq. (4.6), we have

$$x = [1 - \alpha + \beta \ln(T^{\mathrm{s}}(x)T^{\mathrm{p}}/T^{\mathrm{s}}(x))]X \tag{4.7}$$

where α and β are two constants such that $1 \geq f(T^{\mathrm{p}}) \geq 0$. Similarly, for the exponential functional relationship $f(T^{\mathrm{p}}) = \alpha \exp(\beta T^{\mathrm{p}})$, Eq. (4.6) yields

$$x = [1 - \alpha \exp\{\beta T^{\mathrm{s}}(x)T^{\mathrm{p}}/T^{\mathrm{s}}(x)\}]X \tag{4.8}$$

Using Eqs. (4.4) and (4.7), we obtain

$$T^{\mathrm{s}}(x) = \begin{cases} T^{\mathrm{th}}/[\exp(\{\alpha - 1 + (x/X)\}/\beta)], & 1 \geq \exp(\{\alpha - 1 + (x/X)\}/\beta) \geq T^{\mathrm{th}} \\ 1, & \text{otherwise} \end{cases} \tag{4.9}$$

for the logarithmic contact screen, and similarly using Eqs. (4.4) and (4.8), we get

$$T^{\mathrm{s}}(x) = \begin{cases} \beta T^{\mathrm{th}}/[(\ln\{1 - (x/X)\})/\alpha], & 1 \geq \exp \beta^{-1} (\ln\{1 - (x/X)\})/\alpha \geq T^{\mathrm{th}} \\ 1, & \text{otherwise} \end{cases} \tag{4.10}$$

for the exponential contact screen.

The aforementioned contact screen design was formulated with the unrealistic assumption that a hard-clipping recording medium is available for generating the halftone image. But, in practice, very few recording media have hard-clipping characteristics. Accordingly, the uncompensated contact screen, when used, produces gray regions in addition to the usual binary black and white. The precompensation design analysis that follows takes into consideration the actual response of the recording medium using both density and spatial modulations.

For simplicity, the design is limited to only nonsymmetric one-dimensional screens. Assume further that $T^{\mathrm{s}}(0) > T^{\mathrm{s}}(X)$. The optical energy incident on the recording medium at x is given by

$$E(x) = eT^{\mathrm{p}}T^{\mathrm{s}}(x) \tag{4.11}$$

where e is the light intensity incident on the continuous-tone input transparency. Let T^h be the transmittance of the resulting halftone image, which can be related to the incident energy by

$$T^h(E) = \begin{cases} 1, & \text{if } 0 \leq E \leq E_1 \\ 0, & \text{if } E_2 \leq E \leq \infty \end{cases} \tag{4.12}$$

The general nature of $T^h(E)$ is illustrated in Fig. 4.3. On the other hand, the average transmittance T^H of the halftone image may be defined as the average of the displayed transmittance over a single screen cell. T^H can thus be expressed in terms of T^h as

$$T^H = X^{-1} \int_0^X T^h(eT^p T^s(x))dx \tag{4.13}$$

The design problem therefore reduces to determining $T^s(x)$, provided the functional relationship $T^H(T^p)$ and the exact form of $T^h(E)$ are specified. Furthermore, the precompensated screen profile is expected to compensate for the nonbinary nature of the recording medium. The halftone image generated thereby is expected to be binary in nature. It will thus be possible to identify a specific transmittance value $T^{th,eq}$, referred to as the equivalent threshold transmittance, such that

$$T^h(x) = \begin{cases} 1, & \text{if } T^p T^s(x) < T^{th,eq} \\ 0, & \text{if } T^p T^s(x) \geq T^{th,eq} \end{cases} \tag{4.14}$$

Accordingly, T^H may be given by

$$T^H = \begin{cases} 1 - (x/X), & \text{if } T^p T^s(x) \geq T^{th,eq} \\ 1, & \text{if } T^p T^s(x) < T^{th,eq} \end{cases} \tag{4.15}$$

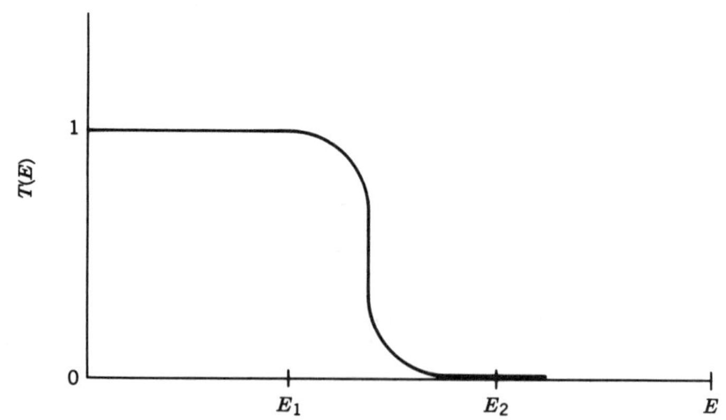

FIGURE 4.3 The general characteristics of T^h versus E.

Furthermore, the screen transmittance is expressed as

$$T^{s}(x) = \begin{cases} T^{\text{th,eq}}, & \text{if } T^{p} \geq T^{\text{th,eq}} \\ 1, & \text{if otherwise} \end{cases} \tag{4.16}$$

Equation (4.16) can thus be used to give

$$\begin{aligned} x \equiv \tau(T^{s}(x)) &= [1 - f(T^{p})]X \\ &= [1 - f(T^{\text{th,eq}}/T^{s}(x))]X \end{aligned} \tag{4.17}$$

for $T^{p} \geq T^{\text{th,eq}}$, where $T^{H} = f(T^{p})$ and τ denotes the inverse function of $T^{s}(x)$. Equation (4.17) can be solved to determine the screen transmittance after having identified $T^{\text{th,eq}}$ and the form of f.

We can rewrite Eq. (4.13) as

$$T^{H} = X^{-1}\left[\int_{0}^{x''} + \int_{x''}^{x'} + \int_{x'}^{X}\right] T^{h}(E)dx \tag{4.18}$$

where

$$eT^{p}T^{s}(x') = E_{1} \tag{4.19}$$

and

$$eT^{p}T^{s}(x'') = E_{2} \tag{4.20}$$

Using Eq. (4.12), we find

$$\int_{0}^{x''} T^{h}(E)dx = 0 \tag{4.21}$$

and

$$\int_{x'}^{X} T^{h}(E)dx = X - x' \tag{4.22}$$

Equation (4.18) therefore reduces to

$$T^{H} = X^{-1}[X - \tau((E_{1})/(eT^{p}))] + X^{-1}\int_{x''}^{x'} T^{h}(E)dx \tag{4.23}$$

Substituting $\Gamma = T^{s}(x)$, we can evaluate the last term of Eq. (4.23) by the method of parts as follows:

$$X^{-1}\int_{x''}^{x'} T^{h}(E)dx = X^{-1}\int_{T^{s}(x'')}^{T^{s}(x')} T^{h}(E)\tau'(\Gamma)d\Gamma$$
$$= X^{-1}\left[\tau\left(\frac{E_1}{eT^{p}}\right) - \int_{T^{s}(x'')}^{T^{s}(x')} eT^{p}T^{h\prime}(E)\tau(\Gamma)d\Gamma\right] \tag{4.24}$$

where $\tau'(\Gamma) = d\tau/d\Gamma$ and $T^{h\prime}(E) = dT^{h}/d\Gamma$. Therefore Eq. (4.23) simplifies to

$$T^{H} = X^{-1}\left[X - \int_{T^{s}(x'')}^{T^{s}(x')} eT^{p}T^{h\prime}(E)\tau(\Gamma)d\Gamma\right] \tag{4.25}$$

$T^{th,eq}$ can now be obtained by introducing Eq. (4.17) into Eq. (4.24). As a first example, let us consider the case of linear mapping $T^{H} = \alpha - \beta T^{p}$, where α and β are two constants such that, for all T^{p}, $1 \geq T^{H} \geq 0$. Accordingly, Eq. (4.24) can be written

$$T^{H} = \alpha - \beta T^{p} = \alpha - \beta T^{th,eq}\, eT^{p}\int_{T^{s}(x'')}^{T^{s}(x')} \{(T^{h\prime}(E))/\Gamma\}d\Gamma \tag{4.26}$$

For linear mapping we can therefore conclude that

$$T^{th,eq} = \left[-\int_{T^{s}(x')}^{T^{s}(x'')} \{(eT^{h\prime}(E))/\Gamma\}d\Gamma\right]^{-1}$$
$$= \left[-\int_{E_1}^{E_2} \{(eT^{h\prime}(E))/E\}dE\right]^{-1} \tag{4.27}$$

and, consequently, the corresponding precompensated screen transmittance is obtained from Eq. (4.17) as

$$T^{s}(x) = \beta\left\{-\int_{E_1}^{E_2} [(eT^{h\prime}(E))/E][\alpha - 1 + (x/X)]\right]^{-1} \tag{4.28}$$

Similarly, for the logarithmic mapping $T^{H} = a - b\ln(T^{p})$, the equivalent threshold transmittance and the precompensated screen transmittance are given respectively, by

$$\ln(T^{th,eq}) = \int_{T^{s}(x')}^{T^{s}(x'')} eT^{p}T^{h\prime}(E)\ln(E/e)d\Gamma \tag{4.29}$$

and

$$T^{s}(x) = \exp\left[-\int_{E_1}^{E_2} T^{h\prime}(E)\ln(E/e)dE - b\{a - 1 + (x/X)\}\right] \quad (4.30)$$

The general nature of Eqs. (4.28) and (4.30) can now be utilized to yield known special cases.

For a familiarly assumed hard-clipping recording medium, the halftone transmittance is given by

$$T^{h}(E) = \begin{cases} 1, & \text{if } E < E^{th} \\ 0, & \text{if } E \geq E^{th} \end{cases} \quad (4.31)$$

where E^{th} is the threshold energy of exposure such that the threshold transmittance $T^{th} = E^{th}/e$. The incorporation of Eq. (4.30) into Eq. (4.27) gives us

$$T^{th,eq} = T^{th} \quad (4.32)$$

for the linear mapping, and therefore

$$T^{s}(x) = \beta T^{th}[\alpha - 1 + (x/X)]^{-1} \quad (4.33)$$

which, for the special case of $\alpha = 0$ and $\beta = -1$, reduces to

$$T^{s}(x) = T^{th}/[1 - (x/X)] \quad (4.34)$$

For the finite-gamma recording medium,

$$T^{h}(E) = \begin{cases} (E/E^{th})^{-g}, & E^{th} \leq E \\ 1, & E < E^{th} \end{cases} \quad (4.35)$$

Precompensated screen transmittance for a linear mapping therefore results in

$$T^{s}(x) = [(g + 1)/g]T^{th}/\{1 - (x/X)\} \quad (4.36)$$

The screen criterion that follows Eq. (4.28) is the most accommodating and is highly precompensating for the true nature of the recording medium, while that following Eq. (4.36) is good only for those recording media whose H-D curves can be approximated by straight lines. Again, Eq. (4.34), which is a special case of Eq. (4.28), represents the transmittance profile of uncompensated screens. All these equations describing the design of nonsymmetric screens are equally applicable for the design of symmetric screens, provided the term $[1 - (x/X)]$ has been replaced with $[1 - (2x/X)]$.

4.3 NONLINEAR OPTICAL PROCESSING

In the previous section, we introduced how a halftone screen can be used to realize pulse-width or pulse-area modulation of an input image. We also identified the transmittance characteristics of the contact screens necessary for realizing arbitrary monotonic transformation. Again, in Chapter 3, we introduced the filters and their usage in optical information processing. As far as optical information processing systems are concerned, it should be borne in mind that up to this point nothing has been done directly to the input image to affect it before it was introduced to the optical system. In this section we explore ideas pertaining to nonlinear optical information processing with the use of contact screens, which are used to preprocess the input image. Accordingly, the intensity at any point of the output plane of the optical information processing system shown in Fig. 4.1 depends on spatial filtering as well as on contact screen preprocessing of the input.

While the halftone technique can achieve a multitude of nonlinear transformations, in this section we focus our attention on a particular optical processing system used to separate noise that is multiplicatively combined with the signal. In many imaging cases, such as x-ray radiograms and images with speckle noises, the images are multiplied by noise. The removal of noise from these images is often desirable or necessary. Logarithmic filtering has been used for the separation of optical images that are multiplied by or convolved with noise in linear homomorphic electronic signal-processing systems in which both the image and noise are assumed to be separated in the spatial frequency domain.

Linear filtering systems are often sufficient to separate a signal and noise that are added to one another. But when the signal and noise are nonadditively combined, a linear system cannot be used to separate them. If the signal is multiplied to noise, the spectra of the signal and noise remain entwined through a convolution operation. This makes signal extraction in a linear system very difficult, if not impossible. This dilemma is resolved in a digital processor by generalizing the concept of a linear system to include a broader class of operations. The generalized system is called a homomorphic system. A system characterized by the operator H is said to be homomorphic if

$$H\{f_1(x) \blacksquare f_2(x)] = H\{f_1(x)\} \blacksquare H\{f_2(x)\} \tag{4.37}$$

and

$$H\{c \cap [f_1(x) \blacksquare f_2(x)]\} = c \cap H\{f_1(x) \blacksquare f_2(x)\} \tag{4.38}$$

for any two arbitrary signals $f_1(x)$ and $f_2(x)$ where $\blacksquare$ denotes a to-be-defined rule of combination and $\cap$ denotes a scalar multiplication. Specifically when $\blacksquare$ represents an addition and $\cap$ represents simple multiplication, the homomorphic system reduces to a conventional linear system. Furthermore, a homomorphic system also includes a system that is invertible, that is, a system for which output is caused by only one input. An invertible system with transformation $\ddot{A}$ and an inverse transformation $\ddot{A}^{-1}$ has the following properties:

$$\ddot{A}\{\ddot{A}^{-1}[f_1(x)] \blacksquare \ddot{A}^{-1}[f_2(x)]\} = f_1(x) \blacksquare f_2(x) \tag{4.39}$$

and

$$\ddot{A}\{c \cap \ddot{A}^{-1}[f(x)]\} = c \cap f \tag{4.40}$$

With a suitable choice of these operations, the system satisfies the generalized principle of superposition.

A homomorphic system can be used to treat the problem of noise T^{n} when it is multiplied with the signal transmittance T^{p}. In particular, the homomorphic system can be represented by a cascade of three operations, as illustrated in Fig. 4.4. The first phase involves transforming a corrupted signal into additive components $\ln(T^{\text{p}})$ and $\ln(T^{\text{n}})$. The second phase involves separating the signal and noise by means of spatial filtering at the Fourier domain. Thereafter, in the third phase, an exponential contact screen processing is used to restore the signal.

The input image $T^{\text{p}} T^{\text{n}}$ can be transformed into a logarithmic halftone image by means of a logarithmic contact screen. Provided the halftone image is placed at the input plane of the optical information processing system of Fig. 4.1, the spectra of $\ln(T^{\text{p}})$ and $\ln(T^{\text{n}})$ appear separated at the Fourier plane. The noise contribution can be spatially filtered and the image recovered at the output can subsequently be processed with an exponential contact screen for restoring the uncorrupted signal. However, if the halftone preprocessing were not done, the Fourier transform operation would have resulted in

$$F\{T^{\text{p}} T^{\text{n}}\} = F\{T^{\text{p}}\} * F\{T^{\text{n}}\} \tag{4.41}$$

where $F\{\cdot\}$ represents the Fourier transform operation and $*$ denotes a convolution operation. It is thus not possible to single out any noise component in the case of linear filtering.

In the following, we discuss the difference between logarithmic and linear filtering in more detail. Consider for simplicity that $T^{\text{n}}(x, y) = T^{\text{n}}(x)$. Obviously,

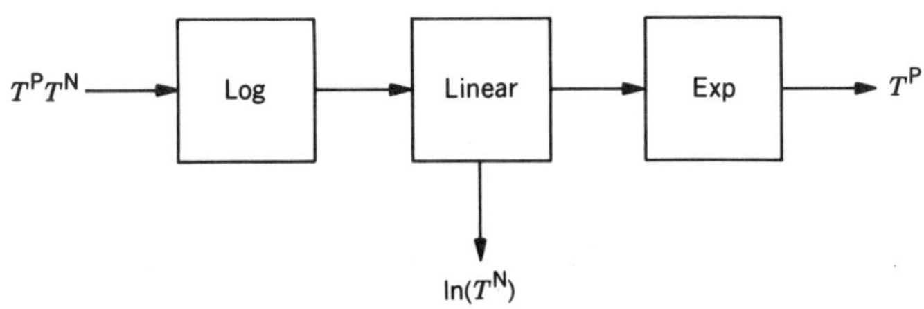

FIGURE 4.4 A typical homomorphic processing operation.

this is not always an oversimplification. For example, in biomedical x-ray imagery, a one-dimensional grid made of lead strips is often used in front of the film-screen cassette to absorb scattered radiation. Such absorbing structures introduce one-dimensional, gridlike multiplicative noise in the radiograph. Provided the noise transmittance function T^n of period X satisfies the conditions for Fourier series expansion, it can be represented by

$$T^n = T_0^{n1}/2 + \sum_{k=-\infty}^{\infty} [T_k^{n1}\cos(kx) + T_k^{n2}\sin(kx)] \tag{4.42}$$

where

$$T_k^{n1} = \frac{2}{X}\int_{-X/2}^{X/2} T^n(x)\cos(2\pi kx/X)dx \tag{4.43}$$

and

$$T_k^{n2} = \frac{2}{X}\int_{-X/2}^{X/2} T^n(x)\sin(2\pi kx/X)dx \tag{4.44}$$

A logarithmic contact screen of period W can be designed, in accordance with Eq. (4.30), such that the average transmittance of the logarithmic halftone image T^H is related to the input transmittance $T(x, y)$ by

$$T^H = [\ln(T) - \ln(T_{min})]/[\ln(T_{max}) - \ln(T_{min})] \tag{4.45}$$

where T_{max} and T_{min} are the maximum and minimum values of $T(x, y)$. For the purpose of generality, we consider that the parallel cell lines of the contact screen are directed along the w direction, which may not necessarily be along either the x or y direction:

$$F(T^H) = F(T^{HP}) + 2\sum_{l=-\infty}^{\infty}\int_{-\infty}^{\infty} T^{HP}(\alpha)\,\mathrm{sinc}(lT^{HP}(\alpha))\cos(2\pi l u_{0w}\alpha) \cdot \exp(-j2u_w\alpha)d\alpha + a_0\delta(u_x, u_y) + \sum_{m=-\infty}^{\infty}(a_m + jb_m)\delta(u_x - mu_{0x}, u_y) \tag{4.46}$$

where

$$a_m = \frac{2}{X}\int_{-X/2}^{X/2} [\ln(T^n(x))/\sigma]\cos(2\pi mx/X)dx \tag{4.47}$$

$$b_m = \frac{2}{X}\int_{-X/2}^{X/2} [\ln(T^n(x))/\sigma]\sin(2\pi mx/X)dx \tag{4.48}$$

$$\sigma = \ln(T_{\text{max}}) - \ln(T_{\text{min}}) \tag{4.49}$$

$$u_{0w} = W^{-1} \tag{4.50}$$

$$u_{0x} = X^{-1} \tag{4.51}$$

$$\text{sinc}(\phi) = \sin(\pi\phi)/\pi\phi \tag{4.52}$$

where u_p is the spatial frequency corresponding to the p axis and T^{HP} is the average transmittance of the logarithmic halftone image obtained using an input of transmittance $T^{\text{p}}(x, y)$. For the case of linear filtering, however, the Fourier transform of the input transmittance $T(x, y)$ becomes

$$F(T^{\text{p}}T^{\text{n}}) = \tfrac{1}{2}\, T_0^{\text{n1}}\, \Gamma(u_x, u_y) + \sum_{k=1}^{\infty} (T_k^{\text{n1}} + jT_k^{\text{n2}})\Gamma(u_x - ku_{0x}, u_y) \tag{4.53}$$

where

$$\Gamma = F(T^{\text{p}}) \tag{4.54}$$

At the Fourier plane, if only the zero-order spectral dot is allowed to pass through a pinhole spatial filter (i.e., when $m = n = 0$), Eq. (4.46) reduces to

$$F(T^{\text{H}})|_{m=n=0} = F(T^{\text{HP}}) + \tfrac{1}{2}\, a_0\delta(u_x, u_y) \tag{4.55}$$

for logarithmic spatial filtering. The corresponding output image has a transmittance proportional to

$$F^{-1}\{F\{T^{\text{H}}\}|_{m=n=0}\} = T^{\text{HP}}(-x, -y) + a_0/2 \tag{4.56}$$

It can be seen that the output transmittance is proportional to the sum of a constant bias term $a_0/2$ and the logarithm of T^{p}. On the other hand, for linear filtering, the zero-order term in Eq. (4.53) is simply

$$F(T^{\text{p}}T^{\text{n}})|_{n=0} = \tfrac{1}{2}\, T_0^{\text{n1}}\Gamma(u_x, u_y) \tag{4.57}$$

The output image accordingly has a transmittance proportional to

$$F^{-1}\{F\{T^{\text{p}}T^{\text{n}}\}|_{n=0}\} = T^{\text{p}}(-x, -y)T_0^{\text{n1}}/2 \tag{4.58}$$

It can be seen that the output transmittance is still proportional to the input signal but is reduced by a factor of $T_0^{\text{n1}}/2$. Clearly, the linearly filtered output is different from the logarithmically filtered output. However, the comparison of the two processes in terms of only zero-order is too idealistic. In practice, it is not easy to isolate the zero-order spectral dot by a pinhole spatial filter of finite aperture. In realistic cases, higher-order terms (say, first orders) cannot totally be eliminated. If, for simplicity, only first orders are present along with the zero-order spectral dot, then the reconstructed transmittance of the output image is proportional to

$$F^{-1}\{F\{T^{\mathrm{H}}\}|_{m=0;n=0,1,-1}\} = T^{\mathrm{HP}}(-x, -y) + a_0/2 + a_1 + jb_1 \quad (4.59)$$

for logarithmic filtering, and that for the linear filtering is given by

$$F^{-1}\{F\{T^{\mathrm{p}}T^{\mathrm{n}}\}|_{n=0,1,-1}\} = T_0^{\mathrm{n}1}T^{\mathrm{p}}(-x, -y)/2 + (T_1^{\mathrm{n}1} + jT_1^{\mathrm{n}2})T^{\mathrm{p}}(-x - X, -y) \quad (4.60)$$

A comparison of Eqs. (4.59) and (4.60) reveals that the logarithmically filtered output contains the logarithm of the signal and a complicated bias term, but the linearly filtered output is proportional to the signal transmittance overlapped with a reduced and displaced duplicate of itself. A similar phenomenon exists if additional higher orders are included in the output. The bias present in the logarithmically filtered output tends to add a constant glare to the logarithm of the signal. On the other hand, the linearly filtered output corresponds to a blurred signal because the signal overlaps with its own displaced replicates. With additional spectral dots, however, the linearly filtered output image deteriorates further because of repeated blurring.

Figure 4.5 shows a comparison of the results of the logarithmically filtered and the homomorphically filtered output of an image multiplied by noises of three spatial frequencies. Both logarithmic and linear filtering are generally successful in removing high-frequency noises. But for noise of frequency less than 0.5 lines/mm, the linearly filtered output usually remains blurred, as expected. For example, consider Figs. 4.6a and 4.6b, which show, respectively, the Fourier spectra of the logarithmic and the corresponding linear processing of an image that has been multiplied with Ronchi-type noise. The figures (from top to bottom) correspond to three different noise frequencies, 2.0, 1.0, and 0.5 lines/mm. The logarithmically obtained spectra can be seen to have additive components, while those corresponding to linearly processed spectra are seen to be localized along one particular frequency axis. In the former, signal information shows up along the vertical axis and the noise information is constrained along the horizontal axis. It can be seen that, for both cases, the noise frequency components along the horizontal axis are located closer to one another with decreasing noise frequency.

It is important to point out that the setup of Fig. 4.1 uses photographic film. Consequently, the filtering process is not only time consuming but also difficult to control because of the clumsy development process involved. Alternatively, however, a real-time recording device such as the liquid crystal light valve (LCLV) can be used to replace the photographic medium.

In this section, we have already used halftone screens for realizing nonlinear optical processing. There is another valid technique, referred to as theta modulation, that can be used for nonlinear filtering operations. In this scheme, illustrated

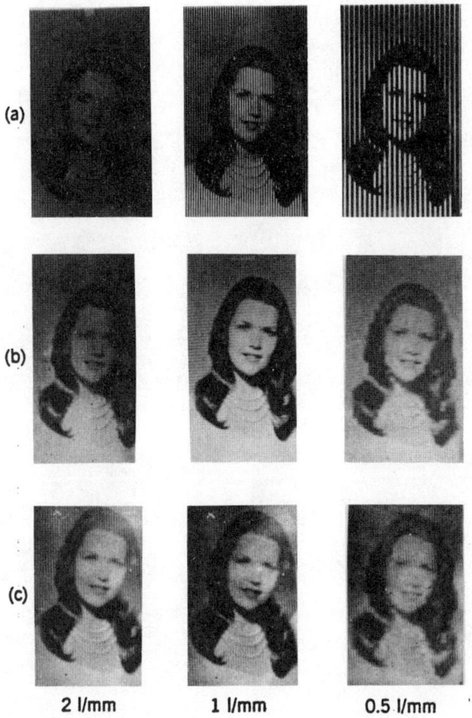

FIGURE 4.5 (a) Input image multiplied with noise; (b) the corresponding logarithmically filtered output; and (c) the corresponding homomorphically processed output.

in Fig. 4.7, the input to be processed is modulated with grating structures. Each particular gray level of the input image is modulated by a grating with a unique angular orientation. When the preprocessed input is introduced into the system, a number of grating orders can be observed at the Fourier domain. Since different gray levels have been modulated by gratings of different angular orientation, each of the Fourier domain grating orders can be associated with a particular gray level. Spatial filtering each diffraction order by a predetermined amount results in an output image with a different set of gray levels. This operation is equivalent to a nonlinear transformation of the gray levels. The technique is rather general enough that it can be used for both monotonic and nonmonotonic nonlinear operations. The scheme, however, lacks a simple easy-to-implement method for preprocessing of the input images.

4.4 ARITHMETIC OPERATIONS

The primary objective of optical computing systems is to make use of photons, instead of electrons, in logical and arithmetic computations. Not surprisingly, the

FIGURE 4.6 Fourier plane corresponding to (a) logarithmic filtering and (b) linear filtering.

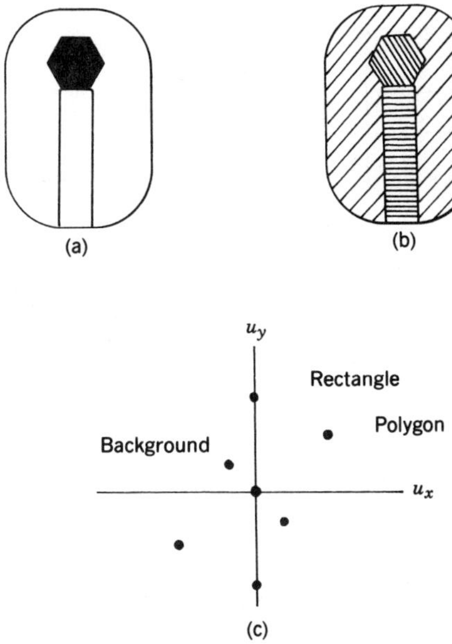

FIGURE 4.7 Theta modulation: (a) input, (b) preprocessed input, and (c) the corresponding Fourier domain characteristics.

Fourier-based analog optical systems are capable of performing selected logic and/or arithmetic operations. Accordingly, in this section, we explore some of these analog optical schemes. It may be noted that optical sources have a large time bandwidth that approaches nearly 1 GHz for some of the laser diodes. In addition, the optical systems provide a large space-bandwidth product.

4.4.1 Addition/Subtraction

Consider two two-dimensional image inputs described by $f_1(x, y)$ and $f_2(x, y)$ that are to be added to one another. An experimental setup such as the one shown in Fig. 4.8 can be used for this purpose. The main light beam is split into two portions by means of a beam splitter. The two beams are subsequently recombined by means of two suitably placed mirrors and a second beam splitter. The two-dimensional inputs are introduced in these two beams. In the absence of any phase shifter, the output corresponds to the sum of the two input images. Addition can also be achieved in the presence of a phase shifter provided the path difference introduced by the phase shifter is an even multiple of the half-wavelength. However, if the phase shifter could be adjusted to provide a path difference equal to odd multiples of the half-wavelength, the output would correspond to the difference of the inputs. Alternatively, this addition/subtraction system can be realized using the system shown in Fig. 4.9. The input images are introduced at the object plane of the coherent adder/subtracter unit. The holographic lens shears and recombines the images at the output. To realize the subtraction operation, the shear lens must be tilted at an angle so that the two optical paths differ by odd multiples of the half-wavelength. This latter system is different from the first in the sense that the two

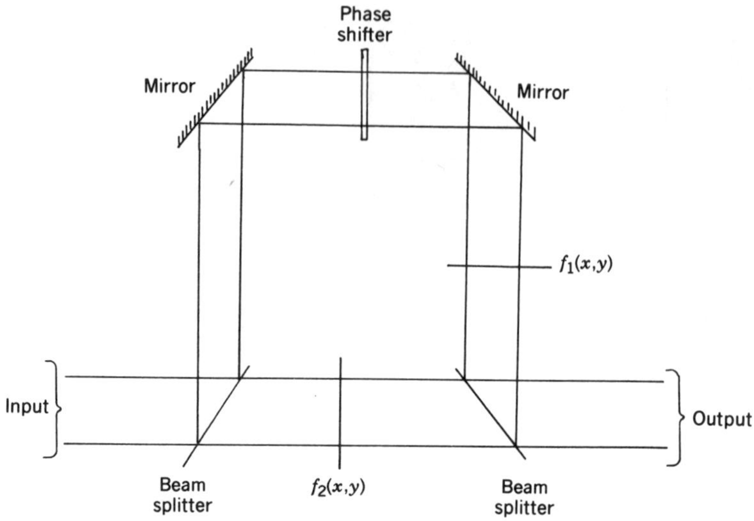

FIGURE 4.8 An optical adder/subtracter using a phase shifter.

images are introduced side by side. Another variation of the system is shown in Fig. 4.10. The input images are illuminated by white light. A red filter is placed next to the upper input image, and a green filter is placed next to the lower input image. With proper alignment of the Wollaston prism, the input images can be combined to produce a pseudocolor difference image at the output. The location(s) where the two images have identical transmittances turn(s) out to be yellow at the output. A red region in the output indicates that the upper image has a higher transmittance at that location than that of the lower image. A green output indicates just the opposite.

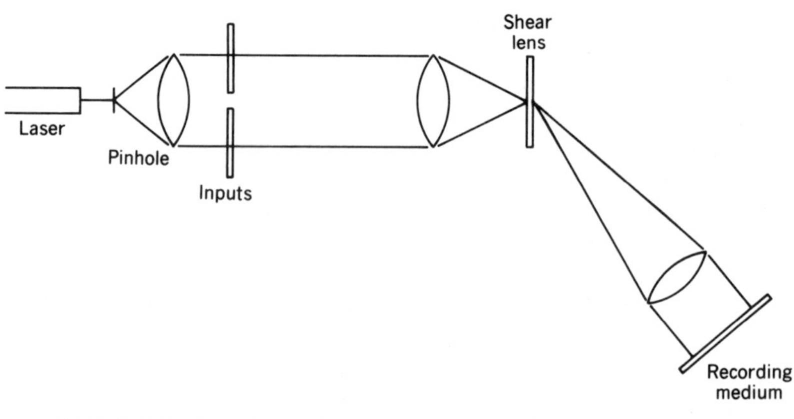

FIGURE 4.9 An optical adder/subtracter using a holographic shear lens.

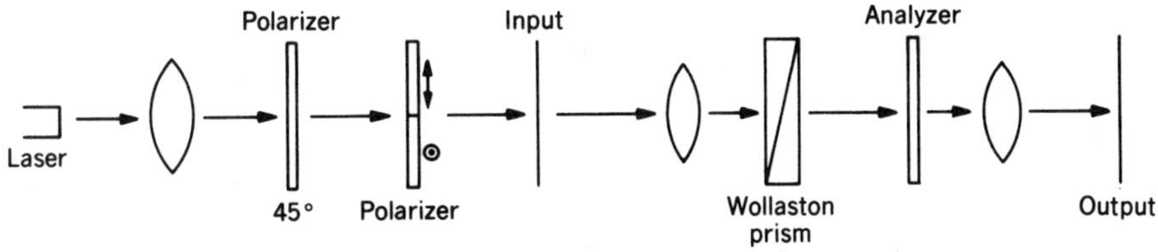

FIGURE 4.10 An optical adder/subtracter using a Wollaston prism.

4.4.2 Multiplication and Division

Multiplication is the simplest of all operations. For input images placed in contact, the transmittance of the combination at any arbitrary point is nothing but the product of the individual transmittances at the same point. However, when images are imprinted on glass plates, they often may not all be placed in contact. Again, even if it were possible, glass plates when slid against each other usually damage easily. For situations such as these, a system like the one in Fig. 4.11 can be used. The lens combination of L_2 and L_3 causes the projection of an inverted real image of $f_1(x, y)$ on $f_2(x, y)$. Finally, the fourth lens projects the product of the two at the output plane. It should be pointed out that both addition and multiplication can be realized using holographic means since signals proportional to the sum and product are both present in the holographic expression of Eq. (3.23).

The division operation can be expected to use the same setup as that of Fig. 4.11. However, for division, $[f_2(x, y)]^{-1}$ should be introduced at the second object plane instead of $f_2(x, y)$. Accordingly, the problem of division reduces to a problem of how to obtain the reciprocal of the transmittance function $f_2(x, y)$. Interestingly, the inverse $f_2(x, y)$ can be expressed as $|f_2(x, y)|^{-2}\{f_2(x, y)\}^*$. This implies that the inverse is obtainable by multiplying $|f_2(x, y)|^{-2}$ and $\{f_2(x, y)\}^*$. The first factor is, however, realizable by a simple photographic process by selecting a film of gamma 2. Again, $\{f_2(x, y)\}^*$ can be realized taking a hologram of $f_2(x, y)$. A multiplication unit such as the one just described can be used to realize $\{f_2(x, y)\}^{-1}$, while a second such unit can provide the result of division.

A class of processors, referred to as discrete, can be generated using the classical optical information processor. These systems, consisting of cylindrical lenses,

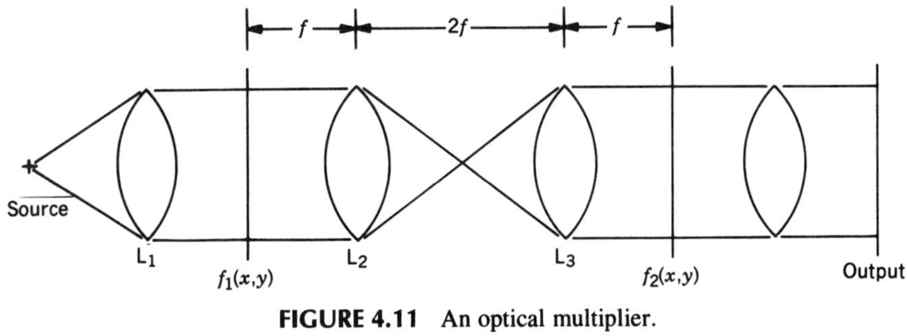

FIGURE 4.11 An optical multiplier.

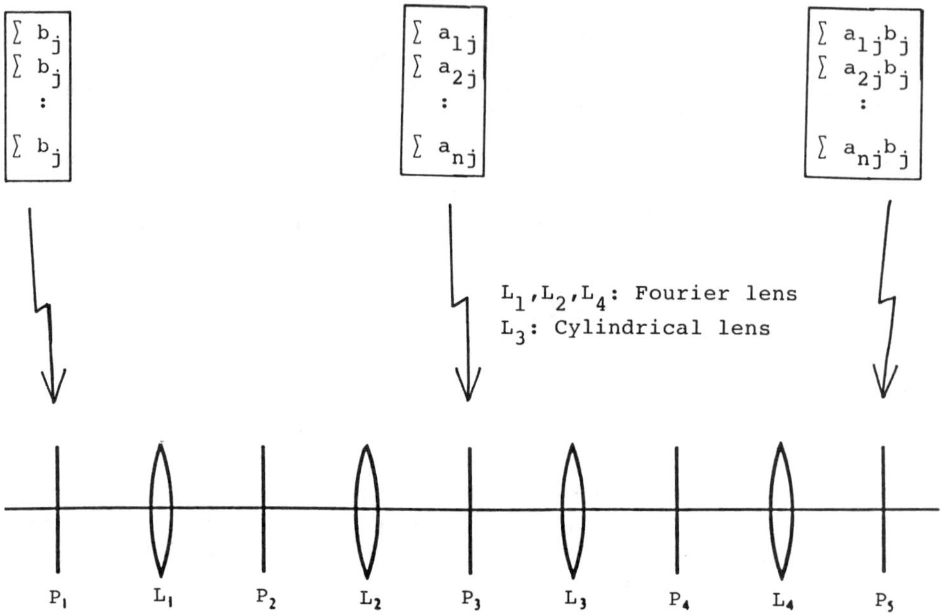

FIGURE 4.12 A parallel matrix-vector multiplier.

are capable of performing sampled operations. While these operations are discrete, they are analog in nature and are subject to the standard limitations on accuracy and dynamic range constraints. Cylindrical lenses that have finite radii of curvature along one of the axes and infinite radii of curvature along the transverse axis are remarkably perfect for simultaneous multiple one-dimensional operations. Consider the matrix-vector multiplier shown in Fig. 4.12, in which the pixels along the vertical columns of the image introduced at P_1 have identical transmittance values. The information of plane P_1 thus corresponds to that of a vector, say, B (of n elements). These columns of light can be superimposed on a matrix mask located at P_3. The information of the matrix mask may correspond to that of a matrix, say, A (of $n \times n$ size). The product C of the stored matrix and the input is then Fourier transformed along only one of the directions by the cylindrical lens L_3, which is then Fourier transformed by the biconvex lens L_4. Consequently, at the output plane P_5, we obtain the column vector of the product C such that its mth row consists of the sum of products $\Sigma a_{mj} b_j$. An array of n detectors placed at the output can pick up the results of this matrix-vector multiplication. While in this chapter the inputs are taken to be fixed, in a subsequent chapter we discuss more about this class of discrete processors to include input elements that are introduced sequentially into the system.

4.4.3 Averaging

Consider the situation where a mask made of an array of pinholes is introduced at the Fourier plane of the system shown in Fig. 4.1. Such linear mask characteristics can be described by a comb function formed of a series of delta functions as

$$\mathrm{Comb}(x/a) = \sum_{n=-\infty}^{\infty} \delta((x/a) - n) = a \sum_{n=-\infty}^{\infty} \delta(x - an) \qquad (4.61)$$

where a is the spacing between the delta functions. Provided an image P has been introduced at the input plane, the system output will be found to contain a large number of relatively displaced input images, as shown in Fig. 4.13a. Figure 4.13b shows the output when two closely positioned images, say, P and Q, were introduced simultaneously at the input. By having varied the separation between P and Q, it is possible to have the output P exactly overlap the adjacent Q output. If the input image Q were now to be replaced by image P, the system output would be equivalent to having P output overlapped with the neighboring P output, as shown in Fig. 4.13c. This concept can be extended to include a nontrivial N number of images at the input while the linear mask can be replaced by a mask having a two-dimensional array of pinholes. The system output will have an array of images with each image, except those on the extreme edges, formed due to addition of

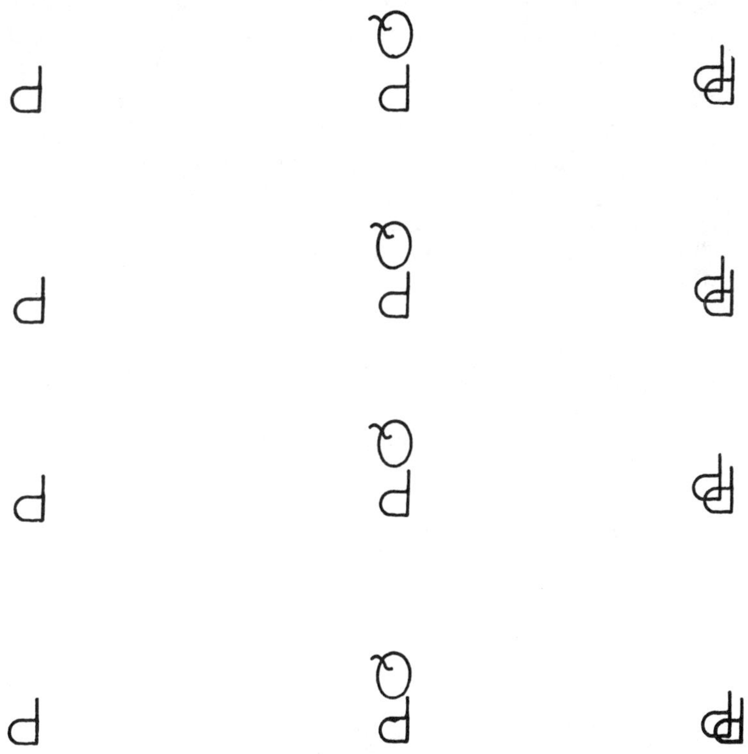

(a) (b) (c)

FIGURE 4.13 Optical system having an array of pinholes at the Fourier plane with (a) single input P, (b) side-by-side input P and Q, and (c) side-by-side input P and P.

the input images. The corresponding output will thus correspond to having taken an arithmetic average of the input images.

4.4.4 Differentiation and Integration

Consider that a one-dimensional input image $g(x)$, for simplicity, is introduced at the input of the optical information processing system shown in Fig. 4.1. The system can be made to take the differentiation or integration of the input, provided an appropriate filter has been introduced at the Fourier plane. In this section, we first consider the case of differentiation with $g^n(x)$ representing the nth derivative with respect to x. The corresponding complex amplitude distribution just beyond the Fourier plane is given by

$$\begin{aligned} T_f(x) &= \int \left(\frac{d^n g(x')}{dx'^n} \right) e^{-j2\pi x' x} dx' \\ &= F\{g(x)\}(j2\pi x)^n. \end{aligned} \tag{4.62}$$

Since $F\{g(x)\}$ is already available on the input side of the filter, the filter function must generate $(j2\pi x)^n$. However, the filter function values range from negative to positive, including zero. Generating a zero value is rather impossible, as that would imply that the filter material would have to have infinite density characteristics. This problem is generally overcome by inserting a blocking filter over the low-frequency components. And, to accommodate the negative values, a phase plate with step characteristics can be used at the Fourier plane.

Figure 4.14 shows the overall scheme for realizing this differentiation filter. Accordingly, except for those frequencies that are covered by the blocking filter, spatial differentiation occurs over the input image information. It should be noted that the differentiation operation when conducted on a two-dimensional input image is able to extract its edges.

The integration operation can likewise be realized. Consider, for example, a one-dimensional filter mask whose transmittance is given by

$$T_f(x) = 2a \operatorname{sinc}(2\pi a x / lf) \tag{4.63}$$

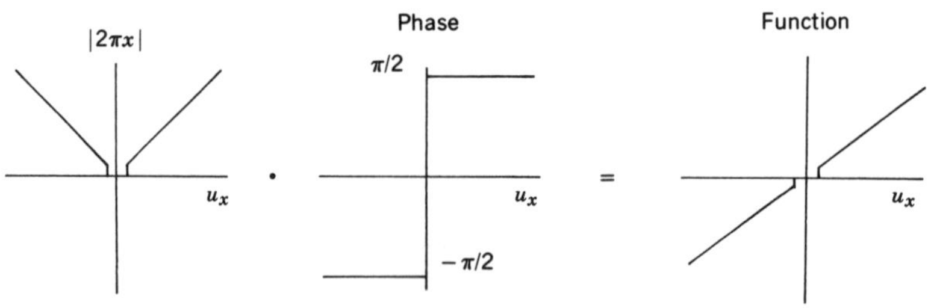

FIGURE 4.14 Realization of a differentiation filter.

where l is the wavelength of light and f is the focal length of the lens. Such a sinc-function-like transmission mask can be realized by optically Fourier transforming a slit, while the negative lobes of the sinc function can be generated using a phase filter. Provided an input image $g(x)$ is introduced at the input plane of the system shown in Fig. 4.1, the system output is found to be given by

$$\begin{aligned} O(x) &= F\{F\{g(x)\}T_f(x)\} \\ &= \frac{\pi}{lf}\left[\int_{-\infty}^{x+a} g(\alpha)d\alpha - \int_{-\infty}^{x-a} g(\alpha)d\alpha\right] \end{aligned} \tag{4.64}$$

Accordingly, the output image consists of both the integral of the input image and a displaced negative integral of the input image. The separation between these two simultaneously occurring output images is determined by the width of the slit used in generating the filter mask.

4.5 PROBLEMS

1. Show that the contact screen characteristics given by

$$T^{s}(x) = T^{th}\{\beta[\{g/(g + n)\}\{\alpha - 1 + (x/X)\}]^{-1}\}^{1/n}$$

satisfy the nonlinear mapping $T^{H} = \alpha - \beta(T^{P})^{n}$, where α and β are two constants such that for all T^{P}, $1 \geq T^{H} \geq 0$, X is the cell period of the screen, T^{th} is the threshold transmittance, and g is the film gamma.

2. For a Ronchi-ruling type one-dimensional noise given by

$$T^{n}(x) = \text{rect}(2x/X) * \text{comb}(x/X)$$

where rect(z) is equal to (a) a when $|z| \leq \frac{1}{2}$ and (b) b when $|z| > \frac{1}{2}$, show that the logarithmically processed output (when both zeroth and first orders are transmitted) has a dc bias, which corresponds to the term $[(\pi + 4)\ln(a) + (\pi - 4)\ln(b)]/2\pi\sigma$, where $1 \geq a \geq b \geq 0$.

3. Show that the homomorphically processed output corresponding to the signal that is multiplied with the Ronchi-ruling type noise transmittance (as described in Problem 2) is indeed proportional to the signal transmittance when the pinhole spatial filtering allows both zeroth and first orders to pass through.

4. Consider a filter mask whose amplitude transmittance is of the form $T_f(x) = A + Bx$, where A and B are controllable constants. Show that this filter can be used to differentiate an input image and that the resultant output corresponds to the superposition of the derivative and the input image.

5. By combining shifted cosine gratings, it is possible to generate a filter mask whose amplitude transmittance is given by $T_f(x) = 1 + \sigma x \sin(nx)$, where $\sigma \rightarrow$

0. Show that this filter mask is able to extract edges of any given image along a direction orthogonal to x.

6. Show that a filter mask whose transmission varies inversely with distance is able to integrate a two-dimensional image introduced at the input plane of Fig. 4.1.

7. Verify Eq. (4.64).

4.6 SUGGESTED READINGS

Armand, A., A. A. Sawchuk, and T. C. Strand, Nonlinear Optical Processing with Halftones: Accurate Predictions for Degradation and Compensation, *Appl. Opt.*, **26,** 1007-1014 (1987).

Armitage, J. D., and A. W. Lohmann, Theta Modulation in Optics, *Appl. Opt.*, **4,** 399-403 (1965).

Dashiell, S. R., and A. A. Sawchuk, Nonlinear Optical Processing: Analysis and Synthesis, *Appl. Opt.*, **16,** 1009–1015 (1977).

Eu, J. K. T., C. Y. C. Liu, and A. W. Lohmann, Spatial Filters for Differentiation, *Opt. Commun.*, **9,** 168–171 (1971).

Goodman, J. W., Operations Achievable with Coherent Optical Information Processing Systems, *Proc. IEEE*, **65,** 29–38 (1977).

Karim, M. A., Fabrication of Precompensated Contact Screens for Finite-Gamma Recording Media, *J. Appl. Photogr. Eng.*, **9,** 100–104 (1983).

Karim, M. A., Realization of Precompensated Monotonic Contact Screens, *Opt. Lett.*, **9,** 527–529 (1984).

Karim, M. A., *Electro-optical Devices and Systems*, PWS-Kent Publishing, Boston, 1990.

Karim, M. A., and H. K. Liu, Linear Versus Logarithmic Spatial Filtering in the Removal of Multiplicative Noises, *Opt. Lett.*, **6,** 207–209 (1981).

Karim, M. A., and H. K. Liu, All-Optical Homomorphic Image-Processing System, *Opt. Lett.*, **7,** 371-373 (1982).

Lizuka, L., *Engineering Optics*, Springer-Verlag, Berlin, 1985.

Peng, F., M. A. Karim, and A. A. S. Awwal, Halftone Image Generation Using Two Carrier Periods, *Opt. Commun.*, **84,** 242–246 (1991).

Reynolds, G. O., J. B. DeVelis, G. B. Parrent, Jr., and B. J. Thompson, *The New Physical Optics Notebook: Tutorials in Fourier Optics*, SPIE, Bellingham, WA, 1989.

Yao, S. K., and S. H. Lee, Spatial Differentiation and Integration by Coherent Optical Correlation Method, *J. Opt. Soc. Am.*, **61,** 474–477 (1971).

Yu, F. T. S., *Optical Information Processing*, Wiley, New York, 1983.

CHAPTER FIVE

Recognition Using Analog Optical Systems

5.1 INTRODUCTION

The previous two chapters have dealt with the computing applications of coherent optical processors, particularly in the areas of spatial filtering, nonlinear operations, and arithmetic. Besides these applications, optical processors are capable of performing another major computing task that is expected to be tied with many of the digital optical computing schemes. In a following chapter, for example, we shall be introducing a digital optical scheme, referred to as symbolic substitution, that is based on the capability of an optical system to recognize two-dimensional input patterns. The Fourier-based coherent optical processors are remarkably suitable for realizing this very important computing task. It must be emphasized that optical pattern recognition activity, though expected to directly impact optical computing, had its own driving motivation, namely, the improvement of target acquisition schemes.

In this chapter, we introduce the key concepts pertaining to optical pattern or character recognition. Several important character recognition filters, such as matched filter, joint Fourier transform filter, phase-only filter, and amplitude-modulated phase-only filter, are introduced. A special class of pattern recognition based on the Mellin transform is also discussed for overcoming scale and rotation issues.

5.2 MATCHED FILTER

Pattern or character recognition basically involves detecting the presence of a target $h(x, y)$ in the input. Consider the function given by

$$\Omega = F\{f(x, y)\}[F\{h(x, y)\}]^* \tag{5.1}$$

The Fourier transform of this composite function is obtained as follows:

$$
\begin{aligned}
F\{\Omega\} &= \int\int F\{f(x, y)\}[F\{h(x, y)\}]^* \exp[-j2\pi(u_x x + u_y y)]du_x\, du_y \\
&= \int\int h^*(\sigma, \mu) f(\sigma - x, \mu - y) d\sigma\, d\mu \\
&= \int\int f(\sigma, \mu) h^*(x + \sigma, y + \mu) d\sigma\, d\mu \qquad (5.2)
\end{aligned}
$$

Thus it is obvious that the Fourier transform of $F\{f(x, y)\}[F\{h(x, y)\}]^*$ is equivalent to a correlation operation between $f(x, y)$ and $h(x, y)$. Consequently, the setup of Fig. 4.1 can be treated as an optical pattern recognition system when $f(x, y)$ is introduced at the input plane and a complex spatial filter characterized by $[F\{h(x, y)\}]^*$ is introduced at the Fourier plane. This particular filter is referred to as the complex matched filter (CMF). If $h(x, y)$ does indeed exist in $f(x, y)$, the output of the system will show a strong correlation peak at a location where the match occurs. Obviously, the character recognition problem reduces to generating a complex spatial filter whose transmittance is proportional to $[F\{h(x, y)\}]^*$.

Equation (3.22) provides the intensity transmittance of a complex spatial filter $T(x, y)$ synthesized using a negative photographic transparency. As a result of having passed through the complex spatial filter placed at the Fourier plane, the field immediately behind the spatial filter is $F\{f(x, y)\}\, T(x, y)$. The output of the coherent optical processor is therefore given by

$$
\begin{aligned}
F\{F\{f(x, y)\}\, T(x, y)\} &= F\{[F\{f(x, y)\}][1 - |H(u_x, u_y)|^2 - A^2]\} \\
&\quad - F\{F\{f(x, y)\} 2AH_0 \cos(2\pi\beta x - \phi)\} \\
&= F\{F\{f(x, y)\}[1 - |H(u_x, u_y)|^2 - A^2]\} \\
&\quad - \int\int h(\sigma, \mu) f(\beta - x - \sigma, -y - \mu) d\sigma\, d\mu \\
&\quad - \int\int f(\sigma, \mu) h^*(x + \sigma + \beta, y + \mu) d\sigma\, d\mu \qquad (5.3)
\end{aligned}
$$

The first term of Eq. (5.3), which manifests itself at the origin of the output plane, is not of much significance in the filtering operation. The second and third terms are, however, of particular interest. The second term corresponds to a convolution of $f(x, y)$ and $h(x, y)$, and the output corresponding to it is centered at the coordinates $(\beta, 0)$ of the output plane. The third term corresponds to the correlation of $f(x, y)$ and $h(x, y)$ and the output corresponding to it is centered at the coordinates $(-\beta, 0)$ of the output plane. The convolution and correlation terms can be separated by choosing a sufficiently high carrier frequency β, that is, by introducing the reference wave at a steeper angle. If $h(x, y)$ is indeed present in $f(x, y)$, the corresponding output at the position of correlation field will show up as a bright identifiable peak.

Consider that the input signal $f(x, y)$ consists of a sequence of n different non-overlapping characters,

$$f(x, y) = \sum_{i=1}^{n} h_i(x, y) \tag{5.4}$$

It can be shown that the magnitude of autocorrelation is generally larger than that of the cross-correlation. The autocorrelation function is given by

$$C_{ii}(\sigma, \mu) = \int\int h_i(x, y)[h_i(x + \sigma, y + \mu)]^* \, dx \, dy \tag{5.5}$$

which at the peak becomes

$$|C_{ii}(0, 0)| = \int\int |h_i(x, y)|^2 \, dx \, dy \tag{5.6}$$

The cross-correlation function, on the other hand, is given by

$$C_{ji}(\sigma, \mu) = \int\int h_j(x, y)[h_i(x + \sigma, y + \mu)]^* \, dx \, dy \tag{5.7}$$

which at the peak becomes

$$|C_{ji}(0, 0)| = \frac{\left|\int\int h_j(x, y)[h_i(x, y)]^* \, dx \, dy\right|^2}{\int\int |h_j(x, y)|^2 \, dx \, dy} \tag{5.8}$$

But since, according to the Schwartz inequality,

$$\left|\int\int h_j h_i^* \, dx \, dy\right|^2 \leq \int\int |h_i|^2 \, dx \, dy \int\int |h_j|^2 \, dx \, dy \tag{5.9}$$

it follows that

$$|C_{ji}(0, 0)| \leq |C_{ii}(0, 0)|. \tag{5.10}$$

Note that when $h_j(x, y)$ is proportional to $h_i(x, y)$, $|C_{ji}(0, 0)|$ is exactly equal to $|C_{ii}(0, 0)|$.

In target recognition, in particular, input is often intermingled with background noise. Consider that the input is an additive mixture of signal $s(x, y)$ and noise $n(x, y)$. The figure of merit, on the basis of which the filter characteristics are optimized, is the signal-to-noise ratio (SNR), which at the origin is given by

$$\mathrm{SNR} = \frac{\left| \int \int H(u_x, u_y) S(u_x, u_y) du_x \, du_y \right|^2}{4\pi^2 \int \int |H(u_x, u_y)|^2 N(u_x, u_y) du_x \, du_y} \tag{5.11}$$

where $H(u_x, u_y)$ and $S(u_x, u_y)$ are the Fourier transforms, respectively, of filter function $h(x, y)$ and $s(x, y)$, and $N(u_x, u_y)$ is the power spectral density of the input noise. But by having defined the noise spectral density as the product of two conjugate factors,

$$N(u_x, u_y) = \tilde{N}(u_x, u_y) \tilde{N}^*(u_x, u_y) \tag{5.12}$$

Eq. (5.11) can be reduced, using Eq. (5.9), to give

$$\begin{aligned} \mathrm{SNR} &\leq \frac{1}{(4\pi)^2} \int\int |[S^*(u_x, u_y)/\tilde{N}^*(u_x, u_y)]|^2 \, du_x \, du_y \\ &\leq \frac{1}{(4\pi)^2} \int\int [|S(u_x, u_y)|^2 / N(u_x, u_y)] du_x \, du_y \end{aligned} \tag{5.13}$$

the equality holding only when

$$H(u_x, u_y) = cS^*(u_x, u_y)/N(u_x, u_y) \tag{5.14}$$

where c is an arbitrary constant. A noise having uniform spectral density over the spatial frequency domain is referred to as white noise. Thus when the additive noise is white, the optimum correlation filter turns out to be our matched filter.

It may be appropriate to note that an optical correlator like the one just described can be realized using either acousto-optic or spatial light modulators. The principal element of an acousto-optic signal processor is a signal input device that serves to convert electric signals to a spatially distributed optical signals. The acousto-optic (AO) devices can be either space-integrating or time-integrating. Figure 5.1a shows a typical acousto-optic space-integrating correlator. The signal $f_2(x, y)$ is stored on a high spatial-frequency carrier in reference transparency that is then correlated with an incoming temporal signal $f_1(t)$ applied to the AO device. A single-order signal diffracted by the reference grating is next detected at the output. As the acoustic signal $f_1(t)$ propagates through the AO device, the effective time delay between the stored reference $f_2(x, y)$ and the incoming signal changes. The time-dependent output is determined by the correlation of the two signals as a function of delay. Figure 5.1b shows the corresponding time-integrating AO correlator. The signal $f_1(t)$ is introduced in the AO device as in the former case, while $f_2(t)$ is introduced via intensity modulation of the light source. A spatial filter is used to image the AO output onto a detector array that is capable of time-integrating the

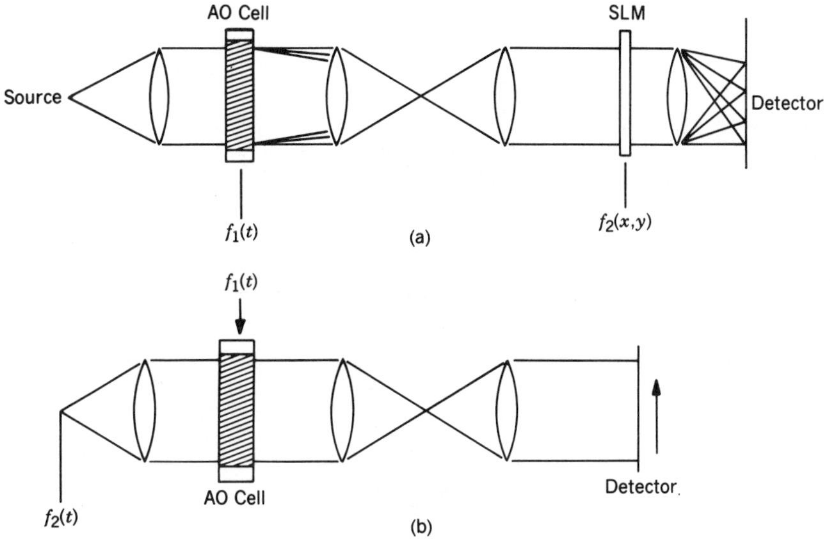

FIGURE 5.1 An acousto-optic correlator: (a) space-integrating and (b) time-integrating.

signals incident upon it. Each detector array accumulates the output incident upon it, but with a different constant relative delay between the two signals.

5.3 JOINT TRANSFORM CORRELATION

The optical pattern recognition scheme that was discussed in the last section requires the fabrication of a complex matched filter. Holographic fabrication of such filters is rather time consuming. Again, alignment of the filter is critical. Thus for real-time operation, the simple Fourier transform based correlator is not too desirable. In this section, we introduce an alternative scheme to overcome this problem by using a joint Fourier transform operation. Accordingly, both signal and the reference are introduced at the first liquid-crystal television (LCTV) of a joint Fourier transform correlator (JFTC), as shown in Fig. 5.2. Subsequently, the interference of the Fourier transforms is obtained in one single step by the CCD camera. The intensity of the interference is transformed again to obtain the correlation between the signal and reference.

Consider that a composite image formed of the input image described by $i(x, y + b)$ and a reference image described by $r(x, y - b)$ is Fourier transformed. Accordingly, the irradiance of the interference of the Fourier transforms is given by

$$\begin{aligned} I(u_x, u_y) = {} & |I(u_x, u_y)|^2 + |R(u_x, u_y)|^2 \\ & + I(u_x, u_y)R^*(u_x, u_y)\exp(-j\phi) \\ & + I^*(u_x, u_y)R(u_x, u_y)\exp(+j\phi) \end{aligned} \tag{5.15}$$

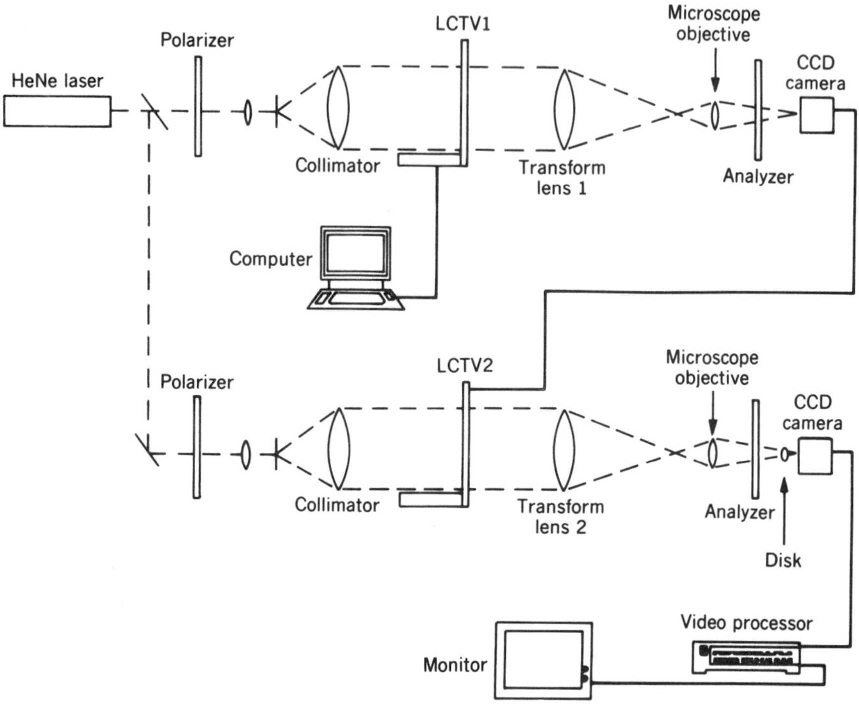

FIGURE 5.2 A joint Fourier transform correlator.

where ϕ is a function of spatial frequency coordinates, wavelength, and focal length. An inverse Fourier transform of this intensity pattern gives

$$\begin{aligned} O(x, y) = {} & a[i(x, y) \circledast i^*(x, y) + r(x, y) \circledast r^*(x, y)] \\ & + b[i(x, y - 2b) \circledast r^*(x, y - 2b) \\ & + i^*(x, y + 2b) \circledast r(x, y + 2b)] \end{aligned} \tag{5.16}$$

where $\circledast$ denotes the correlation operation and a and b are two constants. Equation (5.16) represents a dc term and two correlation terms displaced $2b$ from the center of the dc component.

In Fig. 5.2, the CCD output as represented by Eq. (5.15) is reintroduced at the second LCTV. The second transform lens completes the operation by taking an inverse Fourier transform, which is then displayed on the monitor. Figure 5.3a shows both the input and reference while Fig. 5.3b shows the resulting pattern corresponding to Eq. (5.15). Figure 5.3c shows the autocorrelation output as displayed on the monitor, while Fig. 5.3d shows the same as that of Fig. 5.3c but with the dc term suppressed. Figure 5.4 shows the different stages as in Fig. 5.3 but for the case of cross-correlation.

(a)

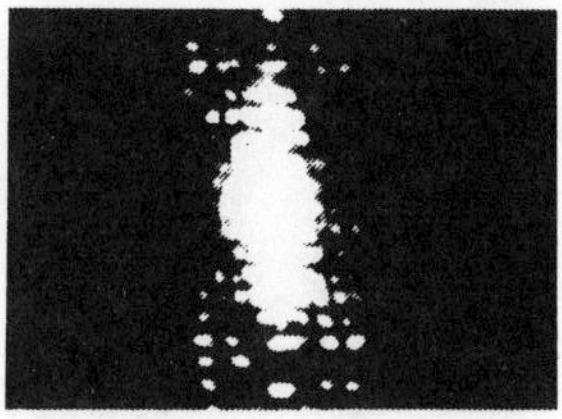

(b)

(c)

(d)

FIGURE 5.3 Autocorrelation in a JFTC: (a) inputs, (b) joint Fourier transform, (c) correlation output, and (d) correlation output with dc term suppressed.

5.4 PHASE-ONLY FILTER

It is known that much of the information of an image can be retrieved from the phase components and the pixel-by-pixel average amplitude of similar images. This information is the basis for pattern recognition using only phase. In fact, for achieving a high optical efficiency in a correlator, neither an amplitude-only filter nor a CMF is a prudent choice.

Let the Fourier transform of the object function $f(x, y)$ be denoted by $F(u_x, u_y) = |F(u_x, u_y)| \exp(j\phi(u_x, u_y))$. A CMF corresponding to this function $f(x, y)$ when introduced at the filter plane is expected to produce the autocorrelation of the input

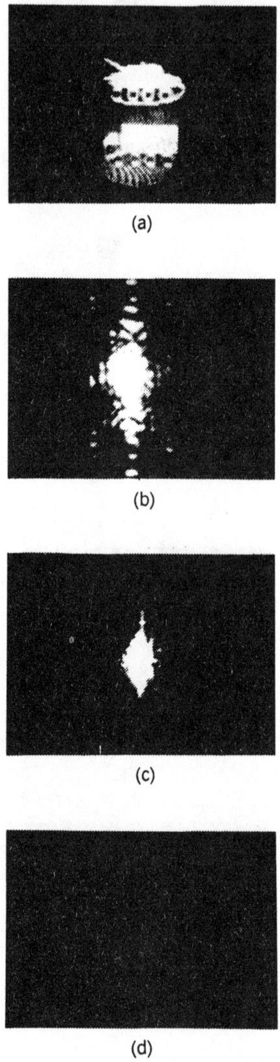

FIGURE 5.4 Cross-correlation in a JFTC: (a) inputs, (b) joint Fourier transform, (c) correlation output, and (d) correlation output with dc term suppressed.

at the output plane. The CMF is thus given by the complex conjugate of the input Fourier spectrum:

$$H_{\mathrm{CMF}}(u_x, u_y) = F^*(u_x, u_y) = |F(u_x, u_y)| \exp[-j\phi(u_x, u_y)] \tag{5.17}$$

An inverse Fourier transform of the product of $F(u_x, u_y)$ and $H_{\mathrm{CMF}}(u_x, u_y)$ results in a convolution of $f(x, y)$ and $f(-x, -y)$, which in turn is equivalent to the autocorrelation of $f(x, y)$. Note that when $|F(u_x, u_y)|$ is unity, H_{CMF} becomes a phase-only filter (POF), which is described by

$$H_{\mathrm{POF}}(u_x, u_y) = \exp(-j\phi(u_x, u_y)) \tag{5.18}$$

The reason why the autocorrelation peak corresponding to a POF-based correlator is sharper becomes apparent when one examines carefully the content of the corresponding Fourier plane. In the case of the autocorrelation corresponding to a CMF-based correlator [with $F(u_x, u_y) \neq 1$], the content of the Fourier domain is generated by a squaring operation given by $|F(u_x, u_y)|^2$. In the case of a POF-based correlator, in comparison, the content of the Fourier plane is directly proportional to $|F(u_x, u_y)|$ only.

It is well known that the squaring operation often produces more harmonics. In other words, a squaring operation shifts the energy from the lower harmonics to one or more of the higher harmonics. Since the total energy must be conserved, the squaring operation reduces the energy of the autocorrelation peak. The POF-based autocorrelation involves no squaring operation, and therefore most of the energy is concentrated at output regions that correspond to the lower harmonics.

The energy of the central autocorrelation peak corresponds to the dc value of the spectrum (at the Fourier plane). Therefore the more the spectrum of the Fourier plane can be flattened the sharper is the correlation peak. In the case of a POF, the autocorrelation peak is proportional to the dc value of $|F(u_x, u_y)|$, and in the case of the CMF, the autocorrelation peak is proportional to the dc value of $|F(u_x, u_y)|^2$. Since, in general, $|F(u_x, u_y)| < 1$, the autocorrelation peak corresponding to a POF is always higher than that corresponding to a CMF.

When compared to both amplitude-only and complex matched filters, a POF is associated with smaller energy absorptance in the filter plane, while it provides a rather narrower and higher autocorrelation peak at the output plane. If one were to measure optical efficiency of POF in terms of energy that is transferred from the input to the output, it amounts to 100% since there is no energy absorption by the filter. It can be inferred that it is the amplitude content of the CMF that is responsible for broadening and simultaneous deterioration of correlation peaks.

5.5 AMPLITUDE-MODULATED RECOGNITION FILTERS

An optical pattern or character recognition operation may consist of matching either amplitudes using an amplitude-only filter, or both amplitudes and phases using a Vander Lugt type CMF, or phases using a POF. The basic Vander Lugt correlation operation indicating a match between two objects involves cancellation of phase and squaring of magnitude at the Fourier plane. In comparison, a POF-based autocorrelation operation results in the cancellation of only the phases associated with the object while the amplitude-only filter squares the amplitude but leaves the phase undisturbed. However, amplitude information, if properly incorporated along with the phase, can also be used to improve the discrimination efficiency. This is the basis for two important amplitude-modulated correlation filters: the amplitude-compensated matched filter (ACMF) and the amplitude-modulated phase-only filter (AMPOF).

ACMF correlates an object with an inverse type of filter. This filter is able to provide larger discrimination (defined as the difference between the autocorrelation and cross-correlation peaks) as well as sharper correlation peaks than that using a POF. Interestingly, inverse filters were dismissed earlier for having poor optical efficiency. In particular, the inverse filter has an operational difficulty in that it becomes indeterminate whenever the function passes through zeros. The AMPOF, on the other hand, effectively uses both amplitude and phase information and, at the same time, avoids the operational difficulty of the inverse filters. Both of these filters are remarkably better in yielding higher correlation discrimination than both CMF- and POF-based correlators.

To assure that the Fourier plane magnitude spectrum is significantly flattened, the characteristics of the POF can be modulated with amplitude. To have a Dirac delta autocorrelation peak, the signal at the Fourier plane would have to be $F^{-1}\{\delta(x, y)\} = 1$, that is, a plane wave. Accordingly, the desired correlation filter is the inverse filter characterized by $1/F(u_x, u_y)$. It is rather difficult to implement an inverse filter optically. Operational difficulty arises when one needs to take an inverse of a value that is less than unity. Thus an inverse operation may necessitate a gain of optical signal. Furthermore, when the value approaches zero, the inverse operation causes an indeterminate condition.

The bottlenecks that are otherwise associated with the inverse filter are overcome in amplitude-modulated recognition filters by decreasing the gain of the filter and deliberately avoiding the indeterminate filter function values. For example, the ACMF function as given by

$$H_{\mathrm{ACMF}}(u_x, u_y) = \begin{cases} \exp[-j\phi(u_x, u_y)], & |H(u_x, u_y)| < T_0 \\ T_0 \exp[-j\phi(u_x, u_y)]/|H(u_x, u_y)|, & |H(u_x, u_y)| > T_0 \end{cases} \tag{5.19}$$

where T_0 is a threshold value, takes the form of a POF where the modulus of the amplitude spectrum is lower than the threshold value and takes the form of an inverse filter when the modulus is higher than the threshold value.

Alternatively, to avoid the absorption of energy at the filter plane and, at the same time, achieve a significantly uniform filter plane spectrum, one can make use of an AMPOF that is characterized by

$$H_{\mathrm{AMPOF}}(u_x, v_y) = \frac{De^{-j\phi(u_x, u_y)}}{|F(u_x, u_y)| + a} \tag{5.20}$$

where $0 < D < 1$ and $|F(u_x, u_y)|_{\min}$ is the minimum value of $|F(u_x, u_y)|$ such that $|F(u_x, u_y)|_{\min} + a > D$. This AMPOF formulation guarantees that the transmittance of the filter is less than unity. For simplicity, the parameter a can be a constant or even a function of u_x and u_y and therefore can be used to either suppress noise or band limit the filter or both. For example, an a priori knowledge of the noise spectrum can be used to formulate $a(u_x, u_y)$ in a way that the noise can be attenuated at the filter plane. When $|F(u_x, u_y)| >> a$ and $D = 1$, the AMPOF approaches a perfect inverse filter. On the other hand, with $a = 0$ and $D << 1$, AMPOF approaches the ACMF.

In AMPOF-based correlators, both amplitude matching and phase matching are effectively employed to yield a narrower and larger autocorrelation peak. When the phases match, the dc value of the spectrum reduces to the dc value of the amplitude spectrum. Again, annihilation of the phase spectrum will result in the concentration of more energy at the autocorrelation peak. But when the filter phase mismatches with the phase of the object, some of the energy will be diverted away from the correlation peak.

To determine the degree of similarity between the object function and the filter, the correlation peak energy is normalized with respect to the total output energy. Both peak and total energies are measurable quantities and consequently the ratio of the two can provide an excellent correlation discrimination. Correlation discrimination quality of a filter is found by determining how much of the energy of the output plane actually appears at the correlation peak.

Figure 5.5a shows the autocorrelation of the character E, for example, for a POF-based correlator, while Fig. 5.5b shows that for the AMPOF-based correlator when both D and a are set to unity. It can be seen that the autocorrelation peak is

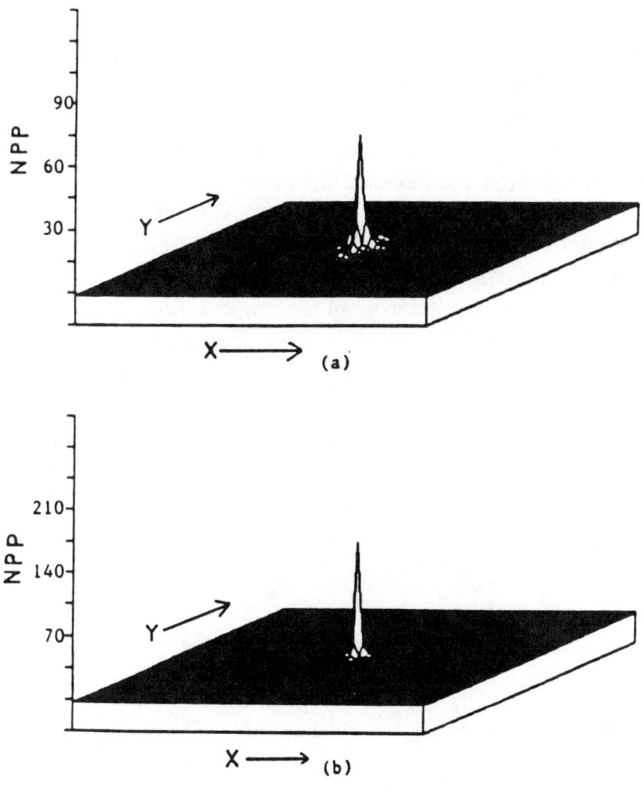

FIGURE 5.5 Autocorrelation of character E using (a) POF and (b) AMPOF.

higher and narrower in the case of the AMPOF-based correlator. Also, the AMPOF-based correlator produces fewer and smaller side lobes at the output plane. As discussed previously, the correlation peak height is normalized with respect to the total energy of the output plane for comparing both autocorrelation and cross-correlation heights for both POF and AMPOF. Table 5.1 lists the important parameters for both autocorrelation and cross-correlation using both the POF and the AMPOF. The normalized peak power (NPP) is obtained by dividing the absolute peak power (APP) with the total output power. Similarly, the normalized power at the side lobes (NPS) is obtained by dividing the peak side-lobe power with the total output power. Note that the NPP for the AMPOF is usually 100% more than that for the POFs. The peak to side lobe intensity ratio for the case of the AMPOF is also significantly larger, indicating that more energy is concentrated at the autocorrelation peak. Furthermore, the AMPOF is able to differentiate better between autocorrelation and cross-correlation.

Figure 5.6 shows four gray-level images (car 1 through car 4), which are also processed in POF- and AMPOF-based correlators but in the presence of Gaussian random noise. Figures 5.7a and 5.7b show the autocorrelation peaks (corresponding to car 3 in the absence of noise) using AMPOF- and POF-based correlators, respectively, while Figs. 5.7c and 5.7d show the corresponding cross-correlation peaks (corresponding to the input car 3 in the absence of noise and filter for car 4). Similarly, Figs. 5.8a and 5.8b show the autocorrelation peaks (corresponding to the input car 3 with noise) using the AMPOF- and POF-based correlators, respectively, while Figs. 5.8c and 5.8d show the cross-correlation peaks (corresponding to the input car 3 with noise and filter for car 4). Figure 5.9 shows the plot of NPP versus the noise standard deviation for both AMPOF- and POF-based correlators. It can readily be concluded that the AMPOF-based correlator provides better correlation discrimination than the POF-based correlator even in the presence of random Gaussian noise.

While LCTV can be used for amplitude modulation, a magneto-optic spatial light modulator (MOSLM) can store the discretized phase components of the filter.

TABLE 5.1 Correlation Results for Noise-Free Case

		POF			AMPOF		
Input	Filter	APP ($\times 10^6$)	NPP	NPS	APP ($\times 10^5$)	NPP	NPS
E	E	1.53	57	3	4.94	134	1
F	F	1.34	62	3	4.60	131	1
G	G	2.03	70	0[a]	6.95	163	0
O	O	1.74	55	0	5.55	144	0
E	F	1.12	41	1	3.64	81	3
F	E	0.96	44	5	3.07	83	2
G	O	1.51	52	0	4.99	105	0
O	G	1.39	44	1	4.14	99	1

[a]The zero value of the side lobes indicates a value between 0 and 1.

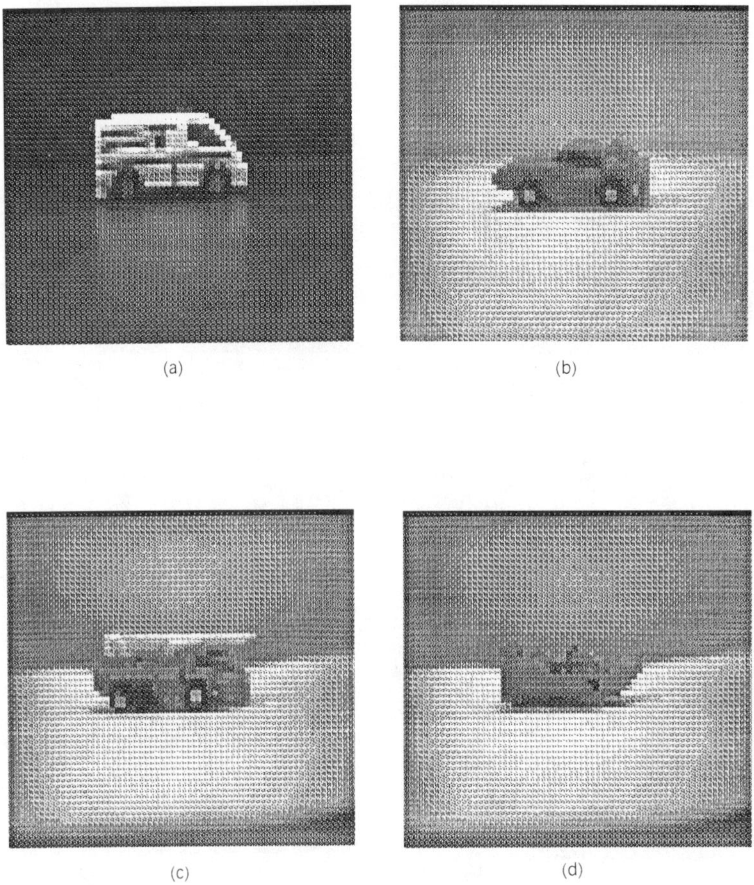

FIGURE 5.6 Four gray-level car images without any noise.

Accordingly, for a real-time pattern recognition task, a combination of computer-controlled LCTV and MOSLM can be used to store the amplitude and phase components, respectively, of the appropriate amplitude-modulated recognition filter.

5.6 GENERALIZED CORRELATION FILTER

In the previous sections, a number of different correlation filters were discussed. Except for the joint Fourier transform correlator, all other correlators actually require the synthesis of a correlation filter. In all these cases, the filter is placed at the Fourier plane and the correlation output is obtained by transforming the product of the filter function and the Fourier transform of the input scene. It is, however, rather difficult to prefer any one of these filters over the others without performing a real comparison. Accordingly, in this section, a new family of correlation filters,

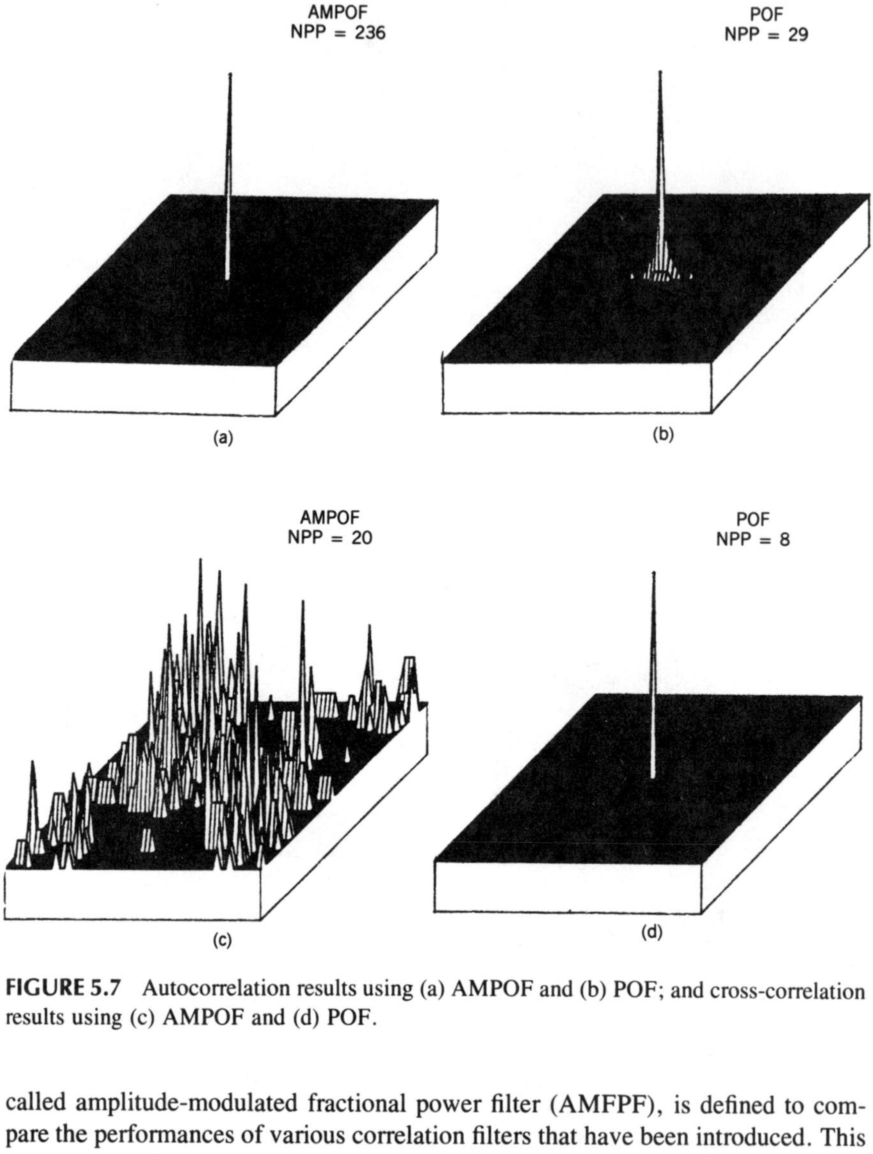

FIGURE 5.7 Autocorrelation results using (a) AMPOF and (b) POF; and cross-correlation results using (c) AMPOF and (d) POF.

called amplitude-modulated fractional power filter (AMFPF), is defined to compare the performances of various correlation filters that have been introduced. This becomes possible because it can be shown that the CMF, inverse filter (IF), POF, ACMF, and AMPOF are all special cases of this more general AMFPF. The design of each of these filters is dictated by a particular set of four parameters used to define an AMFPF.

The generalized AMFPF function may be expressed by

$$F_{\text{AMFPF}}(u, v) = P[k|R(u, v)| + Q]^m \exp[-j\phi(u, v)], \tag{5.21}$$

where $|R(u, v)|$ is the amplitude of the Fourier spectrum $R(u, v)$ of the reference function $r(x, y)$, $\phi(u, v)$ is the phase factor of $R(u, v)$, P and Q are either constants

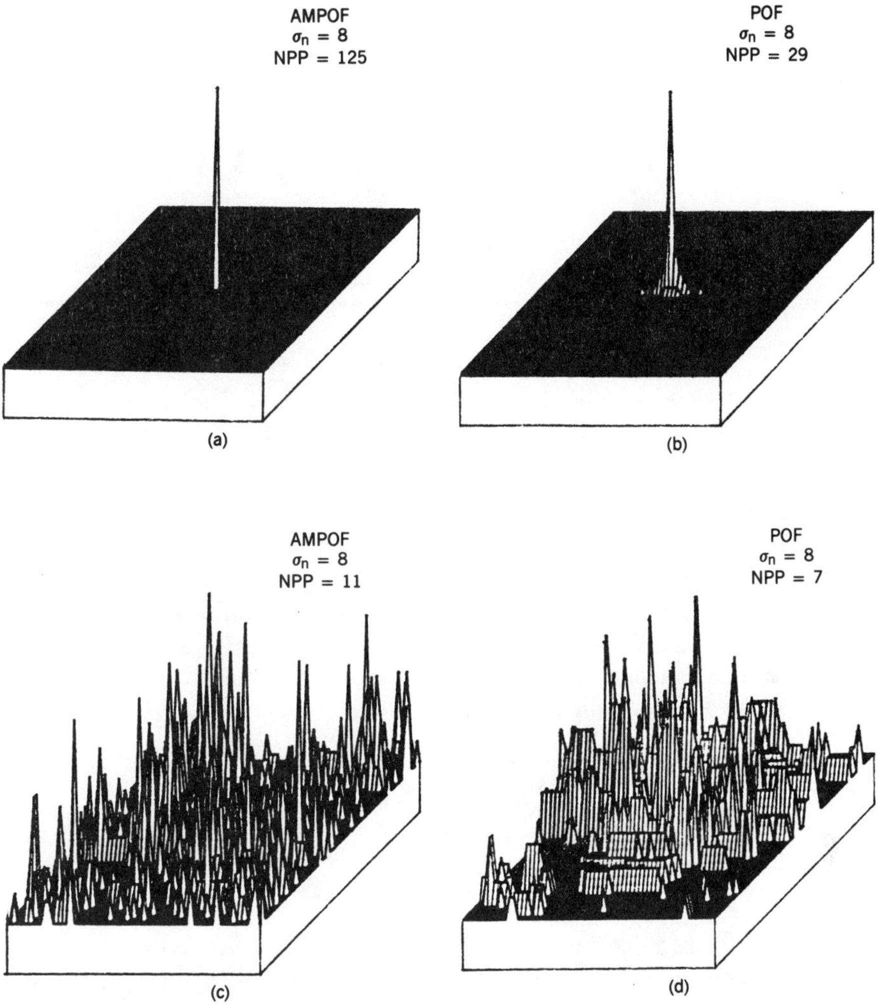

FIGURE 5.8 Autocorrelation results in the presence of noise using (a) AMPOF and (b) POF; and cross-correlation results in the presence of noise using (c) AMPOF and (d) POF.

or functions, and k and m are arbitrary constants. Note that different values of m will result in a completely different correlator. In particular, when $P = 1$, $k = 1$, and $Q = 0$, an AMFPF reduces to a POF, CMF, and IF, respectively, with $m = 0$, 1, and -1. When P is a proper constant, $k = 1$, $Q = 0$, and $m = -1$, and AMFPF reduces to a ACMF. Again, with $k = 1$, $Q \neq 0$, and $m = -1$, the AMFPF becomes an AMPOF.

The AMFPF provides a systematic scheme to choose the right compromise between various correlator performances. For effective comparison, the criteria of optical efficiency (OE), normalized correlation peak power (NCPP), correlation discrimination (CD), and signal-to-noise ratio (SNR) may be employed. As a matter of fact, this evaluation is a tedious process since it is rather difficult to generate

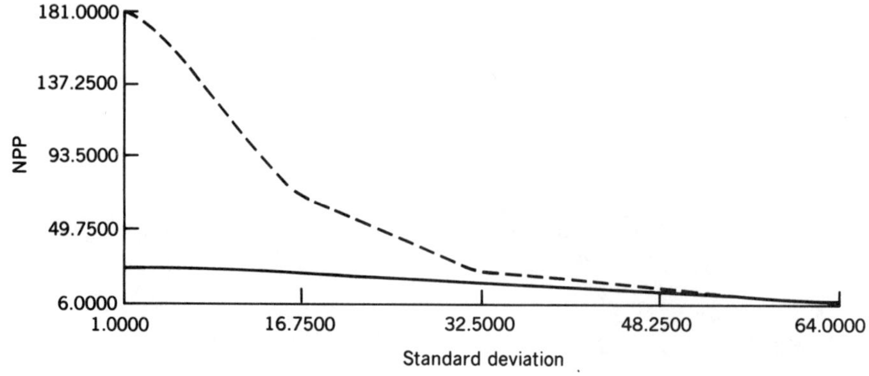

FIGURE 5.9 NPP versus standard deviation of the Gaussian random noise when using POF- and AMPOF-based correlators. Solid curve, POF; dashed curve, AMPOF.

a general analytical expression for comparison. Again, in the case of noise, different noise sources will have different impacts on these four measures. While a particular correlation filter may be good in one of these measures, it may not be very encouraging in other measures. For example, while the POF results in the highest OE, and the CMF produces the largest SNR, the amplitude-modulated correlation filters provide the highest values for both NCPP and CD.

The OE for a recognition filter is given by

$$\mathrm{OE} = e\,\frac{\iint |i(x, y) \circledast f(x, y)|^2 dx\, dy}{\iint |i(x, y)|^2\, dx\, dy} \tag{5.22}$$

where $i(x, y)$ is the input function, $f(x, y)$ is the filter function corresponding to the reference function $r(x, y)$ in the space domain, e is the diffraction efficiency of the recording medium, and $\circledast$ indicates a correlation operation. In a practical correlator, it is desirable to have a high throughput of light to keep the laser power requirements as small as possible.

The second measure NCPP may be given as

$$\mathrm{NCPP} = \frac{|o(0, 0)|^2}{\iint |o(x, y)|^2\, dx\, dy} \tag{5.23}$$

where $o(x, y)$ corresponds to the output at the correlation plane. A higher NCPP value generally corresponds to a narrower correlation peak and thus relates to an increase in the probability of recognition and accuracy of location.

The third characteristic, CD, is a parameter that measures the ability of the correlator to discriminate between two very similar input objects. The correlation discrimination measure may be defined in two ways. The CD measure of the first kind, CD_1, may be expressed as

$$CD_1 = \frac{|i(x, y) \circledast f(x, y)|^2 - |g(x, y) \circledast f(x, y)|^2}{|i(x, y) \circledast f(x, y)|^2} \tag{5.24}$$

where $i(x, y)$ and $g(x, y)$ are input functions such that $i(x, y)$ may include the reference while $g(x, y)$ may not include the reference. CD_1 indicates the difference between the absolute values of autocorrelation peak power (representing a match) and cross-correlation peak power (representing a mismatch), as a fraction of the absolute value of the autocorrelation peak power. On the other hand, the correlation discrimination measure of the second kind, CD_2, may be defined as

$$CD_2 = \frac{NCPP_a - NCPP_c}{NCPP_a} \tag{5.25}$$

where $NCPP_a$ is the NCPP corresponding to the autocorrelation output and $NCPP_c$ is that corresponding to the cross-correlation output. The CD_2 value indicates the sharpness difference between the autocorrelation and cross-correlation peaks as a fraction of the autocorrelation sharpness. Usually, the peak of autocorrelation is sharper than that of cross-correlation. In designing a filter, a larger value for CD_2 is also desirable.

Figure 5.10 shows two binary characters, E (reference) and F (input), and two gray-level images, ambulance (reference) and fire engine (input). For simplicity, we may normalize the maximum amplitude of the AMFPF to unity, and set $e = 1$. Figures 5.11–5.14 show the AMFPF correlator performance results for some specific sets of P, k, Q, and m.

Figure 5.11a shows OE, NCPP, CD_1, and CD_2 values versus m for binary objects in the absence of noise when $P = k = 1$ and $Q = 0$. The best values of NCPP are obtained when $m = -1$, which corresponds to an IF-based correlator. While CD_1 remains almost a constant, the value of CD_2 decreases more than 30 dB as m changes from -2.5 to 2.5. The OE is higher for positive m than for negative m. The best value of OE is 0 dB, which is realized using a POF. Figure 5.11b shows OE, NCPP, CD_1, CD_2, and SNR values versus m when the binary targets are mixed with Gaussian random noise. The best values of NCPP are now shifted to $m = -0.5$. This is because the spectrum of the input signal, which is the reference plus noise, cannot be canceled now by the IF in the Fourier transform plane. However, an approximate uniform distribution is achieved at the Fourier transform plane after having filtered with an AMFPF when $m = -0.5$. When $m = 1$, which corresponds to a CMF, the SNR takes on the highest possible value. On the other hand, the OE, CD_1, and CD_2 trends remain rather unchanged in the presence of noise. Figure 5.12 shows the corresponding performances when gray-level images are used instead of the binary objects. The general tendencies are almost the same as in the case of binary targets, but the CD_1 and CD_2 values are relatively larger.

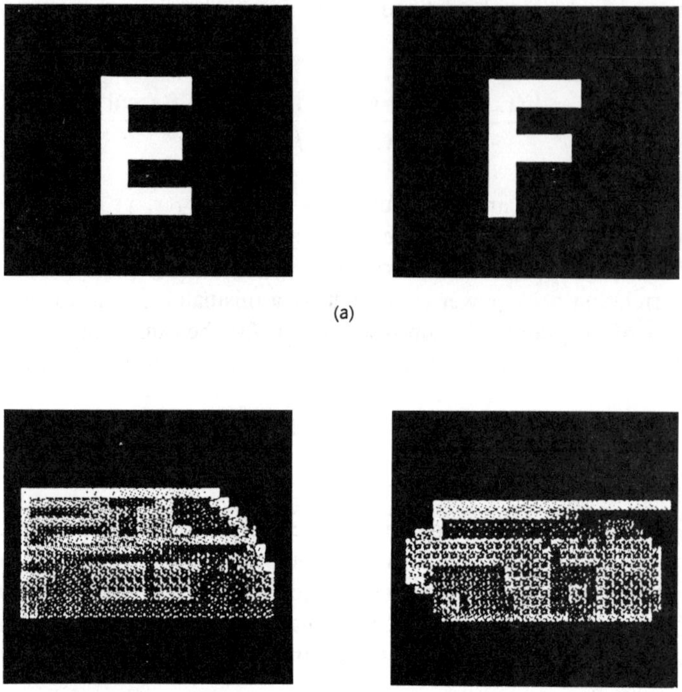

(a)

(b)

FIGURE 5.10 Images for the AMFPF performance simulations: (a) binary characters and (b) gray-level images.

Figures 5.13 and 5.14 show the AMFPF performances, respectively, for binary and gray-level targets but for a large Q, say, 10,000. Accordingly, the case corresponding to $m = -1$ represents one of the AMPOFs. In the absence of noise as well as in the presence of noise, the value of NCPP for an AMFPF decreases with increasing m. With a larger value of Q in these two figures, the OE is improved significantly when m is positive and the CD_2 is improved when m is negative. In general, the SNR increases with m.

The parameters m and $Q(u, v)$ play an important role in determining the AMFPF correlation performances. The parameter Q is sensitive to both NCPP and OE. Increasing the value of Q improves the OE considerably, but at a cost of reduced NCPP value. The parameter k does not influence the correlation performances as significantly as do m and Q. Interestingly, the OE, NCPP, CD, and SNR values in the case of gray-level images are better than those in the case of binary characters.

5.7 MELLIN TRANSFORM BASED CORRELATION

All the correlation schemes that have been discussed so far in this chapter are very sensitive to orientation and scale problems. One obvious way to solve this very

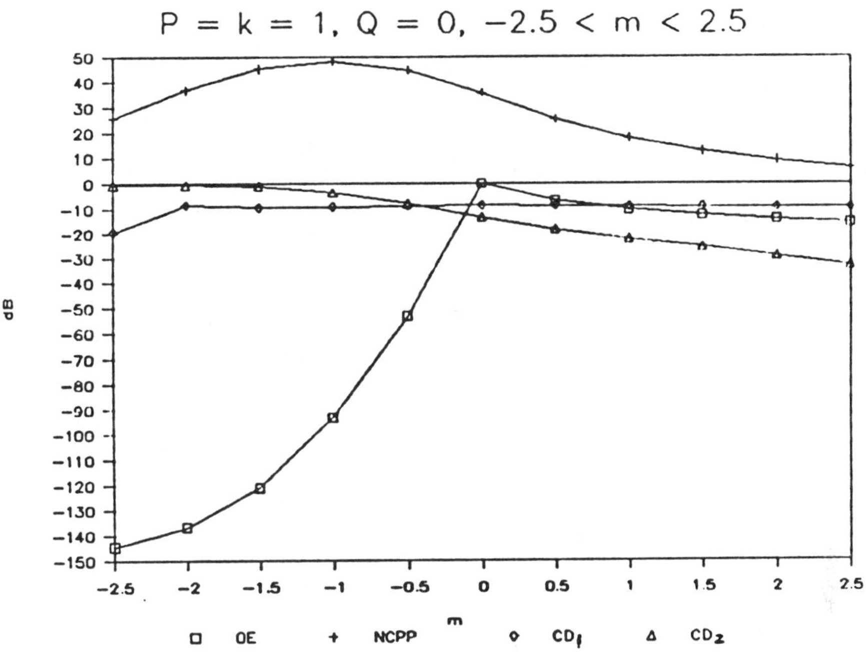

(a)

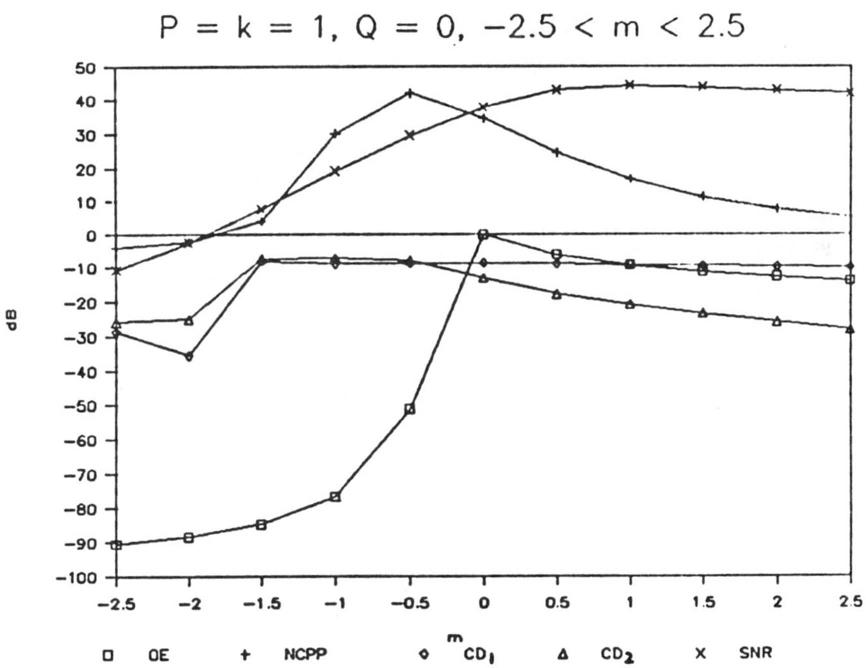

(b)

FIGURE 5.11 AMFPF correlator characteristics when using binary characters with $P = k = 1$ and $Q = 0$: (a) OE, NCPP, CD_1, and CD_2 as a function of m in the absence of noise; (b) OE, NCPP, CD_1, CD_2, and SNR as a function of m in the presence of noise.

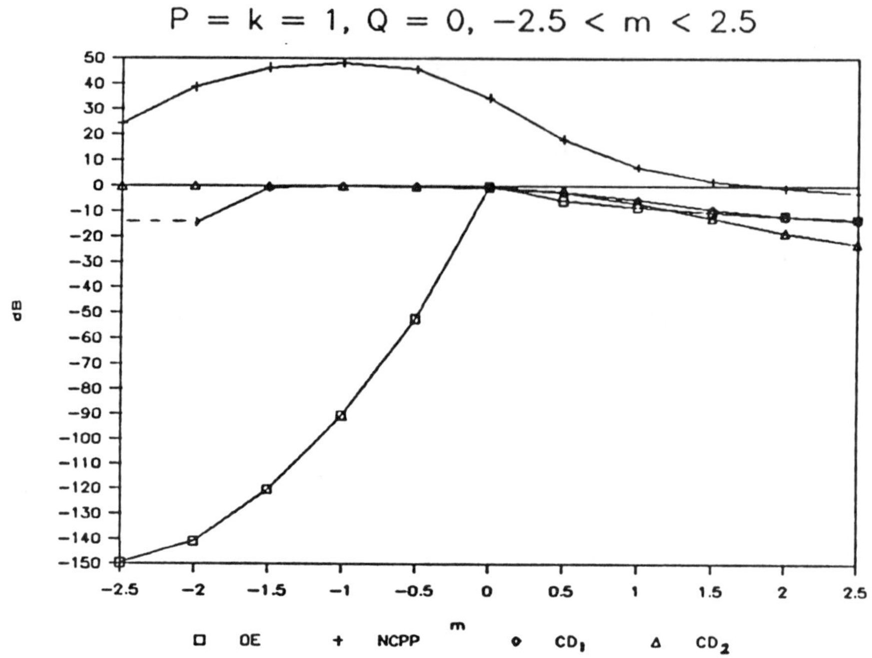

(a)

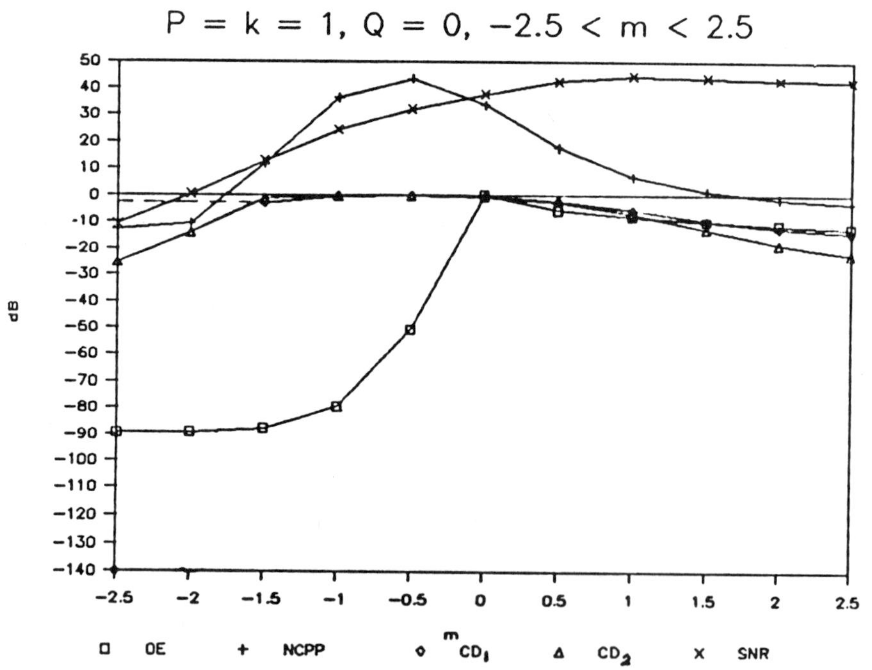

(b)

FIGURE 5.12 AMFPF correlator characteristics when using gray-level images with $P = k = 1$ and $Q = 0$: (a) OE, NCPP, CD_1, and CD_2 as a function of m in the absence of noise; (b) OE, NCPP, CD_1, CD_2 and SNR as a function of m in the presence of noise.

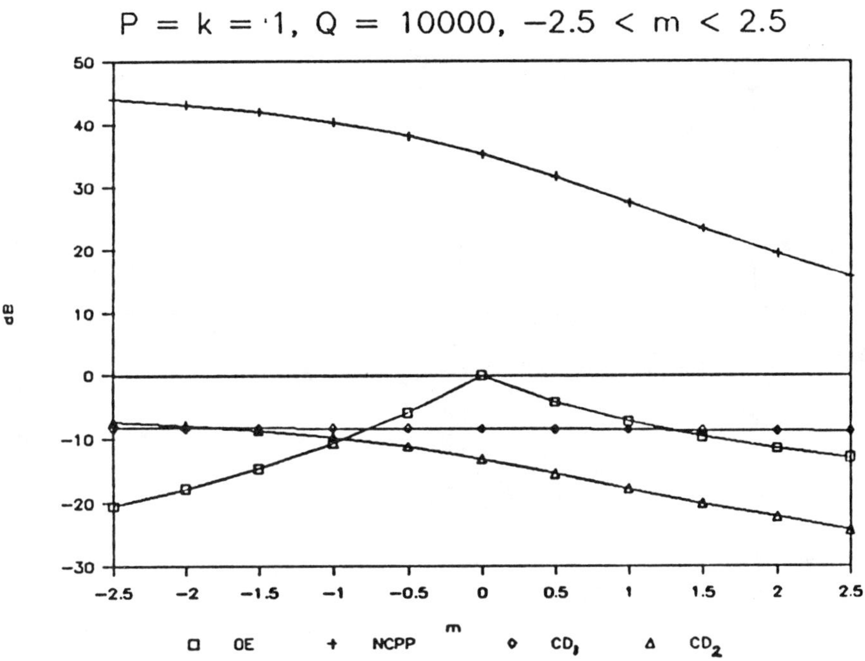

(a)

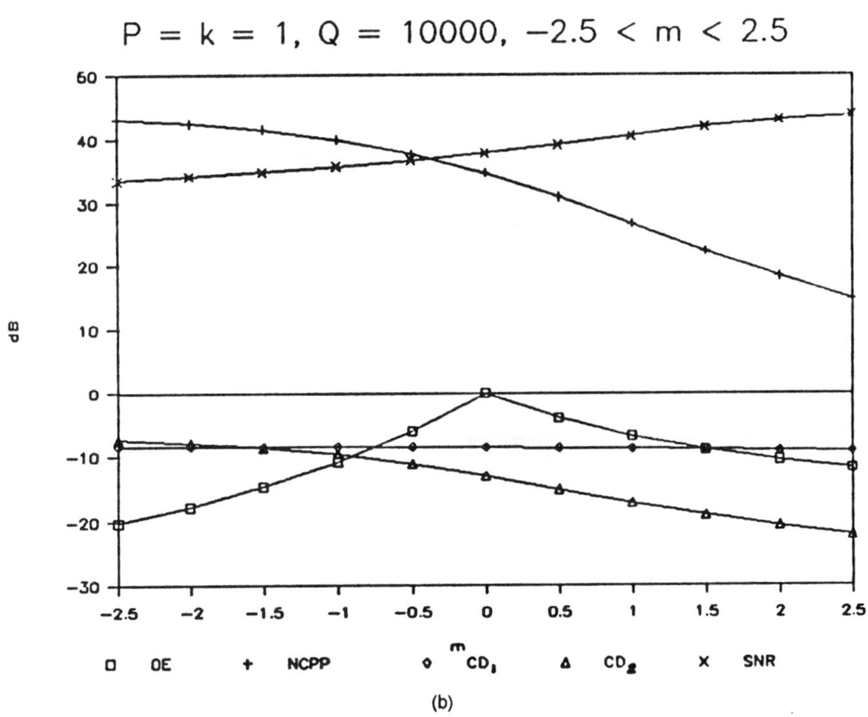

(b)

FIGURE 5.13 AMFPF correlator characteristics when using binary characters with $P = k = 1$ and $Q = 10{,}000$: (a) OE, NCPP, CD_1, and CD_2 as a function of m in the absence of noise; (b) OE, NCPP, CD_1, CD_2, and SNR as a function of m in the presence of noise.

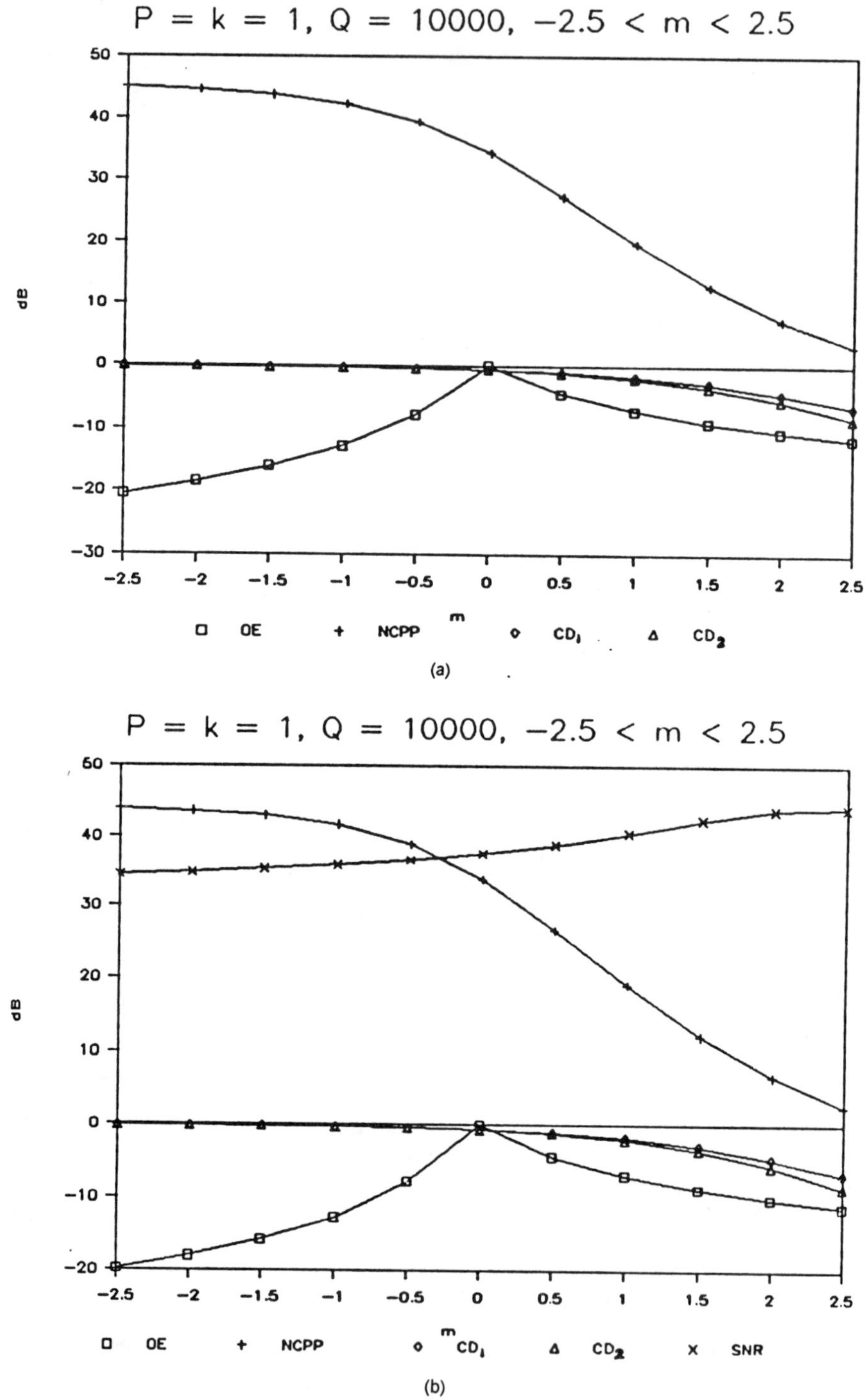

(b)

FIGURE 5.14 AMFPF correlator characteristics when using gray-level images with $P = k = 1$ and $Q = 10{,}000$: (a) OE, NCPP, CD_1, CD_2 as a function of m in the absence of noise; (b) OE, NCPP, CD_1, CD_2, and SNR as a function of m in the presence of noise.

realistic problem is through the use of multiplexed filters. According to this scheme, a composite filter is fabricated from a bank of correlation filters designed for each scale and orientation variation of the reference. Design of such a composite filter using holographic techniques may contribute to the diffraction efficiency problem. Again, if it were to be accomplished using certain spatial light modulators, the complete processing would be rather time consuming. There exists a rather interesting alternative using the Mellin transform operation. However, since this operation is not directly related to Fourier transformation, its optical implementation is not obvious. Accordingly, in this section, we first explore the properties of Mellin transforms and then establish the interrelationship between the Mellin and Fourier transforms.

The two-dimensional Mellin transform along α and β is defined as

$$M(\alpha, \beta) = \int_{-0}^{\infty} \int_{-0}^{\infty} f(\sigma, \mu)\sigma^{-(1+j\alpha)}\mu^{-(1+j\beta)} d\sigma \, d\mu \tag{5.26}$$

The most useful property of the Mellin transform is its scale invariance. Consider that the function $f(\sigma, \mu)$ is replaced with $f(c\sigma, c\mu)$, where c is the scaling constant. Accordingly, the Mellin transform of the scaled function is given by

$$\begin{aligned} M'(\alpha, \beta) &= \int_{-0}^{\infty} \int_{-0}^{\infty} f(c\sigma, c\mu)\sigma^{-(1+j\alpha)}\mu^{-(1+j\beta)} d\sigma \, d\mu \\ &= c^{j(\alpha+\beta)} M(\alpha, \beta) = e^{j\ln(c)} M(\alpha, \beta) \end{aligned} \tag{5.27}$$

Accordingly, we can conclude that $|M'(\alpha, \beta)| = |M(\alpha, \beta)|$. The magnitude of the Mellin transform, however, is strictly position dependent. In other words, $|M\{f(x, y)\}|$ is not equal to, for example, $|M\{f(x - x', y - y')\}|$. The scaling constant c appears as a linear phase factor multiplying the unscaled Mellin transform in the transform space. When used in scale-invariant correlators, a follow-on Fourier transform of $M'(\alpha, \beta)$ will shift the correlation peak by an amount proportional to $\ln(c)$.

The interrelationship between the Mellin and Fourier transforms can be illustrated by making the substitution $\sigma = e^x$ and $\mu = e^y$ in Eq. (5.26). The Mellin transform can then be expressed as

$$\begin{aligned} M(\alpha, \beta) &= \int_{-0}^{\infty} \int_{-0}^{\infty} f(e^x, e^y) \exp[-j(\alpha x + \beta y)] dx \, dy \\ &= F\{f(e^x, e^y)\} \end{aligned} \tag{5.28}$$

The implication of Eq. (5.28) is significant since the Mellin transform of $f(\sigma, \mu)$ is the same as the Fourier transform of $f(e^x, e^y)$, where $\sigma = e^x$ and $\mu = e^y$.

A complex spatial filter described by $M\{f(\alpha, \beta)\} = M(\alpha, \beta) = |M(\alpha, \beta)|e^{-j\phi}$ can be fabricated using holographic interferometric techniques as described by Eq. (3.22). This will involve introducing an object transparency given by $f(e^x, e^y)$ at the input of the recording system. Accordingly, the resulting complex filter will be given by

$$H(\alpha, \beta) = 1 - A^2 - |M(\alpha, \beta)|^2 - 2A|M(\alpha, \beta)| \cos(2\pi\Omega\alpha - \phi(\alpha, \beta)) \quad (5.29)$$

where Ω is a function of wavelength and the tilt angle. Provided that the transparency $f(e^x, e^y)$ has been introduced as an input of the coherent optical processor and that the spatial filter described by Eq. (5.29) has been introduced at its Fourier plane, the system output is found to be

$$\begin{aligned} o(x'', y'') &= (1 - A^2)f(e^x, e^y) - f(e^x, e^y) * f(e^x, e^y) * f^*(e^x, e^y) \\ &\quad - Af(e^x, e^y) * f(e^x + e^{2\pi\Omega}, e^y) \\ &\quad - Af(e^x, e^y) * f^*(-e^x + e^{2\pi\Omega}, - e^y) \end{aligned} \quad (5.30)$$

While the first two terms of Eq. (5.30) are the dc terms, the other two terms are the correlation terms appearing, respectively, at $x'' = 2\pi\Omega$ and $x'' = -2\pi\Omega$.

Figure 5.15 shows a hybrid opto-electronic system for obtaining the Mellin transform of an arbitrary two-dimensional function $f(x, y)$. The first leg of the system takes the Fourier transform $F(u_x, u_y)$ of the input transparency, which is the detected by the detector. Rotation of $f(x, y)$ by any angle results in the rotation of $|F(u_x, u_y)|$ by the same angle. Next, $|F(u_x, u_y)|$ is converted first, to $F(r, \theta)$ by a polar transformation such that $r = (u_x^2 + u_y^2)^{1/2}$ and $\phi = \tan^{-1}(u_y/u_x)$ and,

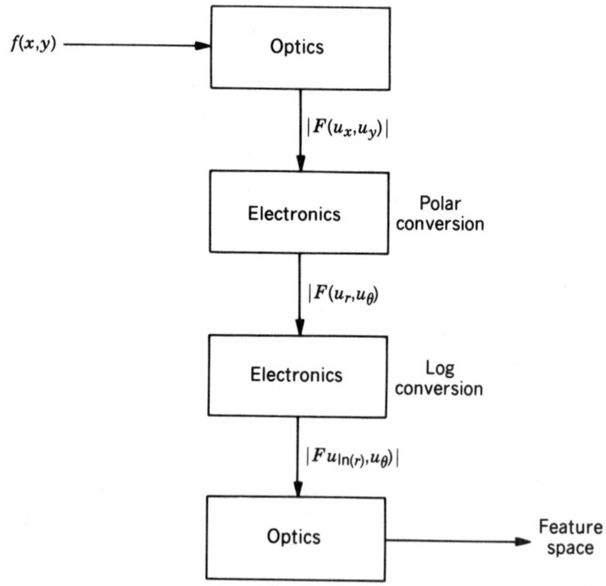

FIGURE 5.15 Scheme for the realization of the Mellin transform.

second, to $F(e^\delta, \phi)$ by substituting $\delta = \ln(r)$. Finally, the Mellin transform is obtained optically by taking a Fourier transform of $F(e^\delta, \phi)$.

5.8 PROBLEMS

1. Consider an input image $f(x - x', y)$ (mixed with additive Gaussian noise), which has been introduced in the joint transform correlator while the reference image is simply $f(x + x', y)$. Evaluate and sketch the resultant light distribution at the output and discuss the correlation peak in terms of the Gaussian noise characteristics.
2. Determine the magnitude of the Mellin transform of a one-dimensional rect function of width 2δ, which is centered at $x = X$, where $\delta < X$. Justify the scale invariance of the Mellin transform by scaling down the width of the rect function by a factor of 2.
3. Show that the Mellin transform of a circular aperture is independent of its radius.

5.9 SUGGESTED READINGS

Abu-Mostafa, Y. S., and D. Psaltis, Recognition Aspects of Moment Invariants, *Trans. IEEE Patten Anal. Machine Intell.*, **PAMI-6,** 698 (1984).

Alam, M. S., A. A. S. Awwal, and M. A. Karim, Improved Correlation Discrimination Using Joint Fourier-Transform Optical Correlator, *Microwave Opt. Technol. Lett.*, **4,** 103–106 (1991).

Arsenault, H. H., Y. N. Hsu, and K. Chalasinska-Macukow, Rotation-Invariant Pattern Recognition, *Opt. Eng.*, **23,** 705–709 (1984).

Awwal, A. A. S., M. A. Karim, and S. R. Jahan, Improved Correlation discrimination Using an Amplitude-Modulated Phase-Only Filter, *Appl. Opt.*, **29,** 233–236 (1990).

Casasent, D., and D. Psaltis, Scale Invariant Optical Transform, *Opt. Eng.*, **15,** 258–261 (1976).

Casasent, D., and D. Psaltis, New Optical Transforms for Pattern Recognition, *Proc. IEEE*, **65,** 77–81 (1977).

Caulfield, H. J., Role of Horner Efficiency in the Optimization of Spatial Filters for Optical Pattern Recognition, *Appl. Opt.*, 4391 (1982).

Davis, J. A., E. A. Merrill, D. M. Cotrell, and R. M. Bunch, Effects of Sampling and Binarization in the Output of the Joint Fourier Transform Correlation, *Opt. Eng.*, **29,** 1094–1100 (1990).

Flannery, D. L., J. S. Loomis, and M. E. Milkovich, Transform-Ratio Ternary Phase Amplitude Filter Formulation for Improved Correlation Discrimination, *Appl. Opt.*, **27,** 4079 (1988).

Gianino, P. D., and J. L. Horner, Additional Properties of the Phase Only Correlation Filter, *Opt. Eng.*, **23,** 695 (1984).

Horner, J. L., Light Utilization in Optical Correlators, *Appl. Opt.*, **21,** 4511 (1982).

Horner, J. L., and P. D. Gianino, Phase Only Matched Filtering, *Appl. Opt.*, **23,** 812 (1984).

Javidi, B., and C. Kuo, Joint Transform Image Correlation Using a Binary Spatial Light Modulator at the Fourier Plane, *Appl. Opt.*, **27,** 663–665 (1988).

Javidi, B., and S. F. Odeh, Multiple Object Identification by Bipolar Joint Transform Image Correlation, *Opt. Eng.*, **27,** 295–300 (1988).

Kaura, M. A., and W. T. Rhodes, Optical Correlator Performance Using a Phase-with-Constrained-Magnitude Complex Spatial Filter, *Appl. Opt.*, **29,** 2587–2593 (1990).

Kumar, B. V. K. V., and Z. Bahri, Phase-Only Filters with Improved Signal to Noise Ratio, *Appl. Opt.*, **28,** 250–257 (1989).

Kumar, B. V. K. V., and L. Hassebrook, Performance Measures for Correlation Filters, *Appl. Opt.*, **29,** 2997–3006 (1990).

Liu, H. K., J. A. Davis, and R. A. Lilly, Optical-Data-Processing Properties of a Liquid-Crystal Spatial Light Modulator, *Opt. Lett.*, **10,** 635 (1985).

Mu, G., X. Wang, and Z. Wang, Amplitude-Compensated Matched Filtering, *Appl. Opt.*, **27,** 3461 (1988).

Oppenheim, A. V., and J. S. Lim, The Importance of Phase in Signals, *Proc. IEEE*, **69,** 529–541 (1981).

Perez, O., and M. A. Karim, An Efficient Implementation of Joint Fourier Transform Correlator Using a Modified LCTV, *Microwave Opt. Technol. Lett.*, **2,** 193–196 (1989).

Shamir, J., H. J. Caulfield, and J. Rosen, Pattern Recognition Using Reduced Information Content Filters, *Appl. Opt.*, **26,** 2311–2314 (1987).

Turin, G. L., An Introduction to Matched Filters, *IRE Trans. Inf. Theory*, **IT-6,** 311–329 (1960).

Vander Lugt, A. B., Signal Detection by Complex Spatial Filtering, *IEEE Trans. Inf. Theory*, **IT-10,** 139–145 (1964).

Yu, F. T. S., *Optical Information Processing*, Wiley, New York, 1983.

Yu, F. T. S., S. Jutamalia, T. W. Lin, and D. Gregory, Adaptive Real-Time Pattern Recognition Using a Liquid Crystal TV Based Joint Transform Correlator, *Appl. Opt.*, **26,** 1370 (1987).

Yu, F. T. S., Q. W. Song, Y. S. Cheng, and D. A. Gregory, Comparison of Detection Efficiencies for Vander Lugt and Joint Transform Correlators, *Appl. Opt*, **29,** 225–232 (1990).

DIGITAL LOGIC

CHAPTER SIX

Number Systems

6.1 INTRODUCTION

An electronic digital system corresponding to a specified function is usually formed by a combination of logic circuit components connected in a specific fashion such that all involved variables (input and output) are designated by only two states. Customarily, electronic devices used in the digital circuit components are both reliable and consistent in their behavior as long as they are maintained in either of these two permissible states. These two states may correspond, respectively, to high- and low-voltage outputs or to closed and open switches. In the digital domain, the combination of these two discrete states is used to represent numbers, characters, codes, and other information.

An understanding of the binary (i.e., two-valued) number system is essential for assessing the characteristics of electronic digital circuits. However, in optics-based logic circuitry, we are not necessarily restricted to using the binary number system. Therefore, in this chapter, in addition to examining the binary number system, we also investigate several other number systems and their interrelationships with one another. Accordingly, we shall then be able to manipulate the circuit inputs and formulate (in Chapter 7) logic circuits for generating output(s) of interest.

6.2 NUMBER REPRESENTATIONS

6.2.1 Decimal Numbers

The number system that we have learned to depend on the most is the base-10 system, also referred to as the decimal number system. This number system is based on the ten Arabic numerals 0, 1, 2, 3, 4, 5, 6, 7, 8, and 9. In fact, use of the base-10 number system has become so habitual that we seldom contemplate the fact that we only use these ten Arabic numerals and can still depict any arbitrary number. This feat is actually accomplished by designating weights to the locations of the digits in a multidigit number. For example, the base-10 number 486.325_{10} can be expressed in polynomial form as

$$(486.325)_{10} = 4 \times 10^2 + 8 \times 10^1 + 6 \times 10^0 + 3 \times 10^{-1} + 2 \times 10^{-2} + 5 \times 10^{-3}.$$

Weights assigned to the integer digits are, respectively, 1, 10, 100, and so on, depending on the position of digits with reference to the decimal point. Again, as we encounter less significant fraction related digits, they are, respectively, assigned the weights 0.1, 0.01, 0.001, and so on.

In the base-R number system, a fixed-point numerical value N may be written

$$(N)_{10} = (a_n a_{n-1} \cdots a_1 a_0 \cdot a_{-1} a_{-2} \cdots a_{-m}) \tag{6.1}$$

where $0 \leq a \leq R - 1$ such that

$$\begin{aligned}(N)_{10} &= a_n R^n + a_{n-1} R^{n-1} + \cdots + a_1 R^1 + a_0 R_0 \\ &\quad + a_{-1} R^{-1} + a_{-2} R^{-2} + \cdots + a_{-m} R^{-m}\end{aligned} \tag{6.2}$$

where the digits a_n through a_0 correspond to the integer portion and the digits a_{-1} through a_{-m} correspond to the fractional portion. The manner in which numbers are represented places subtle and certain fundamental limitations on what types of calculations can be performed on them. One only needs to try performing any computation using Roman numerals and very quickly you will realize the impact and importance of the Arabic number system.

Our discussion so far is limited to decimal numbers of the fixed-point type, implying that the decimal point is fixed. A property of all fixed-point number systems is the interdependence of digits in arithmetic computations. Accordingly, in the addition operation, for example, the carry-out will have to propagate from lower to higher significant digits. Consequently, carry propagation becomes a fundamental limitation for many high-speed digital electronic addition schemes. Advances in VLSI, however, have increased the feasibility of floating point arithmetic elements. In general, a floating point number consists of a pair of fixed-point numbers, a mantissa M, and an exponent E. Accordingly, a floating point number N is given by

$$N = MR^E \tag{6.3}$$

where R is the radix of the floating point system. Typically, the exponent is an integer, while the mantissa is a combination of integer and fraction parts.

6.2.2 Base-2^n Number Systems

In theory, R can be almost any integer; however, it may not be wise to make it too large. In particular, R is equal to 2 when electronic digital circuits are involved. But as we shall see later, a base-2 (binary) number consisting of a combination of

1's and 0's may be too large a string. Each binary digit is called a bit. The number of bits in the binary representation poses no problem since in most electronic digital systems the operations take place in parallel. For the sake of ease of writing or of data entry, however, a binary number is often expressed in terms of its base-R equivalent when $R = 2^n$ and then entered for processing. In practice, only two values of n (i.e., 3 and 4) are used for this purpose. The corresponding number systems are, respectively, base-8 (octal) and base-16 (hexadecimal) systems. The binary numbers necessitate a longer sequence of digits than either of the other two systems. Accordingly, in this section we investigate binary, octal, and hexadecimal numbers. Table 6.1 shows the corresponding numbers in these three number systems. A scrutiny of this table should assure us that a mechanized representation of a base-10 number would be simplest in the base-2 number system since it requires only two values. In such a case, the electronic processor will have to account for only two attributes, such as high and low or on and off.

The privilege of using the binary number system is that there are many devices, even optical ones, that exhibit two stable states. One of these stable states can be designated by the value 1, while the other is designated 0. In solid state devices, for example, the bits are represented by voltages. In general, we specify a distinct range of voltages as logic 1 and another span of voltages as logic 0. The two allowed ranges are isolated by a forbidden range of voltages. This separation, in general, provides for accuracy in the device output.

Conversion between binary, octal, and hexadecimal equivalents is rather straightforward, as will be shown later, since all their bases are members of the

TABLE 6.1 Equivalence Table

$(N)_{10}$	$(N)^2$	$(N)_8$	$(N)_{16}$
0	0	0	0
1	1	1	1
2	10	2	2
3	11	3	3
4	100	4	4
5	101	5	5
6	110	6	6
7	111	7	7
8	1000	10	8
9	1001	11	9
10	1010	12	A
11	1011	13	B
12	1100	14	C
13	1101	15	D
14	1110	16	E
15	1111	17	F
16	10000	20	10
⋮	⋮	⋮	⋮
226	11100010	342	E2

same family. However, converting a number of any arbitrary base to its equivalent in another arbitrary base may not be so easy. For such difficult conversions, two particular conversion schemes need to be studied: the conversion of a base-R number (where $R \neq 10$) to its base-10 equivalent and the conversion of a base-10 number to its base-R equivalent (where $R \neq 10$). Of these two, the former conversion directly follows Eq. (6.2). Each of the base-R digits is first multiplied by the respective R^j value, where j represents the respective positional value of the integer between n and $-m$. The products are subsequently added to yield the equivalent decimal number. However, the base-10 to base-R conversion involves two distinct processes: the conversion of the integer component and that of the fraction component.

To convert a decimal integer to its equivalent base-R integer, we must determine each of the integer coefficients (a_n through a_0) of Eq. (6.1). The base-10 integer, DI, can be expressed as

$$(DI)_{10} = (((\cdot \cdot \cdot (((a_nR + a_{n-1})R + a_{n-2})R + a_{n-3}) \cdot \cdot \cdot)R + a_1)R + a_0) \tag{6.4}$$

Therefore, when both sides of this equation are divided by R, the least significant coefficient, a_0, is obtained as the remainder. Subsequent divisions of the corresponding quotients by R yield the remaining coefficients, a_1, a_2, and so on, until all the coefficients have been identified.

The conversion of decimal fractions into their equivalent base-R fractions requires the evaluation of the fraction coefficients a_{-1} through a_{-m}. Whenever the base-10 fraction is multiplied by R, the integer part of the resulting product is a_{-1}. If only the generated fractional part is now multiplied again by R, the integer part of the resulting product would produce the next coefficient, a_{-2}. This process of multiplication by R can be carried out until all the other coefficients, a_{-3} through a_{-m}, have been identified. This conversion recipe is elucidated by Example 6.1. Unlike the integer conversions, however, conversions between decimal and binary fractions do not necessarily result in convergence. Instead, an infinite string of digits may be required for the equivalent number. This is complicated further by the fact that fractions like $(\frac{1}{3})_{10}$ are themselves inexact. Conversions between the binary, octal, and hexadecimal systems, however, are without such problems.

To maintain at least the same level of accuracy of the given d-digit decimal fraction, the equivalent binary fraction must be worked through at least b bits, where

$$b = I(3.32d + 1) \tag{6.5}$$

and $I(x)$ is the integer value of x such that

$$(10)^{-d} \leq (2)^{-b}. \tag{6.6}$$

Similarly, when a b-bit binary fraction is converted to its decimal fraction equivalent, the minimum number of decimal fractions is given by

$$d = I(0.3b + 1). \tag{6.7}$$

Example 6.1 Convert $(537.32)_{10}$ to its base-2 equivalent number.

Solution Proceeding first with the integer part, we obtain

		Remainder	
2	537		
2	268	1	Least significant
2	134	0	
2	67	0	
2	33	1	
2	16	1	
2	8	0	
2	4	0	
2	2	0	
2	1	0	
2	0	1	Most significant

Therefore $(537)_{10} = (1000011001)_2$. Again,

$$
\begin{aligned}
0.32 \times 2 &= 0.64 \longrightarrow 0 \quad \text{Most significant}\\
0.64 \times 2 &= 1.28 \longrightarrow 1\\
0.28 \times 2 &= 0.56 \longrightarrow 0\\
0.56 \times 2 &= 1.12 \longrightarrow 1\\
0.12 \times 2 &= 0.24 \longrightarrow 0\\
0.24 \times 2 &= 0.48 \longrightarrow 0\\
0.48 \times 2 &= 0.96 \longrightarrow 0\\
0.96 \times 2 &= 1.92 \longrightarrow 1 \quad \cdots
\end{aligned}
$$

Thus

$$
\begin{aligned}
(537.32)_{10} &= (1000011001.01010001 \cdots)_2\\
&= (1000011001.0101000)_2
\end{aligned}
$$

since $b = I(3.32 \times 2 + 1) = 7$.

Our discussion so far was limited to converting a decimal number to its equivalent nondecimal number and vice versa. When converting a particular nondecimal number to its equivalent in another nondecimal number system, say, from octal to binary, the octal number may first be converted to its equivalent decimal number, which is then converted to its binary equivalent. This two-step process is time consuming. However, the bases of some of the nondecimal number systems are related. For example, the bases 2, 8, and 16 can be obtained by raising 2 to the powers of 1, 3, and 4, respectively. This observation allows for easy conversion between the binary, octal, and hexadecimal number systems using the substitution scheme of Table 6.2. This can be explained as follows. When a given binary number is grouped in fours, starting from the binary point while scanning both to the left and to the right, the equivalent hexadecimal number is obtained by directly substituting each group of 4 bits by its hexadecimal equivalent. Similarly, when converting a hexadecimal number to its binary equivalent, each of the hexadecimal digits needs to be replaced by the corresponding group of 4 bits. The rule for converting a binary number to its octal equivalent and vice versa follows the same principle except that groups of 3 bits are used in place of the groups of 4 bits. The corresponding substitution rules are also listed in Table 6.2.

Example 6.2 Find the base-2 and base-16 equivalents of $(574.62)_8$.

Solution Using Table 6.2, we find

$$\begin{aligned}(574.62)_8 &= (101\ 111\ 100\ .\ 110\ 010)_2\\ &= (0001\ 0111\ 1100\ .\ 1100\ 1000)_2\\ &= (17C.C8)_{16}\end{aligned}$$

It must be emphasized that adding or deleting zeros from the extreme left of the integer part and from the extreme right of the fraction part leaves the number unchanged.

TABLE 6.2 Substitution Equivalence Between Binary and Octal and Between Binary and Hexadecimal Numbers

Octal	→	Binary	Hexadecimal	→	Binary
0		000	0		0000
1		001	1		0001
2		010	2		0010
3		011	3		0011
4		100	4		0100
5		101	5		0101
6		110	6		0110
7		111	7		0111
			8		1000
			9		1001
			A		1010
			B		1011
			C		1100
			D		1101
			E		1110
			F		1111

It is important to note that even though the microprocessors are trained to process only binary numbers, the microprocessor users are not always required to enter binary numbers. Such an option is desirable. It is much easier to remember a hexadecimal number like B70EE5 or an octal number 55607345, for example, instead of the rather lengthy binary number 101101110000111011100101. This simplification is accomplished by using switches that are coded in groups of three or four to facilitate the use of octal and hexadecimal numbers, respectively, However, in such a case, an additional circuit will have to be included to convert the actual input to the corresponding binary code.

6.2.3 Negative Number Representation

In fixed-point number systems, numbers are represented using n bits: a sign bit (the most significant bit of the number, a_{n-1}) and $n - 1$ data or magnitude bits (a_{n-2} through a_0). However, there are three common ways to represent base-R negative fixed-point numbers:

1. *Sign and Magnitude Scheme* A negative number is obtained from the corresponding positive number by keeping the magnitude unchanged while the sign bit is changed from 0 to $R - 1$.
2. *Signed (R − 1)'s Complement Representation* The negative number is obtained by subtracting each digit of the corresponding positive number from $R - 1$.
3. *Signed R's Complement Representation* The negative number is determined by adding a 1 to the least significant position of the corresponding $(R - 1)$'s complement negative number.

In all three representations, the sign bit of a positive number is expressed by a 0 and that of the negative number is represented by $R - 1$. This form provides for explicit representation of sign. When only base-2 numbers are concerned, the $(R - 1)$'s and R's complement representation schemes lead to what are commonly referred to as 1's complement and 2's complement numbers, respectively.

A decimal number (0 09852) when complemented gives $(1\ 09852)_{10}$ in sign and magnitude representation, $(1\ 90147)_{10}$ in 9's complement scheme, and $(1\ 90148)_{10}$ in 10's complement representation. Similarly, the binary positive number $(0\ 0101101)_2$, for example, when complemented yields $(1\ 0101101)_2$ in sign-and-magnitude representation, $(1\ 1010010)_2$ in signed 1's complement representation, and $(1\ 1010011)_2$ in signed 2's complement representation.

For the binary number system, in particular, the rules for determining the complements can be further simplified. For finding 1's complement of a binary number, every digit is complemented by changing each 1 to a 0 and each 0 to a 1. Interestingly, the 2's complement can be obtained by scanning the corresponding positive number from the right to the left and complementing all bits that are encountered after the first found 1. For the fixed point binary numbers, the complement number N_c is represented as

$$N_c = \begin{cases} (1 - 2a_{n-1}) \sum_{j=0}^{n-1} a_j 2^{(j-n+1)}, & \text{sign and magnitude} \\ \sum_{j=0}^{n-2} (a_j - a_{n-1}) 2^{j-n+1}, & \text{1's complement} \\ -a_{n-1} + \sum_{j=0}^{n-2} a_j 2^{j-n+1}, & \text{2's complement} \end{cases} \tag{6.8}$$

It is interesting to note that the 2's complement technique is the only means by which both +0 and −0 have the same representation. Table 6.3 compares 4-bit fractional fixed-point numbers in all three of the complement representations. Another significant difference between the 2's complement representation and the other two representations is their behavior under truncation. Both of these considerations are why almost all digital electronic computers use 2's complement representation for data processing.

6.2.4 Multiple-Valued Number System

Even though electronic computing has almost succeeded in keeping itself away from anything but the binary number system, questions have continually been asked about use of the binary number system. These questions have become more probing since the concept of optical computing has begun to lay the groundwork for competing all-optical as well as optoelectronic computing technologies. With the binary number system, there is an inefficient use of interconnect in the computers. This issue need not be ignored since the major area of the VLSI chip is wasted in interconnection. Traditional approaches taken to increase the logic density and capacity of VLSI include increased interconnection through multilayer conductors

TABLE 6.3 Comparison of 4-Bit Fractional Fixed-Point Numbers

Number	Sign and Magnitude	1's Complement	2's Complement
$+\frac{7}{8}$	0 111	0 111	0 111
$+\frac{3}{4}$	0 110	0 110	0 110
$+\frac{1}{2}$	0 100	0 100	0 100
$+\frac{1}{4}$	0 010	0 010	0 010
+0	0 000	0 000	0 000
−0	1 000	1 111	0 000
$-\frac{1}{4}$	1 010	1 101	1 110
$-\frac{1}{2}$	1 100	1 011	1 100
$-\frac{3}{4}$	1 110	1 001	1 010
$-\frac{7}{8}$	1 111	1 000	1 001

and stacked device structures. For example, a base-4 signal on a single wire carries twice the information carried by a binary signal. In addition, such multiple-valued logic (with $R > 2$) may be useful in reducing the wiring between the chips and thus solve the pin-out problem, where an insufficient number of pins restricts the flow of information. Furthermore, multiple-valued circuits are expected to be more compact.

In the previous section, we discussed two multiple-valued number systems, namely, base-8 and base-16, besides the decimal. It must be emphasized, however, that the use of the aforementioned two number systems in an electronic computer is marginal. The number crunching is still done in binary while these base-2^n numbers are used solely for the convenience of data entry.

While it is almost a nonissue to name the single-variable binary functions, it is not so in the case of multiple-valued logic systems. A base-R system has R possible outputs for each of R possible input states. This amounts to R^R functions of a single base-R variable. In the case of binary, the number of possible functions is four, namely, identity, complement, and the constants 0 and 1. For a simple base-3 (referred to also as either trinary or ternary), the number of such functions become 27. A base-4 number contributes to a total of 256 such functions of which only four are constants. Trying to give names to the remaining 252 functions can lead to real confusion.

The trinary number system is relatively more optimal in terms of storage complexity and interconnection. We examine this particular scheme, as an example of multiple-valued number systems, in detail. Generally, there are two particular sets of numbers that can be used for the trinary number system: a positive set (consisting of 0, 1, and 2) and a symmetric set (consisting of $\bar{1}$, 0, and 1), where $\bar{1}$ refers to -1. Using Eq. (6.2) and either set, an unsigned trinary number N can be represented; by using the symmetric set, it is possible to represent a signed number. For example, the decimal number 345 can be represented as $(110210)_3$ using the positive trinary numbers or $(111\bar{1}10)_3$ using the symmetric trinary numbers. The negative number $(-345)_{10}$, on the other hand, can be represented by only $(\bar{1}\bar{1}\bar{1}1\bar{1}0)_3$ in the symmetric trinary number system. Interestingly, we can see that $(-345)_{10} = (\bar{1}\bar{1}\bar{1}1\bar{1}0)_3$ can be generated from $(345)_{10} = (111\bar{1}10)_3$ by replacing each 1 with $\bar{1}$ and each $\bar{1}$ with 1.

6.2.5 Modified Signed Digit Number System

Modified signed-digit (MSD) number representation is an alternate means of representing both positive and negative numbers capable of carry-free addition and subtraction with both fixed-point as well as floating point capability. For simplicity, only the fixed-point representation is addressed here. MSD numbers have three members ($\bar{1}$, 0, and 1) like those of the symmetric trinary system, but its positional weights follow the convention of binary numbers. In fact, MSD is a subset of the signed-digit representation with radix $R > 2$ and members forming the set of $\{-a, \cdots, -1, 0, 1, \cdots, a\}$. This signed-digit representation is a redundant number system since it allows for more than one representation of the numbers. The value of R is 2 for the MSD number system. For example, using 6-bit trinary precision, $(11)_{10}$ can be represented as follows:

$$
\begin{aligned}
(11)_{10} &= (01\bar{1}1\bar{1}1)_{\text{MSD}} = 2^4 - 2^3 + 2^2 - 2^1 + 2^0, \\
&= (10\bar{1}\bar{1}10)_{\text{MSD}} = 2^5 - 2^3 - 2^2 + 2^1, \\
&= (001011)_{\text{MSD}} = 2^3 + 2^1 + 2^0.
\end{aligned}
$$

Obviously, the sign information is carried along with this representation. MSD negative number is the complement of the MSD positive number. This should be obvious from our previous discussion of symmetric trinary numbers. Accordingly, one can show that $(-11)_{10}$ can be given by either $(0\bar{1}1\bar{1}1\bar{1})_{\text{MSD}}$, or $(\bar{1}0 1 1\bar{1}0)_{\text{MSD}}$, or $(00\bar{1}0\bar{1}\bar{1})_{\text{MSD}}$.

6.2.6 Residue Number System

All the number systems considered so far are linear, positional, and weighted. In the binary system, for example, the weights of the positions are $2^{n-1}, \cdots, 2^2, 2^1$, and 2^0. In comparison to all these number systems, the residue number system is an unweigthed system. It uses positional bases that are relatively prime (having no common factor) to each other, which are referred to as moduli. Residue number systems are inherently integer systems by their very definition. An integer N can be represented by a p-tuple $\{N_1, N_2, \cdots, N_p\}$, where $N_j = |N|_a$ (read N mod a) is the positive remainder that is obtained by dividing N with each of the moduli a.

To convert a conventionally weighted number to the residue system, the residues of the number with respect to each of the positional moduli are determined. For any given set of relatively prime moduli $a_1, a_2, \cdots, a_p$, the set of residues is unique for a dynamic range of

$$M = a_1 \times a_2 \times \cdots \times a_p \tag{6.9}$$

Consider, for example, a set of three moduli 2, 3, and 7. These moduli cover a dynamic range of $2 \times 3 \times 7 = 42$. For instance, the residues corresponding to 31 are, respectively, 1, 1, and 3 since 31 can be expressed as $2 \times 15 + 1$, $3 \times 10 + 1$, and $7 \times 4 + 3$. Table 6.4 lists the numbers 0 through 41 and their residues to bases 2, 3, and 7. All these sets of residues uniquely identify a number. For example, the set of residues $\{1, 2, 1\}$ identifies 29 just as uniquely as binary 11101.

The important, advantage of the residue number system, as will be shown later, is the absence of carry-outs between the columns during addition and subtraction. It therefore allows for high-speed computation even when the involved numbers are large. This is contrary to what usually takes place in the case of the weighted numbers when the computation time increases with the increase of the value of the involved numbers.

TABLE 6.4 Residues to Bases 2, 3, and 7

Number	Residue 2	3	7	Number	Residue 2	3	7	Number	Residue 2	3	7
0	0	0	0	14	0	2	0	28	0	1	0
1	1	1	1	15	1	0	1	29	1	2	1
2	0	2	2	16	0	1	2	30	0	0	2
3	1	0	3	17	1	2	3	31	1	1	3
4	0	1	4	18	0	0	4	32	0	2	4
5	1	2	5	19	1	1	5	33	1	0	5
6	0	0	6	20	0	2	6	34	0	1	6
7	1	1	0	21	1	0	0	35	1	2	0
8	0	2	1	22	0	1	1	36	0	0	1
9	1	0	2	23	1	2	2	37	1	1	2
10	0	1	3	24	0	0	3	38	0	2	3
11	1	2	4	25	1	1	4	39	1	0	4
12	0	0	5	26	0	2	5	40	0	1	5
13	1	1	6	27	1	0	6	41	1	2	6

The residue number system as defined thus far only manipulates positive numbers. Thus to handle carry-free subtraction operation, complement coding is used to represent negative numbers. The signed residue system is realized by designating numbers less than $M/2$ as positive and those equal to or greater than $M/2$ as negative. A number $N \geq M/2$ is treated as $N - M$, such that

$$(N - M)\text{mod } M = N \text{ mod } M \tag{6.10}$$

and the complement of N mod M is

$$(\overline{N \text{ mod } M}) = (M - N)\text{mod } M \tag{6.11}$$

In residue representation, where $N = \{N_1, N_2, \cdots, N_p\}$ such that $N_j = N$ mod a_j, the complement of N (represented as $\overline{N}$) is the complement of $\{N_1, N_2, \cdots, N_p\}$. Then

$$\overline{N_j} = (a_j - N_j)\text{mod } a_j \tag{6.12}$$

and

$$\overline{N} = \{\overline{N_1, N_2, \cdots, N_p}\} = \{\overline{N}_1, \overline{N}_2, \cdots, \overline{N}_p\} \tag{6.13}$$

since $a_j - N_j$ and $-N_j$ are congruent mod a_j. In the 2,3,7 residue system, $M = 42$. Accordingly, the dynamic range of the corresponding signed residue number system spans from -21 through $+20$. While 0 through 20 are treated as positive numbers, the numbers 21 through 41 are taken to represent, respectively, the neg-

TABLE 6.5 Signed Residue Numbers for the Bases 2, 3, and 7

	Residue				Residue				Residue		
Number	2	3	7	Number	2	3	7	Number	2	3	7
0	0	0	0	14	0	2	0	−14	0	1	0
1	1	1	1	15	1	0	1	−13	1	2	1
2	0	2	2	16	0	1	2	−12	0	0	2
3	1	0	3	17	1	2	3	−11	1	1	3
4	0	1	4	18	0	0	4	−10	0	2	4
5	1	2	5	19	1	1	5	−9	1	0	5
6	0	0	6	20	0	2	6	−8	0	1	6
7	1	1	0	−21	1	0	0	−7	1	2	0
8	0	2	1	−20	0	1	1	−6	0	0	1
9	1	0	2	−19	1	2	2	−5	1	1	2
10	0	1	3	−18	0	0	3	−4	0	2	3
11	1	2	4	−17	1	1	4	−3	1	0	4
12	0	0	5	−16	0	2	5	−2	0	1	5
13	1	1	6	−15	1	0	6	−1	1	2	6

ative numbers −21 through −1. The scheme to find the representation of a negative number is therefore straightforward. The number $(17)_{10}$, for example, in the signed residue number system is represented by $\{1,2,3\}$. The number $(-17)_{10}$'s residue representation is therefore $\{2-1 \bmod 2,\ 3-2 \bmod 3,\ 7-3 \bmod 7\}$ or $\{1,1,4\}$. Table 6.5 shows the corresponding signed residue representation of the numbers −21 through +20.

6.2.7 Logarithmic Number System

We have seen that in order to accommodate large dynamic range using the residue number system, for example, large numbers of large moduli are necessary. This problem of large dynamic range is often encountered by using instead a signed logarithmic number system. Like the residue number system, it also allows for high-speed computation but for relatively low precision computing systems. In this scheme, the number is represented by listing a sign bit and the logarithm of the absolute value of the said number. The sign bit s is generally a 0 for all numbers equal to or greater than 0, otherwise it is a 1. The binary logarithm of the magnitude of the number X is given by

$$L = \begin{cases} \log(|wX|), & \text{if } |X| > 1/w \\ 0, & \text{if } |X| \le 1/w \end{cases} \tag{6.14}$$

where w is a scaling factor to ensure that $L \ge 0$. The signed logarithm representation SL_x of the number is thus given by $(1-2s)2^L$. The number X may be found from SL_x by separating s and L from it and then applying the expression

$$X = (1-2s)2^L/w \tag{6.15}$$

In general, this number system is able to achieve a very large dynamic range with small word size. A parameter known as relative dynamic range and defined as the difference between the most positive and the most negative numbers divided by the step size for small signals may be used to make this comparison. The relative dynamic range of an n-bit 2's complement number system is 2^{2n-1}, whereas it is $2^m - 1$ for an k-bit signed logarithm number system where $m = 2^{k-1}$. Accordingly, a 6-bit signed logarithm number has a larger relative dynamic range than a 16-bit 2's complement number.

6.3 CODES

Even though digital electronic computers process only binary numbers, several special binary codes have been developed for representing numbers. There are occasions when it is difficult for the user to convert a decimal number to its binary equivalent, for example, in real-time. Binary codes are designed to address this issue of user convenience. A minimum of n bits may be used to code 2^n distinct quantities in binary. Since there is no constraint on the maximum number of bits, if efficient codes are not essential, there could be many possibilities for binary coding.

Consider the decimal number 54.25. We can express it in binary as 110110.01 or each of the individual decimal digits can be coded separately to give 0101 0100. 0010 0101. This alternative means of expressing the number in binary form requires few additional bits but it is significantly more handy and useful since the coded version could readily be obtained. This form of representation is referred to as the binary coded decimal (BCD) code. The BCD code, also referred to as the 8-4-2-1 code, makes use of the binary system in that the 4-bit binary representations of only the ten decimal digits are used. Accordingly, the 4-bit combinations 1010 through 1111 are not used in BCD.

Table 6.6 shows a few of the other codes that might be used to code each decimal digit to form a binary coded number. These codes are referred to as weighted because each bit position is assigned a weight. The XS3 (excess-3) code

TABLE 6.6 Some Weighted Binary Codes

Decimal	BCD (8421)	XS3	4221	5043210
0	0000	0011	0000	0100001
1	0001	0100	0001	0100010
2	0010	0101	0010	0100100
3	0011	0110	0011	0101000
4	0100	0111	1000	0110000
5	0101	1000	0111	1000001
6	0110	1001	1100	1000010
7	0111	1010	1101	1000100
8	1000	1011	1110	1001000
9	1001	1100	1111	1010000

is very similar to the BCD code, except that the binary value of each XS3 code is three more than the corresponding BCD value. One advantage of this particular code over the BCD code is that an XS3 code can be complemented simply by taking its 1's complement. For example, 4 in XS3 is given by 0111, whereas its complement 5 (=9 − 4) is represented in XS3 as 1000, which can be obtained readily by taking 1's complement of 0111.

The 5043210 code listed in Table 6.6 is a 2-of-7 code, meaning that in each code group two of the seven bits are 1. Codes such as this one can be used for detecting errors that may be introduced during data processing. The codes listed in Table 6.6 are just a sample of the very many in use. In fact, coding theory has become a separate discipline. Because of the many coding possibilities, devices such as encoders and decoders are used in digital systems for translating one code to another.

Codes are devised not necessarily for only binary numbers. They are encountered in the case of multivalued numbers as well. For example, it may be more difficult to realize an optical multistable switch than its binary equivalent. This may be reason enough to code trinary numbers using binary representations. Using a scheme referred to as pulse-position coding, each of the trinary digits may be represented by a combination of two binary digits. For example, one may decide to code the positive trinary numbers as follows:

$$\begin{aligned} 0_3 &= 0,\ 0 \\ 1_3 &= 1,\ 0 \text{ or } 0,\ 1 \\ 2_3 &= 1,\ 1 \end{aligned} \tag{6.16}$$

while the symmetric trinary numbers may be coded by the following combinations:

$$\begin{aligned} \bar{1}_3 &= 0,\ 1 \\ 0_3 &= 0,\ 0 \text{ or } 1,\ 1 \\ 1_3 &= 1,\ 0 \end{aligned} \tag{6.17}$$

The binary coded positive trinary number $(110210)_3 = 345_{10}$ can be represented as either (110110, 000100) or (000100, 110110). Similarly, the equivalent binary coded symmetric trinary number $(111\bar{1}10)_3$ can be represented as either (111010, 000100) or (111011, 000101).

6.4 ARITHMETIC OPERATIONS

6.4.1 Binary Arithmetic

To realize binary subtraction, multiplication, and division, one needs to be able to perform addition and take either 1's or 2's complement. The common decimal

TABLE 6.7 Rules for Binary Addition

Input			Output	
Augend	Addend	Carry-In	Carry-Out	Sum
0	0	0	0	0
0	0	1	0	1
0	1	0	0	1
0	1	1	1	0
1	0	0	0	1
1	0	1	1	0
1	1	0	1	0
1	1	1	1	1

arithmetic rules apply equally well to the binary arithmetic as well. Table 6.7 lists the addition operation rules in the form of a truth table. The addition of binary numbers may be described in terms of the two input variables, addend and augend, at the least significant position. At the remaining bit positions, however, an additional variable, carry-in, will be necessary. For all the bit positions, however, there can be only two outputs, the carry-out and the sum. The sum output is a 1 when only one input is a 1 or when all the three inputs are equal to 1. The carry-out is a 1 if two or more of the three inputs are equal to 1.

The rules for binary subtraction may be derived directly from decimal subtraction as shown in Table 6.8. Subtraction involves three input variables, the minuend, the subtrahend, and the borrow-in, while the output variables are the borrow-out and the difference.

Subtraction as summarized in Table 6.8 may appear to be an additional burden for a machine trained in addition. Subtraction of a number B from a number A is equivalent to performing an addition of A and the complement of B. Again, since multiplication uses repeated addition and division uses repeated subtraction, perhaps the computing machine needs to be equipped with the understanding of only the addition process as it is alone capable of all possible arithmetic operations.

TABLE 6.8 Rules for Binary Subtraction

Input			Output	
Minuend	Subtrahend	Borrow-In	Difference	Borrow-Out
0	0	0	0	0
0	0	1	1	1
0	1	0	1	1
0	1	1	0	1
1	0	0	1	0
1	0	1	0	0
1	1	0	0	0
1	1	1	1	1

The idea central to such expectation, however, depends on our ability to handle complemented binary numbers.

The binary addition of two numbers, at least one of which is a negative number represented in 2's complement form, can be carried out as follows. All bits including the sign bits (the most significant bit) are added and the carry-out of the sign bit is discarded. If the resultant sign bit is a 1, the 2's complement sum is negative; otherwise, it is interpreted as positive provided the sum hasn't resulted in any overflow. Again, if at least one of the two operands is a negative number represented in 1's complement form, the carry-out of the sign bit is added to the result. If the resulting sign bit is a 1, the resulting 1's complement sum is negative; otherwise, it is positive.

Example 6.3 (a) Add $(14)_{10}$ and $(-6)_{10}$, and (b) subtract $(11)_{10}$ from $(4)_{10}$.

Solution For part (a), we have

$$\begin{aligned} (-6)_{10} &= (1\ 111001)_2 \quad \text{in signed 1's complement} \\ &= (1\ 111010)_2 \quad \text{in signed 2's complement} \end{aligned}$$

Therefore, either

$$\begin{array}{rrl} (14)_{10} = & 0\ 001110 & \\ & + & \\ (-6)_{10} = & 1\ 111001 & \text{1's complement} \\ \hline & 10\ 000111 & \\ & 1 & \\ \hline & 0\ 001000 & (8)_{10} \end{array}$$

or

$$\begin{array}{rrl} (14)_{10} = & 0\ 001110 & \\ & + & \\ (-6)_{10} = & 1\ 111010 & \text{2's complement} \\ \hline & 0\ 001000 & = (8)_{10} \end{array}$$

where the carry has been discarded.

For part (b), we have

$$(-11)_{10} = (1\ 10100)_2 \quad \text{in signed 1's complement}$$
$$= (1\ 10101)_2 \quad \text{in signed 2's complement}$$

Therefore, either

$$\begin{array}{rrl} (4)_{10} = & 0\ 00100 & \\ & + & \\ (-11)_{10} = & 1\ 10100 & \text{1's complement} \\ \hline & 01\ 11000 & \\ & 0 & \\ \hline & 1\ 11000 & = (-7)_{10} \end{array}$$

or

$$\begin{array}{rrl} (4)_{10} = & 0\ 00100 & \\ & + & \\ (-11)_{10} = & 1\ 10101 & \text{2's complement} \\ \hline & 01\ 11001 & = (-7)_{10} \end{array}$$

It is appropriate to introduce an alternative form of binary addition. This scheme, referred to as symbolic substitution, consists of n substitution steps so that the sum is finally obtained without actually adding the two given n-bit numbers. First, the jth pair of the addend (a_j) and augend (b_j) bits are replaced by a pair of intermediate augend (at the $(j + 1)$th position) and addend (at the jth position) bits according to the rules of Table 6.9 while the new values of both a_0 and b_{n-1} are set equal to 0. It must be noted that the entries of Table 6.9 are the same as those of

TABLE 6.9 Substitution Rules for Binary Addition

$a_j b_j$	$a_{j+1} b_j$
00	00
01	01
10	01
11	10

2-bit addition with the replaced set of numbers corresponding to carry-out and sum, respectively. This substitution process is repeated n times until the augend becomes equal to the final sum. The substitution-based addition scheme is illustrated in Example 6.4.

Example 6.4 Use the substitution scheme for adding 21_{10} and 13_{10}.

Solution The substitution rules can be applied in parallel and four iterations are required due to the propagation of carries. Note that when the fourth iteration is completed, no further iterations will change the result.

Initial	$0010101 = 21_{10}$
	$0001101 = 13_{10}$
First iteration	$0001010 = 10_{10}$
	$0011000 = 24_{10}$
Second iteration	$0010000 = 16_{10}$
	$0010010 = 18_{10}$
Third iteration	$0100000 = 32_{10}$
	$0000010 = 2_{10}$
Fourth iteration	$0000000 = 0_{10}$
	$0100010 = 34_{10}$

It is important to realize that the binary number system is a weak candidate for the substitution-based addition scheme. The substitution-based iteration operations are inherently parallel but the complete process involves too many iterations. Thus the overall process involving binary numbers is not attractive enough for high-speed computation. However, we shall find later that the MSD-based substitution scheme is able to overcome this apparent bottleneck.

6.4.2 BCD Arithmetic

The BCD form may be used also to perform arithmetic operations of numbers using the standard binary adders. But since the BCD scheme does not make use of six of the sixteen different 4-bit combinations, the corresponding arithmetic leads to two specific problems: (1) the use of binary adders results in 4-bit combinations that may not represent allowable BCD codes without further modification; and (2) and 9's complement of BCD values is not readily obtained from the BCD code. The latter problem can be solved, however, by the use of the XS3 code, in which each decimal digit is represented by a binary number 3 greater than the decimal digit.

The XS3 code has a distinct advantage, which can be explained as follows. When ordinary binary addition is performed on two decimal digits represented in XS3 code, their binary sum produces a carry bit from only those digit positions that would also have produced a carry if the numbers were represented in the base-10 system. Again, subtraction of XS3 numbers involves a scheme that is similar to the subtraction of binary numbers. In fact, XS3 subtraction involves adding XS3 minuend to the complement of XS3 subtrahend. Interestingly, the 9's complement of a XS3 number is readily obtained by complementing each bit of the corresponding XS3 code.

Example 6.5 Add the decimal numbers 2644 and 2474 using (a) XS3 and (b) BCD codes.

Solution (a) For the XS3 code,

Carries:	1100	1	1	0	0
	2644	0101	1001	0111	0111
	2474	0101	0111	1010	0111
	5118	1011	0001	0001	1000

(b) For the BCD code,

Carries:	1100	0	0	0	0
	2644	0010	0110	0100	0100
	2474	0010	0100	0111	0100
	5118	0100	1010	1011	1000

In Example 6.5, the carries only in the addition of XS3 numbers occur at the same digit positions as in decimal addition. Note that the XS3 addition shown above has resulted in an undefined number 0001 and similarly the corresponding BCD addition has resulted in two undefined numbers 1010 and 1011. In the case of the XS3 addition, all the resulting XS3 numbers are incorrect, while in the case of BCD addition, all but the least significant digits are incorrect. In either case, it is obvious that additional processing will be necessary to convert coded decimal digits to their correct XS3 or BCD forms.

6.4.3 MSD Arithmetic

Addition of two n-digit MSD numbers (digit 0 through digit $n - 1$) can be carried out in several ways. The first scheme to be discussed consists of three steps. The overall process is relatively free of carry propagation. While the first two steps basically involve substituting each pair of input bits with a particularly unique pair of intermediate bits, the final step involves regular addition of the pairs of intermediate bits. The steps and the necessary operations are carried out in a manner such that when the numbers are finally added, no carry is generated. Thus the pairs of MSD numbers that will have to be generated by the first two steps will include the combinations 1 and $\bar{1}$, $\bar{1}$ and 1, 0 and 1, 0 and $\bar{1}$, 1 and 0, and $\bar{1}$ and 0.

For each pair of the given augend and addend input bits (a_j and b_j, respectively), a pair of first-order transfer and weight bits (T_j and W_j, respectively) are first determined as per the rules of Table 6.10. In the second step, each pair of T_{j-1} and W_j bits are replaced by a pair of second-order transfer and weight bits (T'_{j-1} and W'_{j-1}, respectively) for all $n > j \geq 1$. Finally, in the third step, the second-order weight bit W'_j and the second-order transfer bit T'_{j-1} are added to give the sum bit S_{j+1} for all $n - 1 > j \geq 1$. The remaining three bits of sum S_0, S_1, and S_n are, respectively, W_0, W'_0, and the sum of T'_{n-2} and T_{n-1}.

The MSD subtraction is accomplished, as in binary, by adding the complement of the subtrahend. Interestingly, determining the complement of a MSD number is very simple. It is accomplished by replacing 1 with $\bar{1}$ and $\bar{1}$ with 1.

Example 6.6 Realize (a) $27_{10} + 15_{10}$ and (b) $27_{10} - 15_{10}$, using MSD arithmetic.

Solution (a) There can be several MSD representations for both 27_{10} and 15_{10}. In this example, we take them to be

$$27_1 = 1110\bar{1} \quad \text{and} \quad 15_{10} = 1000\bar{1}$$

Therefore $27_{10} + 15_{10}$ can be determined as follows:

$$\begin{array}{lll}
a & 1110\bar{1} & = 27_{10} \\
b & 1000\bar{1} & = 15_{10} \\
T & 1110\bar{1} & \\
W & \;0\bar{1}\bar{1}00 & \\
T' & 0000 & \\
W' & \;10\bar{1}\bar{1} & \\
S & 110\bar{1}\bar{1}0 & = (42)_{10}
\end{array}$$

(b) Now subtracting 15_{10} from 27_{10} is equivalent to adding 27_{10} and -15_{10}. But $15_{10} = \bar{1}0001$; therefore the difference can be obtained as follows:

$$\begin{array}{lll}
a & 1110\bar{1} & = 27_{10} \\
b & \bar{1}0001 & = -15_{10} \\
T & 01100 & \\
W & \;0\bar{1}\bar{1}00 & \\
T' & 0000 & \\
W' & \;10\bar{1}0 & \\
S & 010\bar{1}00 & = (12)_{10}
\end{array}$$

An alternative form of MSD addition/subtraction can be carried out in just two steps. In the first step, the pairs of input MSD bits are replaced by pairs of transfer and weight bits as per the rules of Table 6.11. We see that for some of the cases, the values of the less significant input bits play an important role. In the final step, the pairs of T_{j-1} and W_j bits are added to give the carry-free sum.

TABLE 6.10 Substitution Rules for MSD Addition

$a_j b_j$	$\longrightarrow$	$T_j W_j$	$T_{j-1} W_j$	$\longrightarrow$	$T'_{j-1} W'_{j-1}$
00		00	00		00
01		$1\bar{1}$	01		01
$0\bar{1}$		$\bar{1}1$	$0\bar{1}$		$0\bar{1}$
11		10	11		10
10		$1\bar{1}$	10		01
$1\bar{1}$		00	$1\bar{1}$		00
$\bar{1}1$		00	$\bar{1}1$		00
$\bar{1}0$		$\bar{1}1$	$\bar{1}0$		$0\bar{1}$
$\bar{1}\bar{1}$		$\bar{1}0$	$\bar{1}\bar{1}$		$\bar{1}0$

Example 6.7 Obtain (a) $12_{10} + 15_{10}$ and (b) $12_{10} - 15_{10}$ using the two-step MSD addition scheme.

Solution (a) For $12_{10} + 15_{10}$, we obtain

$$
\begin{aligned}
a &= 1100 \\
b &= 1111 \\
T_j &= 1111 \\
W_j &= \;\;00\bar{1}\bar{1} \\
S &= \;\;\;\;1110\bar{1} = 27_{10}
\end{aligned}
$$

(b) For $12_{10} - 15_{10}$, we obtain

$$
\begin{aligned}
a &= 1100 \\
b &= \bar{1}\bar{1}\bar{1}\bar{1} \\
T_j &= 00\bar{1}0 \\
W_j &= \;\;001\bar{1} \\
S &= \;\;\;\;00\bar{1}1\bar{1} = -3_{10}
\end{aligned}
$$

TABLE 6.11 Substitution Rules for Two-Step MSD Addition

$A_j B_j$	$A_{j-1} B_{j-1}$	$T_j W_j$
11		10
10	Both positive	$1\bar{1}$
01	Otherwise	$0\bar{1}$
00 $0\bar{1}$ $\bar{1}1$		00
$0\bar{1}$	Both positive	$0\bar{1}$
$\bar{1}0$	Otherwise	$\bar{1}1$
$\bar{1}\bar{1}$		$\bar{1}0$

6.4.4 Residue Arithmetic

The primary advantage of using the residue number system is the absence of carries during both addition and multiplication of numbers. Arithmetic operations are performed on the corresponding residues without any interaction between the residues. Again since $X \bmod a - Y \bmod a = (X - Y) \bmod a$, the subtraction operation is also relatively straightforward. The guiding theorems for conducting addition, subtraction, and multiplication are given by

$$X \pm Y = \{[X \pm Y] \bmod a_k, [X \pm Y] \bmod a_{k-1}, \cdots, [X \pm Y] \bmod a_0\} \quad (6.18)$$

$$XY = \{XY \bmod a_k, XY \bmod a_{k-1}, \cdots, XY \bmod a_0\} \quad (6.19)$$

It must be noted that the computed result is generally valid as long as it hasn't exceeded the allowable dynamic range.

Operations such as sign detection, overflow detection, and division are not very easy using the residue number system. Algorithms exist for such purposes, but they are generally very complex, so much so that the advantage of using the residue number system may become questionable.

Example 6.8 Show the following operations using residue arithmetic and bases 3, 5, and 7: (a) 21 + 13; (b) 17 − 38; (c) 46 − 21; (d) 22 × 4; and (e) 68 + 53.

Solution

(a)

$$\begin{array}{rcll} 21 & \longrightarrow & \{0, 1, 0\} & \\ & & + & \\ 13 & \longrightarrow & \{1, 3, 6\} & \\ \hline & & \{1, 4, 6\} & \text{which is the residue of 34} \end{array}$$

(b)

$$\begin{array}{rcll} 17 & \longrightarrow & \{2, 2, 3\} & \\ & & + & \\ -38 & \longrightarrow & \{1, 2, 4\} & \text{using signed residue scheme} \\ \hline & & \{0, 4, 0\} & \text{which is the residue of } -21 \end{array}$$

(c)

$$\begin{array}{rcll} 46 & \longrightarrow & \{1, 1, 4\} & \\ & & + & \\ -21 & \longrightarrow & \{0, 4, 0\} & \text{using signed residue scheme} \\ \hline & & \{1, 0, 4\} & \text{which is the residue of 25} \end{array}$$

(d)

$$\begin{array}{rcll} 22 & \longrightarrow & \{1, 2, 1\} & \\ & & \times & \\ 4 & \longrightarrow & \{1, 4, 4\} & \\ \hline & & \{1, 3, 4\} & \text{which is the residue of 88} \end{array}$$

(e)

$$\begin{array}{rcll} 68 & \longrightarrow & \{2, 3, 5\} & \\ & & + & \\ 53 & \longrightarrow & \{2, 3, 4\} & \\ \hline & & \{1, 1, 2\} & \text{which is the residue of 16} \end{array}$$

This latter result is invalid since the true result 121 falls outside the dynamic range 0 through 104.

6.4.5 Signed Logarithmic Arithmetic

It is easier to comprehend first the multiplication and division operations in the logarithmic number system. This follows from the fact that the logarithm of the product of several factors is given by the sum of their logarithms. Considering that the product of two numbers X and Y were to be determined, the logarithmic representation of the product is given by

$$\begin{aligned} L_{XY} &= \log(|wXY|) \\ &= \log(|w^2XY|) - \log(w) \\ &= \log(|wX|) + \log(|wY|) - \log(w) \\ &= L_X + L_Y - \log(w) \end{aligned} \tag{6.20}$$

where L_X and L_Y are the logarithmic representations of X and Y, respectively, and the sign is determined by modulo 2 addition of the signs of the factors X and Y. In a manner similar to that of Eq. (6.20), the quotient of the two numbers X and Y can be obtained from

$$L_{X/Y} = \log(|wX|) - \log(|wY|) + \log(w) \tag{6.21}$$

Addition of numbers follows directly from both Eqs. (6.20) and (6.21). Since the sum of X and Y can be given by $S = X(1 + Y/X)$, the sum of two numbers can be expressed in the form of a product as well. Depending on the signs of X and Y, the logarithmic representations of the sum and difference can be found, respectively, to be

$$L_s = \begin{cases} L_X + f(L_Y - L_X), & \text{if } L_X \geq L_Y \\ L_Y + f(L_X - L_Y), & \text{if } L_Y > L_X \end{cases} \tag{6.22}$$

and

$$L_d = \begin{cases} L_X + g(L_Y - L_X), & \text{if } L_X \geq L_Y \\ L_Y + g(L_X - L_Y), & \text{if } L_Y > L_X \end{cases} \tag{6.23}$$

where $f(z) = \log(1 + 2^z)$ and $g(z) = \log(1 - 2^z)$.

6.5 PROBLEMS

1. Convert the base-R numbers listed below to their base-10 equivalents.
   (a) $(1010101.10111)_2$ **(b)** $(10101110.0011101)_2$
   (c) $(231230.3211)_3$ **(d)** $(347462.4342)_8$
   (e) $(456703.7641)_8$ **(f)** $(64A3200.11A)_{16}$
   (g) $(12122.12112)_3$
2. Convert the base-10 numbers listed below to their equivalent numbers in base 2, 3, 4, 5, 8, and 16.
   (a) 012546.203 **(b)** 4315162.672 **(c)** 43227160.0210
   (d) 4230202.03 **(e)** 11092.0911 **(f)** 4321.123
3. Truncate the 6-bit fractional fixed-point numbers (between 1 and -1) to form (a) 2-bit, (b) 3-bit, (c) 4-bit, and (d) 5-bit fractions and then plot the truncated values against the actual values of the fractions for all three forms of binary representations. Summarize the result.
4. Plot the signed logarithm values of numbers between -10 and 10 for three different values of the scale factor. Show that the signed logarithm is a monotonic function and find its relationship with the scaling factor w. Justify that a dead band of width $2/w$ exists around zero.

5. Show the following arithmetic operations using both two-step as well as three-step MSD addition scheme.
(a) $1433_{10} + 542_{10}$ (b) $1534_{10} - 2435_{10}$
(c) $1232_{10} + 124_{10} - 3256_{10}$ (d) $3241_{10} + 413_{10} - 23_{10}$

6. Conduct the following operations using a residue number system of bases 3, 5, and 9.
(a) $46_{10} + 53_{10}$ (b) $-42_{10} - 19_{10}$
(c) $59_{10} - 42_{10}$ (d) $12_{10} \times 7_{10}$
(e) $23_{10} + 91_{10} + 31_{10}$ (f) $8_{10} + 53_{10} - 24_{10}$

7. Derive Eqs. (6.22) and (6.23).

6.6 SUGGESTED READINGS

Abraham, G., Multiple-Valued Logic for Optoelectronics, *Opt. Eng.*, **25,** 3–13 (1986).

Cody, W. J., Jr., Static and Dynamic Numerical Characteristics of Floating Point Arithmetic, *IEEE Trans. Comput.*, **C-22,** 598 (1973).

Dao, T. T., and D. M. Campbell, Multiple-Valued Logic: An Implementation, *Opt. Eng.*, **25,** 14–21 (1986).

Eichmann, G., Y. Li, and R. R. Alfano, Optical Binary Coded Ternary Arithmetic and Logic, *Appl. Opt.*, **25,** 3113–3121 (1986).

Garner, H. L., Theory of Computer Addition and Overflows, *IEEE Trans. Comput.*, **C-27,** 297 (1978).

Gaylord, T. K., and M. M. Mirsalehi, Truth-Table Look-up Processing: Number Representation, Multilevel Coding, and Logical Minimization, *Opt. Eng.*, **25,** 22–28 (1986).

Gaylord, T. L., M. M. Mirsalehi, and C. C. Guest, Optical Digital Truth Table Look-up Processing, *Opt. Eng.*, **24,** 48 (1985).

Hurst, S. L., Multiple-Valued Threshold Logic: Its Status and Its Realization, *Opt. Eng.*, **25,** 44–55 (1986).

Johnson, E. L., and M. A. Karim, *Digital Design: A Pragmatic Approach*, PWS-Kent Publishing, Boston, 1987.

Smith, K. C., Multiple-Valued Logic: A Tutorial and Appreciation, *Computer*, 17–27 (April 1988).

Swartzlander, E. E., Merged Arithmetic, *IEEE Trans. Comput.*, **C-29,** 946 (1980).

Swartzlander, E. E., Digital Optical Arithmetic, *Appl. Opt.*, **25,** 3021–3032 (1986).

Swartzlander, E. E., Jr., D. V. S. Chandra, H. T. Nagle, Jr., and S. A. Starks, Sign/Logarithm Arithmetic for FFT Implementation, *IEEE Trans. Comput.*, **C-24,** 1238 (1975).

Waser, S., and M. J. Flynn, *Introduction to Arithmetic for Digital Systems Designers*, Holt, Rinehart and Winston, New York, 1982.

CHAPTER SEVEN

Logic Elements and Operations

7.1 INTRODUCTION

In Chapter 6, we studied a variety of different number systems as well as a number of different arithmetic operations using the different number systems. What we have identified so far are the operations but we haven't yet identified any devices suitable for these operations. However, while digital electronics have delivered devices suitable for only binary operations, there aren't any significant devices that could handle the nonbinary number systems. The potential for optical technology perhaps lies in other number systems. But in order to develop any such capability, it is necessary to understand how similar arithmetic and logic operations are realized using electronics.

The operations of digital electronic devices are based on several basic logic operations. An understanding of these operations is vital for both designing and analyzing digital electronic circuits. These logic functions operate on variables that are allowed to assume only two values, 1 or 0.

The concept of logic operations and devices (also referred to as logic gates) is used to develop strategies for defining an arbitrary logic function of multiple variables, for designing logic circuits, and for reducing the complexities of a logic circuit using both algebraic and graphical means. These discussions are followed by an introduction to several special types of logic devices, some of which are possible candidates for optical implementation.

7.2 BASIC LOGIC OPERATIONS

To describe any arbitrary logic function of multiple input variables, only three basic logic functions are necessary. They are AND, OR, and NOT logic functions. There are four other logic functions that can be derived from these three and they are also considered in this section. While only NOT logic is a single-input operation, the remaining logic operations deal with two or more inputs. The operations of these elementary functions or any other complex function are customarily explained by means of truth tables. In general, a truth table lists all possible variable value combinations and the corresponding values of the output(s).

7.2.1 AND Function

The function $f(A, B) = A \cdot B$ describes an AND logic operation of two inputs A and B if the output is a 1 when both A and B are 1 and is a 0 otherwise. For simplicity, the dot used for expressing the logical AND operation may be omitted. In any case, an AND operation should not be confused with the multiplication operation. The definition of two-input AND logic, however, may be expanded to include more than two input variables. For a multi-input device, the AND output is a 1 as long as all the inputs are 1 and 0 otherwise.

An AND function is best illustrated by an electrical circuit having multiple switches and a lamp all in series as shown in Fig. 7.1a. In this circuit, an open switch corresponds to logic 0 and a closed switch corresponds to logic 1. The lamp lights up, indicating an output of 1, when all the switches are closed. For all other switch combinations, the lamp remains turned off. Figure 7.1b shows the truth table and Fig. 7.1c shows the logic symbol of a two-input AND gate.

7.2.2 OR Function

The logical OR function of two input variables A and B, expressed mathematically as $f(A, B) = A + B$, defines a function that is a 1 when either A or B or both are 1. The logical sign + stands for an OR operation and it should not be misunderstood to represent an addition operation. The definition of OR logic may also be extended to include more than two input variables. For a multi-input logical OR device, the output is a 0 as long as all the inputs are 0 and a 1 otherwise.

The OR function can be understood from an analogy of the circuit shown in Fig. 7.2a. This electrical circuit has multiple switches connected in parallel and a lamp in series to the combination of switches. The lamp lights up, indicating an output of 1, as long as at least one of the switches remains closed and it remains

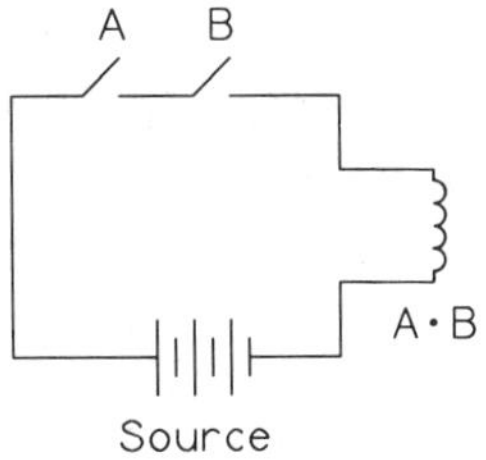

(a)

A	B	$A \cdot B$
0	0	0
0	1	0
1	0	0
1	1	1

(b)

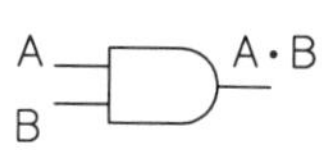

(c)

FIGURE 7.1 A two-input AND function described in terms of (a) an analogous circuit, (b) a truth table, and (c) a logic symbol.

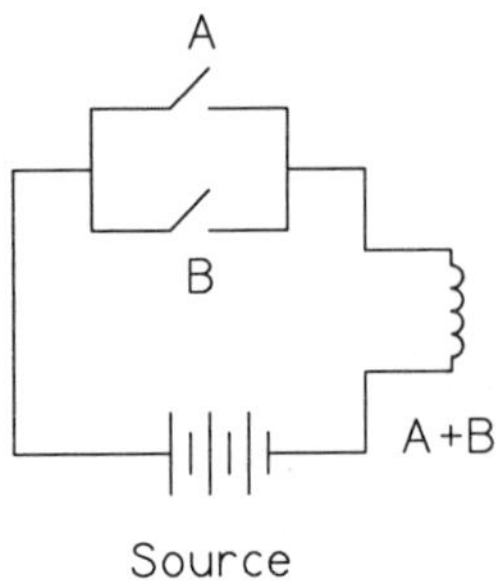

(a)

A	B	$A + B$
0	0	0
0	1	1
1	0	1
1	1	1

(b)

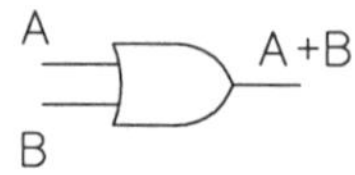

(c)

FIGURE 7.2 A two-input OR function described in terms of (a) an electrical analogy, (b) a truth table, and (c) a logic symbol.

turned off if all the switches are open. Figure 7.2b shows the truth table and Fig. 7.2c shows the logic symbol for a two-input logical OR function.

7.2.3 NOT Function

The NOT logic expressed as $f(A) = \overline{A}$ defines a single-input function of A, which is a 1 when the input is a 0, otherwise it is a 0. The bar on input A indicates the operation and is called interchangeably the NOT, or complement, of A. Figure 7.3 shows the truth table and the logic symbol of the NOT function.

7.2.4 NAND Function

The NAND logic function $f(A, B) = \overline{A \cdot B} = \overline{AB}$ of two inputs A and B defines a logic operation that is a complement of the corresponding AND function. In NAND logic therefore, the output is a 0 only when all the inputs are 1, otherwise it is a 1. A NAND logic function can be realized by a series combination of an AND function followed by a NOT function. Its truth table and logic symbol are shown in Fig. 7.4.

7.2.5 NOR Function

The function $f(A, B) = \overline{A + B}$ of two inputs A and B defines a NOR logical function that is the complement of the OR function. The NOR output is a 1 only when all the inputs are 0, and a 0 otherwise. The NOR logic is thus equivalent to

A	$\overline{A}$
0	1
1	0

(a)

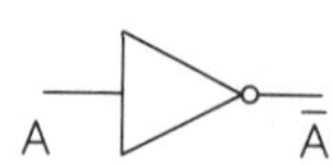

(b)

FIGURE 7.3 NOT function: (a) truth table, and (b) logic symbol.

A	B	$\overline{A \cdot B}$
0	0	1
0	1	0
1	0	0
1	1	0

(a)

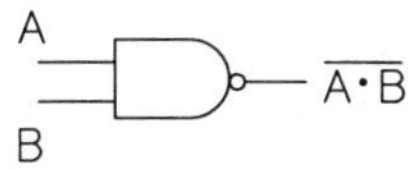

(b)

FIGURE 7.4 A two-input NAND function: (a) truth table and (b) logic symbol.

A	B	$\overline{A + B}$
0	0	1
0	1	0
1	0	0
1	1	0

(a)

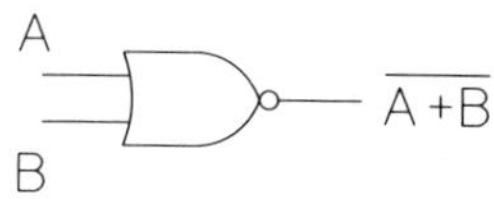

(b)

FIGURE 7.5 A two-input NOR function: (a) truth table and (b) logic symbol.

a series combination of an OR function followed by a NOT function. The truth table and logic symbol for a two-input NOR function are given in Fig. 7.5.

7.2.6 EXCLUSIVE-OR and EXCLUSIVE-NOR Functions

The function $f(A, B) = A \oplus B$ of two inputs A and B defines an EXCLUSIVE-OR (X-OR for short) logic operation. Its output is a 1 if an odd number of its inputs are 1, otherwise it is 0. The EXCLUSIVE-NOR (X-NOR for short) logic is just the complement of X-OR logic. Figure 7.6 shows the truth table and the logic symbols for both of these logic functions. It is important to note that such logic gates are used in tracking odd or even combinations of inputs.

Any arbitrary function can be realized using a combination of AND, OR, and NOT functions. Interestingly, however, either only NAND functions or only NOR functions can be used to realize any arbitrary logic function. This is why NAND and NOR logic functions are commonly referred to as universal logic functions. Consider, for example, the NAND-based logic circuit of Fig. 7.7. The nature of the logic outputs at d can be determined after having determined the outputs at points a, b, and c.

In order to interpret the characteristics of this three-level NAND logic circuit, we need to be knowledgeable only with the NAND function. The outputs at a, b, c, and d can be found by successively employing NAND logic as follows:

X	Y	a	b	c	d
0	0	1	1	1	0
0	1	1	1	0	1
1	0	1	0	1	1
1	1	0	1	1	0

A	B	$A \oplus B$	$\overline{A \oplus B}$
0	0	0	1
0	1	1	0
1	0	1	0
1	1	0	1

(a)

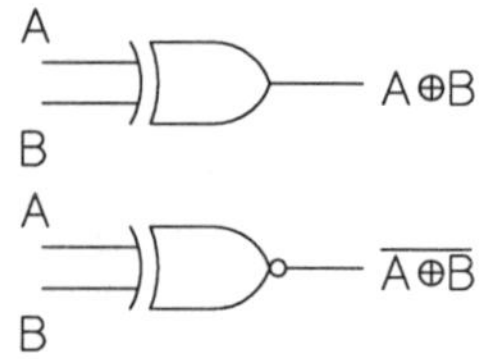

(b)

FIGURE 7.6 Two-input X-OR and X-NOR functions: (a) truth table and (b) logic symbols.

The output d may be compared with the X-OR output column of Fig. 7.6a. We find, accordingly, that the circuit of Fig. 7.7 functions as an X-OR logic. This equivalency points to the fact that there are more than one valid solution for almost any digital circuit. While one solution may require too many logic gates, others may need only a few. Finding the optimal logic circuit is a cherished goal for all logic circuit designers.

7.3 LOGIC FUNCTION FORMULATION

As shown earlier, truth tables are important tools that allow the designer to examine the combination of all possible inputs and the corresponding outputs. The truth table write-up becomes the first step in *designing* a logic function, simple or complex. For a binary logic function of n variables, the number of rows of the corresponding truth table is 2^n. The entries in the output section of the truth table generally follow the requirements indicated by the word statement of the problem.

Consider a device that takes in three inputs, A, B, and C. The logic circuit is to output a 1 when most of the inputs are 1. The logic circuit can be completely specified by considering each possible input combination. A truth table such as that in Table 7.1 can be used to accomplish this. The minterm list shows the corresponding AND combination of the input variables that would yield a 1. Again, the maxterms are the OR combinations of the variables that would give a 0. Thus corresponding to the input combinations $A = 0$, $B = 1$, and $C = 0$, the minterm is $\overline{A}B\overline{C}$ and the maxterm is $A + \overline{B} + C$. In fact, it can be shown that $\overline{A}B\overline{C}$ is the complement of $A + \overline{B} + C$. The minterms are sometimes conveniently referred to by the decimal equivalents of the input variables. Accordingly, minterm $\overline{A}B\overline{C}$, for example, is described by its decimal equivalent 2.

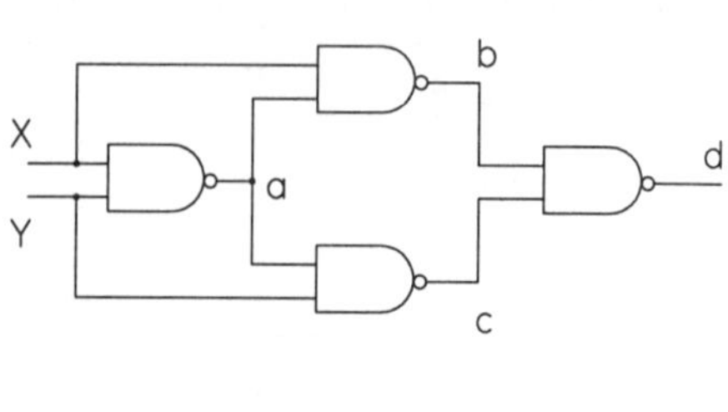

FIGURE 7.7 A NAND-only logic circuit.

TABLE 7.1 An Arbitrary Truth Table

Decimal Equivalents	A	B	C	Output	Minterms	Maxterms
0	0	0	0	0	$\overline{A}\overline{B}\overline{C}$	$A + B + C$
1	0	0	1	0	$\overline{A}\overline{B}C$	$A + B + \overline{C}$
2	0	1	0	0	$\overline{A}B\overline{C}$	$A + \overline{B} + C$
3	0	1	1	1	$\overline{A}BC$	$A + \overline{B} + \overline{C}$
4	1	0	0	0	$A\overline{B}\overline{C}$	$\overline{A} + B + C$
5	1	0	1	1	$A\overline{B}C$	$\overline{A} + B + \overline{C}$
6	1	1	0	1	$AB\overline{C}$	$\overline{A} + \overline{B} + C$
7	1	1	1	1	ABC	$\overline{A} + \overline{B} + \overline{C}$

The function $f(A, B, C)$ can be determined either as a function of all input combinations that yield an output of 1 or as the complement of a function of all input combinations that yield an output of 0. The first type of logic function, referred to as the sum-of-products (SOP) form, is obtained by ORing the minterms that cause the output to be a 1. Accordingly, the function is given by

$$\begin{aligned} f(A, B, C) &= \overline{A}BC + A\overline{B}C + AB\overline{C} + ABC \\ &= \sum m(3, 5, 6, 7) \end{aligned}$$

where 3, 5, 6, and 7 are the decimal equivalents of the minterms that cause the function to be a 1. The second form of the logic function, referred to as the product-of-sums (POS) form, however, is obtained by ANDing the maxterms corresponding to the 0 outputs. The function is thus also given by

$$\begin{aligned} f(A, B, C) &= (A + B + C)(A + B + \overline{C})(A + \overline{B} + C)(\overline{A} + B + C) \\ &= \prod M(0, 1, 2, 4) \end{aligned}$$

where 0, 1, 2, and 4 are the decimal equivalents of the maxterms involved. The circuits that implement the said function are shown in Fig. 7.8.

The logic circuit designer is usually concerned with issues such as the cost and speed of the logic circuit. An SOP-based circuit usually is more cost effective than the corresponding POS-based circuit whenever the number of 1's in the truth table output column is less than or equal to the number of 0's. Otherwise, the POS-based logic circuit is more cost effective. The voting function, however, can be reduced further using Boolean algebra to give

$$f(A, B, C) = AB + BC + AC$$

We may verify that the simplified function will be a 1 for the same input values as the more complex function. While each of the logic circuits of Fig. 7.8 uses a total of 8 gates, the circuit corresponding to this reduced function will involve only

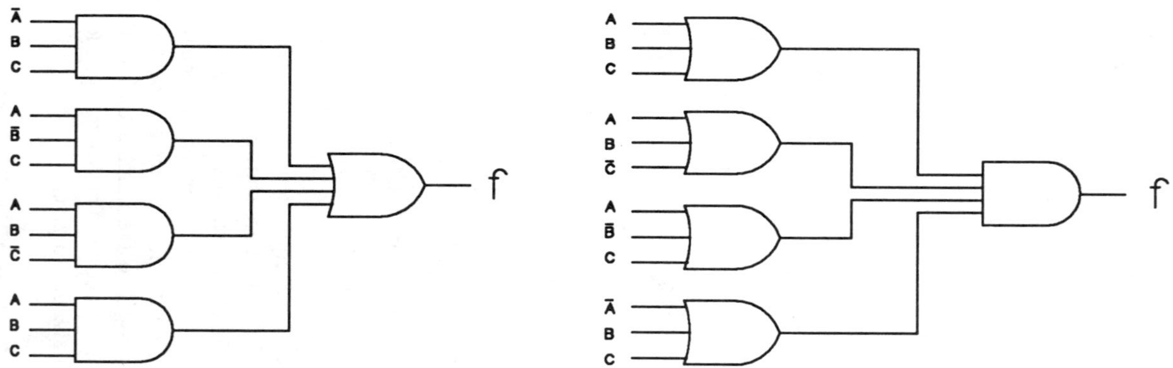

(a) (b)

FIGURE 7.8 The voting logic circuit: (a) SOP type and (b) POS type.

3 two-input AND gates and 1 three-input OR gate. The next section on Boolean algebra will familiarize the reader with the techniques pertaining to such logic function simplification.

7.4 BOOLEAN ALGEBRA

Boolean algebra, first introduced by George Boole, is useful for simplifying and manipulating logic expressions. Interestingly, the formulation of Boolean algebra relies on a set of postulates, known as Huntington's postulates. Each postulate has a dual, such that given a logic expression that is valid, the dual of the same logic expression is also valid. The dual of a logic expression is usually obtained by substituting all 0's with 1's, all 1's with 0's, all ANDs with ORs, and all ORs with ANDs. For example, the dual of the earlier mentioned logic function $\overline{A}BC + A\overline{B}C + AB\overline{C} + ABC = AB + BC + AC$ is

$$(\overline{A} + B + C)(A + \overline{B} + C)(A + B + \overline{C})(A + B + C) = (A + B)(B + C)(A + C)$$

Boolean algebra involves a finite set S of two or more variables and the three basic binary operators AND, OR, and NOT. Huntington's postulates are listed as follows.

Postulate 1 For all x in set S, (a) $x + 0 = x$ and (b) $x \cdot 1 = x$.

Postulate 2 $x + y = y + x$ for all x and y in set S.

Postulate 3 $x(y + z) = xy + xz$ for all x, y, and z in set S.

Postulate 4 $x + (y + z) = (x + y) + z$ for all x, y, and z in set S.

Postulate 5 For every element x in the set it has a complement $\overline{x}$ such that $x + \overline{x} = 1$ and $x\overline{x} = 0$.

The Boolean theorems listed below are based on the aforementioned postulates, for all, x, y, . . . , and z in set S.

Theorem 1 $x + x = x$.

Theorem 2 $x + xy = x$.

Theorem 3 If $x + y = y$ and $xy = y$, then $x = y$.

Theorem 4 If $x + y = 1$ and $xy = 0$, then $x = \overline{y}$.

Theorem 5 $x = \overline{\overline{x}}$.

Theorem 6 Any SOP or POS function of $f(x, y, \ldots, z)$ can be complemented by changing the ORs to ANDs, the ANDs to ORs, and complementing each of the variables.

Theorem 7 $x + \bar{x}y = x + y$.

Theorem 8 $xy + \bar{x}z + yz = xy + \bar{x}z$.

Theorem 9 $xy + xz = (x + z)(x + y)$.

Example 7.1 For a three-input logic circuit show that

$$\sum m(3, 5, 6, 7) = \prod M(0, 1, 2, 4)$$

Solution We proceed as follows:

$$\begin{aligned}
\prod M(0, 1, 2, 4) &= (A + B + C)(A + B + \bar{C})(A + \bar{B} + C)(\bar{A} + B + C) \\
&= (AA + AB + A\bar{C} + BA + BB + B\bar{C} + CA + CB + C\bar{C}) \\
&\quad \cdot (A\bar{A} + AB + AC + \bar{B}\bar{A} + \bar{B}B + \bar{B}C + C\bar{A} + CB + CC) \\
&= (A + AB + A\bar{C} + B + B\bar{C} + AC + BC)(AB + AC + \bar{A}\bar{B} \\
&\quad + \bar{B}C + \bar{A}C + BC + C) \\
&= [A(1 + B + \bar{C} + C) + B(1 + \bar{C} + C)][AB + \bar{A}\bar{B} + C(A \\
&\quad + \bar{B} + \bar{A} + B + 1)] \\
&= (A + B)(AB + \bar{A}\bar{B} + C) \\
&= AAB + A\bar{A}\bar{B} + AC + BAB + B\bar{A}\bar{B} + BC \\
&= AB + AC + BC \\
&= AB(C + \bar{C}) + AC(B + \bar{B}) + BC(A + \bar{A}) \\
&= ABC + AB\bar{C} + ABC + A\bar{B}C + ABC + \bar{A}BC \\
&= \bar{A}BC + A\bar{B}C + AB\bar{C} + ABC \\
&= \sum m(3, 5, 6, 7)
\end{aligned}$$

7.5 MINIMIZATION OF FUNCTION USING K-MAP

One of the objectives of the logic circuit designer is to keep the number of logic gates and number of their inputs to a minimum. As described earlier, the reduction of the logic function is realized using algebraic manipulations, but this is often tedious and at times the designer is not certain if the manipulations have yielded the optimal logic circuit. An easier as well as faster minimization scheme involves plotting minterms on a two-dimensional map. This mapping technique, referred to as Karnaugh map (K-map for short), is used to obtain optimized two-level functions with very little effort. In digital electronic circuit design, it may be noted that the concern for circuit minimization isn't any more critical because of the availability of read-only memory (ROM)-like devices.

The K-map based graphical scheme allows one to position the minterms of a function in a two-dimensional map, such that the minterms that are different only in one of the inputs are positioned as neighbors. The map corresponding to an n-variable function consists of 2^n cells arranged in an ordered two-dimensional array—either square or rectangular. Each of the cells on the map corresponds to one of the 2^n possible minterms (having decimal values ranging from 0 to 2^{n-1}) of the function in question. The possible K-maps for functions of up to six variables are shown in Fig. 7.9. The K-map variable values are listed in the order of the gray code so that as one moves from cell to cell from left to right or from top to bottom, only one input variable changes. Note that the K-map cells on the extreme right column are adjacent to those on the extreme left column. The same is true between the corresponding cells of the top and bottom rows. Figure 7.10 shows examples of some of the possible groups in a four-variable K-map, for example.

The K-map based minimization scheme consists of the following steps:

1. Identify the cells for the minterms of the function.
2. Group minterms having adjacencies.
3. Determine the variables common within each of the groups.
4. Generate a minimized SOP expression for the function.

First, a 1 is entered in each one of the cells corresponding to the minterms that form the function. A 0 is entered in each of the remaining cells. In an allowable group, it must be possible to start at a cell and travel from cell to cell, with only one variable changing, through each cell in the group without passing through any cell twice and return to the starting cell. Each one of the 1's on the map must be included in at least one group. For a nontrivial n-variable logic function, the largest group size is 2^{n-1}, while the rest of the group sizes can be $2^{n-2}, 2^{n-3}, \ldots, 2^0$.

The minimized SOP logic function can be determined by the following steps.

Step 1 The 1's that cannot be grouped with any other 1's are identified. Each of these isolated 1's will contribute to an n-variable AND term to the minimized SOP logic function.

Step 2 Identify 1's in groups of 2^1. Each of these pairs will contribute to an $(n - 1)$-variable AND term to the minimized SOP function.

Step 3 Identify 1's in groups of 2^2. Each of these groups of four will contribute to an $(n - 2)$-variable AND term to the minimized SOP function.

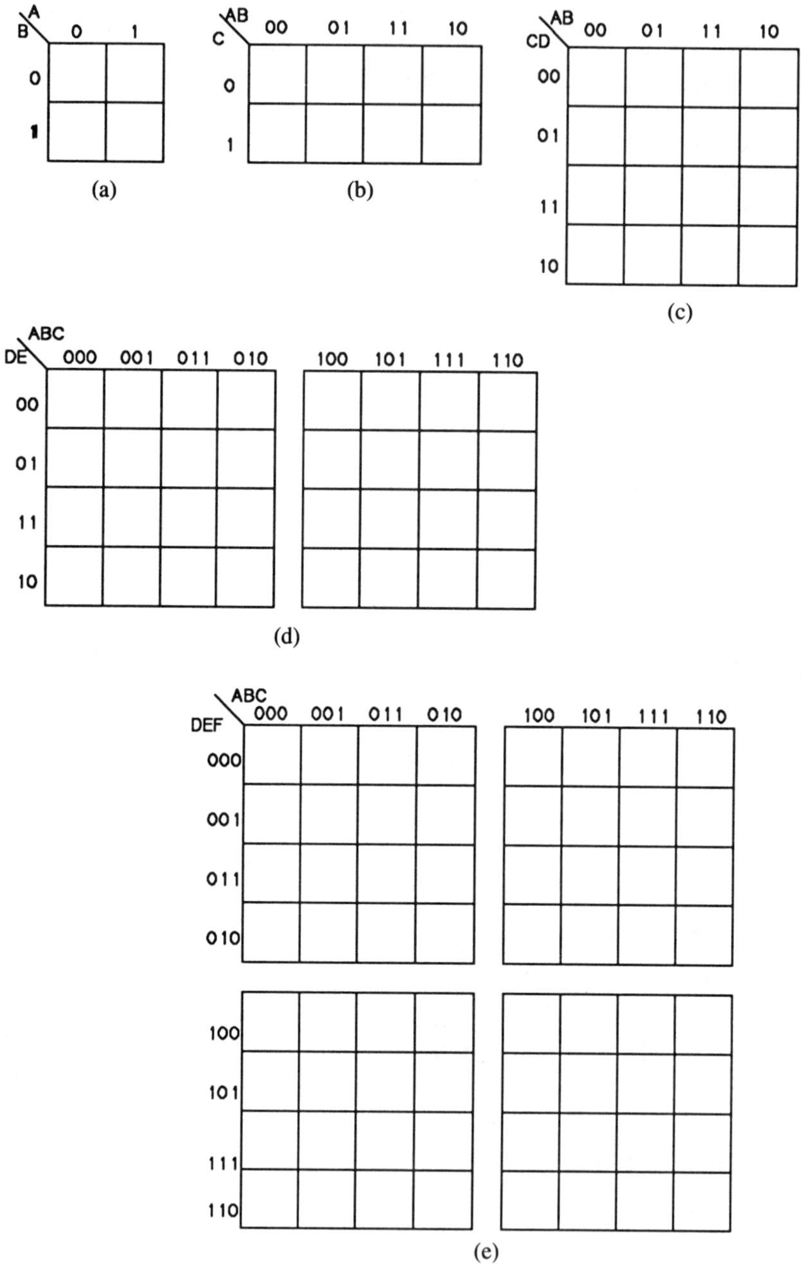

FIGURE 7.9 Karnaugh maps (K-maps) for a function of (a) two variables (b) three variables, (c) four variables, (d) five variables, and (e) six variables.

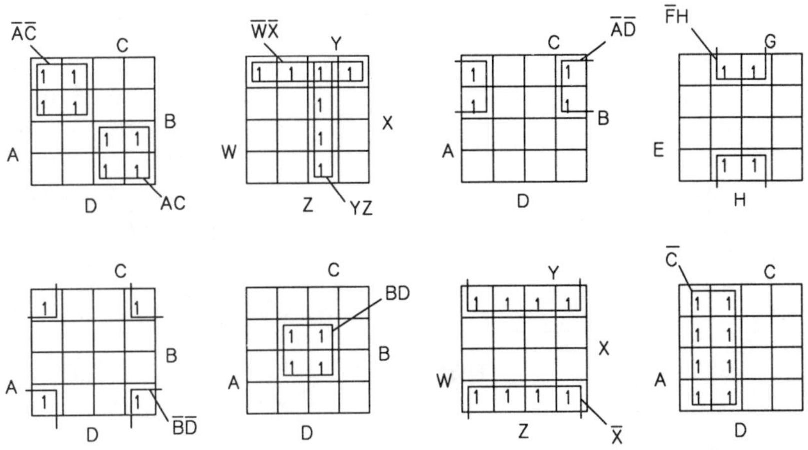

FIGURE 7.10 Sample groupings in a four-variable K-map.

Step 4 Continue the grouping process until group sizes of 2^{n-1} have been considered.

The steps outlined above are postponed as soon as all the 1's have been included in a group. Next, the smaller groups, if there are any, are eliminated if they are enclosed completely within a larger group. For each of the groups, literals, which are constant within the group, are used to form an ANDed term. In general, for an n-variable function, a group size of 2^k would contribute to an AND term of $n - k$ literals, as shown in Fig. 7.10.

The five-variable and six-variable K-maps use four-variable K-maps as building blocks. Consider, for example, the first 16 five-variable minterms expressed in binary form. We can see that the most significant digit in all 16 minterms is a 0, while that of the other 16 minterms is a 1. Similarly, the six-variable minterms can be divided into four groups of 16. Each group of 16 is distinguished from the others in terms of the first two bits. Accordingly, a five-variable K-map is formed of two four-variable K-maps and a six-variable K-map is formed of four four-variable K-maps. Note that the minterms appearing at the same location on the two halves are logically adjacent. For the purpose of grouping, the two halves may be assumed to be lying on top of each other. For a six-variable K-map, for example, the right-half and the left-half as well as the top-half and the bottom-half are considered to be adjacent.

Example 7.2 Find the minimized five-variable logic function that is able to detect the presence of a prime number at the input.

Solution This five-variable prime indicating circuit function can be expressed as

$$f(A, B, C, C, E) = \sum m(1, 2, 3, 5, 7, 11, 13, 17, 19, 23, 29, 31)$$

The function can be plotted on a five-variable K-map as in Fig. 7.11.

DE \ ABC	000	001	011	010
00	0	0	0	0
01	1	1	1	0
11	1	1	0	1
10	1	0	0	0

100	101	111	110
0	0	0	0
1	0	1	0
1	1	1	0
0	0	0	0

FIGURE 7.11

The groupings are

(1, 2, 5, 7) yielding $\overline{A}\overline{B}E$
(1, 2, 17, 19) yielding $\overline{B}\overline{C}E$
(3, 7, 19, 23) yielding $\overline{B}DE$
(2, 3) yielding $\overline{A}\overline{B}\overline{C}D$
(3, 11) yielding $\overline{A}\overline{C}DE$
(5, 13) yielding $\overline{A}C\overline{D}E$
(29, 31) yielding ABCE

Therefore the reduced function is given by

$$f(A, B, C, D, E) = \overline{A}\overline{B}E + \overline{B}\overline{C}E + \overline{B}DE + \overline{A}\overline{B}\overline{C}D + \overline{A}\overline{C}DE + \overline{A}C\overline{D}E + ABCE$$

It should be noted that another choice could have been considering the groups (13, 29) and (23, 31) instead of the groups (5, 13) and (29, 31). In that case, the reduced function would have the AND terms $BC\overline{D}E$ and $ACDE$ instead of $\overline{A}C\overline{D}E$ and $ABCE$.

Often it may be more easy to implement the POS form of a function. Fortunately, the K-map can be used to derive the POS form of the minimized function as well. In that case, a standard K-map is generated but only 0's are grouped instead of the 1's. The 0's are grouped following the same rules that were used earlier in the grouping of 1's. The resulting SOP form corresponds to the complement of the reduced function. When this SOP expression is thus complemented, it results in the reduced POS function.

It is also important to realize that there can be logic functions for which output values for all possible input conditions may not be specified. Practical constraints may exclude a particular group of minterms from occurring. This makes their occurrence in the function strictly optional. These minterms are referred to as the "don't care" minterms. For example, the BCD inputs are expressed in four bits (say, A, B, C, and D) but only 10 of the possible 16-bit combinations are considered valid. The six combinations, 1010, 1011, 1100, 1101, 1110, and 1111, are simply not used in BCD. Constraints of this type often become very helpful since the don't care minterms may be included in the K-map as either a 0 or a 1 depending on which leads to the minimum logic function. Thus the logic function $f(A, B, C, D) = \Sigma m(0, 2, 4, 6, 8)$, for example, can safely be redefined also as including any or all of the minterms 10, 11, 12, 13, 14, and 15. A minterm list for a function with don't cares has the form $f = \Sigma m(\cdot \cdot \cdot) + d(\cdot \cdot \cdot)$, where the don't cares are listed within $d(\cdot \cdot \cdot)$. Thus $f(A, B, C, D)$ can be expressed as $\Sigma m(0, 2, 4, 6, 8) + d(10, 11, 12, 13, 14, 15)$. The don't cares are entered in the K-map with a $-$ sign. While grouping, only those don't cares are grouped which help in forming larger groups of the minterms. Such don't cares are considered as 1's while the remaining don't cares are treated as 0's. Groups of only don't cares are never used.

As we shall find out next, some of the minimized logic functions lead to logic circuits that are unreliable. In general, when the values of the circuit inputs are changed, some or all of the outputs may change in response. However, any output that changes more than once to a single input variable change is symptomatic of a circuit hazard. Such erratic output changes are referred to as "glitches" and they often result from the fact that all practical logic gates have inherent propagation delays.

The hazard outputs are often momentary and therefore do not pose serious problems in a purely combinational network. However, if these outputs were to drive circuits sensitive to $1 \rightarrow 0$ and $0 \rightarrow 1$ transitions, as in a sequential circuit, the hazards may lead to disastrous consequences.

There can be two types of hazards: static and dynamic. A static hazard refers to an error condition that results in a momentary change of output in response to an input change. Circuits with static hazards finally settle at the old steady-state value. In comparison, a dynamic hazard is an error condition that occurs when, in response to an input change, the output changes more than once before approaching the new steady-state value. These hazard outputs are shown in Fig. 7.12.

There are two fundamental sources that contribute to hazards. A *function induced hazard* is caused by the function itself, while a *logic induced hazard* is caused by the particular logic network implementing the function. A function induced hazard is usually eliminated by restricting only one input to change at a time.

An example of a circuit with a logic induced static hazard is shown in Fig. 7.13. The K-map minimization of logic function $f(A, B, C) = \Sigma m(3, 4, 6, 7)$, shown in Fig. 7.13a, leads to the circuit of Fig. 7.13b since the reduced function may be given by $f(A, B, C) = A\overline{C} + BC$. The circuit of Fig. 7.13b has been obtained by grouping the 1's of the corresponding K-map. In order to obtain this logic circuit, the minterms (4, 6) and (3, 7) were grouped as pairs. Accordingly, the output must be a 1 as long as the minterms corresponding to the current input conditions remain within the cells 3, 4, 6, and 7. Consider now the timing diagram of Fig. 7.13c, corresponding to the case when both A and B have been set at 1 but

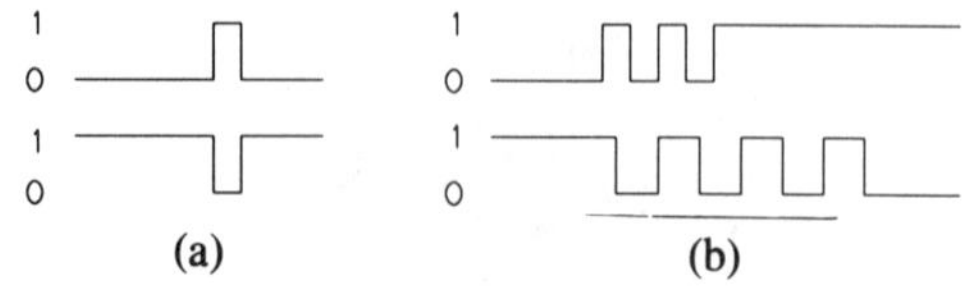

FIGURE 7.12 Hazards: (a) static type and (b) dynamic type.

the input C is suddenly changed from 1 to 0. This corresponds to the input condition changing from $ABC = 111$ to $ABC = 110$. During this input transition the function moves from the group represented by BC to the group represented by $A\overline{C}$, that is, from cell 7 to cell 6. However, the change in C requires that the signal goes through a NOT logic gate that, because of its inherent capacitance, contributes to a finite propagation delay between input changes and corresponding output response. Thus, in practice, C and $\overline{C}$ are both momentarily 0, which forces the output to be 0 for a time equal to the gate-delay time. This momentary drop in the function output, as shown by the timing diagram of Fig. 7.13c, is the static hazard.

Consider now the circuit shown in Fig. 7.13d, which includes an additional AND gate corresponding to a pair of minterms covering minterms 6 and 7. Addition of this AND gate essentially eliminates the hazard condition, which is evident from the timing diagram of Fig. 7.13e. As is evident from this example, it is not difficult to identify hazards. They can be recognized by noting the K-map entries where one group is exited and another is entered with no other group covering the cells during the time a transition is made. Again, as illustrated by the example of Fig. 7.13, this type of hazard is removed simply by including additional AND term(s) to cover the transition from one group to another.

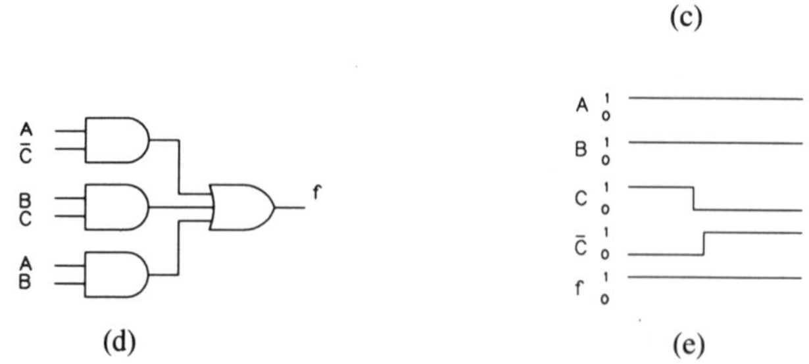

FIGURE 7.13 (a) K-map, (b) the corresponding circuit, (c) timing diagram showing the presence of a hazard, (d) the corrected logic circuit, and (e) a timing diagram showing the removal of a hazard.

Note that there will be situations when a circuit must have their hazard conditions removed and other situations when they may simply be ignored. For example, if the variable change causing the hazard never occurs or if it is determined that the erratic outputs will not significantly affect the operations, circuit hazards may be left untreated.

7.6 QUINE–McCLUSKEY TABULAR REDUCTION

The K-map based minimization scheme is not particularly suited for the logic function involving more than six variables. It often becomes very difficult to visually keep track of too many input variables. Also, it is possible to come up with more than one valid minimization because of the different ways the maps may be interpreted. Again, because the K-map scheme involves a visual process, it cannot easily be programmed for solution by a digital computer. To overcome these problems, a particular minimization algorithm, known as the Quine–McCluskey (Q-M) tabular reduction technique, is used. It is reasonably straightforward but time consuming.

The Q-M technique uses a systematic method to conduct an exhaustive search to determine all possible combinations of logically adjacent cell entries and it then directs the designer to an optimal solution. Accordingly, it is possible to write a software program pertaining to the Q-M method that allows the computer aided design of logic circuits. The Q-M technique can also be used effectively to optimize the logic functions for circuits that have common input variables but multiple output functions.

The Q-M tabular reduction technique may be summarized by the following steps.

Step 1 The n-variable minterms (also referred to as zero cubes) are converted to their equivalent binary form and then grouped in accordance with the total number of 1's in their binary representations.

Step 2 All possible groupings of pairs of zero-cubes are identified. These groups are referred to as one-cubes since the resulting AND term has $n - 1$ literals. Two zero-cubes may be grouped only if the numbers of 1's present in their binary representations differ by a 1. Moreover, the minterms must differ by 2^m, where m is any integer smaller than n, and the zero-cube having less 1's must be smaller than the one having more 1's.

Step 3 Next, all possible groupings of one-cubes are identified to form what is referred to as two-cubes (with the AND term having $n - 2$ literals). In forming two-cubes from one-cubes, the one-cubes must also have the same common literal. This process is continued until no more higher cubes can be formed.

Step 4 The cubes that could not be used in the formation of higher order cubes are identified. These cubes are called prime implicants (PIs).

Step 5 The PIs that are necessary to provide a minimum covering of all the minterms are identified next.

Step 6 The AND term corresponding to each of these PIs are ORed next to obtain the minimized SOP expression for the function.

The following example illustrates the algorithm in detail.

Example 7.3 Use the Q-M tabular procedure to find the PIs of the Boolean function given by

$$f(A, B, C, D) = \sum m(0, 2, 4, 5, 7, 8, 10, 13, 15)$$

Solution The zero-cubes are first classified as shown in Fig. 7.14.

Next, the one-cubes are formed from the zero-cubes. Starting from the top, we see that 0 and 2 can be combined to form a valid one-cube given by 0,2(2) since $0 < 2$, and $2 - 0 = 2$. The AND term corresponding to this pair is $\overline{A}\overline{B}\overline{D}$ (with $3 = 4 - 1$ literal) since $\overline{A}\overline{B}\overline{D}$ is common between $\overline{A}\overline{B}C\overline{D}$ (representing 2) and $\overline{A}\overline{B}\overline{C}\overline{D}$ (representing 0). This one-cube is listed in the top section of the next column, as shown in Fig. 7.15. For bookkeeping purposes, an x is written on the left side of 0 and also on the right side of 2. Continuing similarly, the zero-cubes 0 and 4 form the one-cube 0, 4(4) corresponding to the AND term $\overline{A}\overline{C}\overline{D}$ while the zero-cubes 0 and 8 form the third one-cube 0, 8(8) corresponding to the AND term $\overline{B}\overline{C}\overline{D}$. This completes matching the top two sections of the zero-cube column. A line is then drawn below the one-cubes just completed to separate the one-cubes with no 1's from those with one 1.

We next group the zero-cubes of the second section with those of the third section. Between the second and third sections of the zero-cube column three one-cubes are possible. They are 2 and 10 giving 2,10(8); 4 and 5 giving 4,5(1); and 8 and 10 giving 8,10(2). In all these cases the difference between the minterms is

Minterms	Binary representation	Number of 1's
0	0000	0
2	0010	1
4	0100	1
8	1000	1
5	0101	2
10	1010	2
7	0111	3
13	1101	3
15	1111	4

FIGURE 7.14

Zero-Cube	One-Cube
xxx 0	0,2(2)
x 2 x	0,4(4)
x 4 x	0,8(8)
x 8 x	
xx 5 x	2,10(8)
10 xx	4,5(1)
x 7 x	8,10(2)
x 13 x	5,7(2)
15 xx	5,13(8)
	7,15(8)
	13,15(2)

FIGURE 7.15

always a positive power of 2. In addition, the minterm from the upper section is smaller than the one from the lower section. Each time a group is formed a check mark is placed on the left side of the upper section minterm and one is placed on the right of the lower section minterm. This process is continued until no more one-cubes can be formed. The number of x's on the left side of a column should equal the number of x's on the right side of the same column.

Next, the two-cubes are determined from the one-cubes as shown in Fig. 7.16. A valid two-cube may be formed by combining two one-cubes from the neighboring sections only if the entries within the parenthesis are the same and the minterms of one one-cube differ from the corresponding minterms of the other one-cube by a power of 2.

Note that the one-cubes 0,2(2) and 8,10(2) may be combined to form the two-cube 0, 2, 8, 10(2,8) since $8 - 0 = 10 - 2 = 2^3$ and both have 2 within their brackets. Note that 0, 8(8) and 2, 10(8) also produce the same two-cube. All these

One-Cube	Two-Cube
x 0,2(2)	0,2,8,10(2,8)
0,4(4)	. . .
x 0,8(8)	
2,10(8) x	5,7,13,15(2,8)
4,5(1)	
8,10(2) x	
x 5,7(2)	
x 5,13(8)	
7,15(8) x	
13,15(2) x	

FIGURE 7.16

one-cube, however, are checked off. As a general rule, the formation of every m-cube corresponds to m pairs of $(m - 1)$-cubes. Next, a line is drawn to separate this particular two-cube. The second and the third sections of the one-cube column are searched next. We see that there isn't a single possible two-cube. We draw another line to indicate an empty section of the two-cube column. However, on studying the third and fourth one-cube section entries, we find that the one-cubes 5, 7(2) and 13, 15(2), or equivalently 5, 13(8) and 7, 15(8), can be combined to give 5, 7, 13, 15(2, 8) since $13 - 5 = 15 - 7 = 8$ and $7 - 5 = 15 - 13 = 2$. In both cases the same number is present within the brackets. This completes the list of all possible one-cubes. The two-cubes are investigated in similar fashion. In this example, no combination is possible since there aren't any neighboring two-cube entries.

All unchecked cubes are the PIs. Accordingly, the PIs for this function are:

0, 2, 8, 10(2, 8)
5, 7, 13, 15(2, 8)
0, 4(4)
4, 5(1)

At the end of the Q-M reduction process, the final minimized SOP expression is obtained by ORing the AND terms obtained from each of the PIs. Accordingly, for Example 7.2, the reduced function would be given by $\overline{B}\overline{D} + BD + \overline{A}\overline{C}\overline{D} + \overline{A}BC$. Note that if hazards are not an issue it may be possible to further minimize the function. If we were to include only those PIs that are essential and enough to cover all the minterms, some of the PIs may be eliminated from the SOP expression. This feat is accomplished by using an implication table. The rules for constructing the implication table are summarized as follows.

1. All minterms of the original function are used as the column headings while all PIs are used as the row labels.
2. In each of the row, checkmark x's are entered to indicate which of the minterms are covered by the PI in question.
3. The table entries are examined next for locating the essential PIs. An essential PI is one that is the only cover for one or more minterms. All minterms that are covered by an essential PI are then removed from active consideration.
4. The secondary essential PIs are determined from the remaining PIs by making judicious choices. A particular PI is preferred over another if it covers more of the remaining minterms or it is higher in order than the other.

This process is continued until all the minterms have been considered.

Example 7.4 Obtain the most reduced version for the function of Example 7.3.

Solution Figure 7.17 shows the implication table for the function of Example 7.3.

PI	0	2	4	5	7	8	10	13	15
* *a* 0,2,8,10(2,8)	x	x				x	x		
* *b* 5,7,13,15(2,8)				x	x			x	x
c 0,4(4)	x		x						
d 4,5(1)			x	x					
Covering PI	a or c	a	c or d	b or d	b	a	a	b	b

FIGURE 7.17

In Fig. 7.17, both of the two-cubes (indicated by *) are the essential PIs since *a* is the only PI covering minterms 2, 8, and 10 and *b* is the only PI covering minterms 7, 13, and 15. To complete the cover, either *c* or *d* may be chosen as the secondary essential PI. There is no particular disadvantage in choosing *c* over *d*. The reduced form of the function is therefore obtained as follows:

$$f(A, B, C, D) = a + b + c = \overline{B}\overline{D} + BD + \overline{A}\overline{C}\overline{D}$$

or equivalently

$$f(A, B, C, D) = a + b + d = \overline{B}\overline{D} + BD + \overline{A}BC$$

While using K-maps, it was shown that the presence of don't cares usually provides an opportunity for reducing a function. The same opportunity exists when using the Q-M technique. If the to-be-reduced function includes don't cares, the tabular reduction is carried out exactly as shown in Examples 7.3 and 7.4 with only one important difference. The don't cares are included in all the steps but in the implication table. In doing so, cubes formed exclusively of don't cares are not considered as PIs.

7.7 UNIVERSAL LOGIC GATES

The realization and the subsequent minimization of the logic function is not the end of logic circuit design. Both K-map and Q-M techniques lead to reduced logic functions. While the SOP form of the reduced function leads to an AND-OR type of logic circuit, the POS form of the reduced function leads to an OR-AND type of logic circuit. But digital integrated circuits, in particular, have practical limitations such as fan-in and fan-out problems that may affect the implementation of logic circuits. More importantly, integrated circuits are more frequently available in the NAND and NOR forms than in the AND and OR forms. In fact, both NAND and NOR gates are easily realized with electronic components, and therefore either of these two gates can serve as the basic module in the fabrication of all other logic gates. Therefore NAND and NOR logic gates are referred to as universal logic gates since almost any logic function can be realized using only either of these gates.

Either NAND or NOR gates can easily be manipulated to obtain any arbitrary logic function. This follows directly from Boolean theorems according to which the NAND and NOR operators are duals of each other. The dual logic expression is generally obtained by replacing every OR with AND, AND with OR, 0 with 1, and 1 with 0. The basic dual properties pertaining to these universal logic gates are as follows:

$$
\begin{aligned}
\overline{a \cdot 0} &= 1 & \qquad \overline{a + 1} &= 0 \\
\overline{a \cdot 1} &= \bar{a} & \qquad \overline{a + 0} &= \bar{a} \\
\overline{a \cdot a} &= \bar{a} & \qquad \overline{a + a} &= \bar{a} \\
\overline{a \cdot b} &= \bar{a} + \bar{b} & \qquad \overline{a + b} &= \bar{a}\bar{b} \\
\overline{\bar{a} + \bar{b}} &= a + b & \qquad \overline{\bar{a} \cdot \bar{b}} &= ab \\
\overline{\overline{a \cdot b}} &= ab & \qquad \overline{\overline{a + b}} &= a + b
\end{aligned}
$$

All standard logic functions may thus be generated using the aforementioned properties of NAND and NOR logic. Figure 7.18 shows the logic circuits for various functions using only NAND or NOR logic functions.

Figure 7.18 shows how NAND and NOR gates may be cascaded to form each of the logic functions, NOT, AND, OR, X-OR, and X-NOR. Since either NANDs or NORs may be used to implement all the logic operations, designers often prefer to use only NANDs or only NORs in order to decrease costs. The easiest way to accomplish designing circuits using either only NANDs or only NORs would be to substitute every one of the logic operations with the corresponding NAND/NOR equivalent circuit. This technique is referred to as the brute force method. However, it is important to note that besides the NAND and NOR, a yet-to-be-defined multiple-output logic gate, referred to as the Fredkin gate, also behaves like a universal gate. In the next three subsections, we shall see how these universal logic gates can be used to design complex logic circuits.

7.7.1 Logic Function Using Only NANDs/NORs

Any arbitrary SOP logic function may be designed using two levels of NAND logic gates. This is achieved in two steps:

1. The given logic function is first complemented by complementing the AND terms and replacing the OR operation with an AND operation.
2. The function is finally recovered by complementing the complemented function obtained in step 1.

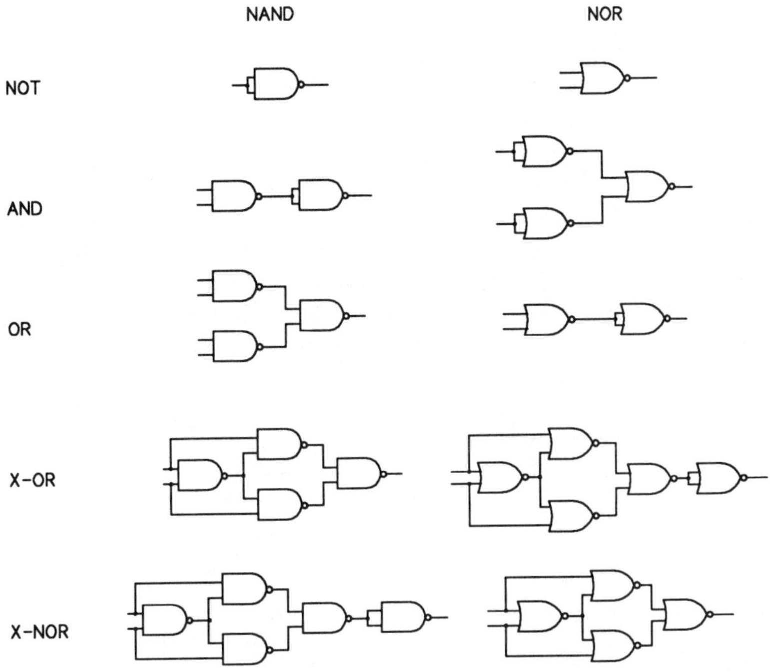

FIGURE 7.18 Logic functions using NANDs and NORs.

For example the SOP function $f = A\overline{B} + \overline{A}C$ can be complemented to give $\overline{f} = \overline{A\overline{B}}\;\overline{\overline{A}C}$ which in turn may be complemented to give $f = \overline{\overline{A\overline{B}}\;\overline{\overline{A}C}}$. Consequently, the function may be realized using three two-input NAND gates provided both $\overline{A}$ and $\overline{B}$ are available as inputs.

The function can also be realized using only NOR gates, provided it is available in the equivalent POS form. The design process involves the following steps:

1. The function is plotted on a K-map with the complemented functions obtained by grouping the zeros.
2. Each of the AND terms are next transformed into NOR form using DeMorgan's theorem.
3. The NOR based function is finally obtained by complementing the complemented function.

The function $f = A\overline{B} + \overline{A}C$, again for illustration, can be plotted on a K-map from which we can determine $\overline{f} = \overline{A}\overline{C} + AB$ by grouping the 0's. The expression for $\overline{f}$ can be rewritten finally as $\overline{A + C} + \overline{\overline{A} + \overline{B}}$ using DeMorgan's theorem, and therefore f becomes $\overline{\overline{A + C} + \overline{\overline{A} + \overline{B}}}$. The function can thus be realized using three two-input NOR gates provided both $\overline{A}$ and $\overline{B}$ are available.

7.7.2 Logic Function Using Fredkin Gates

In all conventional logic gates, the number of input lines is far larger than the number of output lines. Consequently, some of the information coming into the gate is lost. For example, it is often difficult to retrieve the inputs from the output information. A Fredkin gate is a specially designed universal gate that overcomes this limitation. These gates are capable of performing all conventional logic operations while it also preserves original information. In the purely electronic domain, except for reasons of information loss, it is meaningless to design circuits using Fredkin gates since reliable NAND and NOR gates are easily fabricated. However, there are real possibilities for implementing these gates optically and accordingly it is worthwhile to study how an arbitrary logic circuit can be designed using only Fredkin gates.

The basic Fredkin gate, shown in Fig. 7.19 is a three-input three-output device. One of these three binary inputs, Z_i, serves as a selector, which determines the operation of the gate on the other two inputs, X_i and Y_i. The device functions as follows:

$$Z_o = Z_i \tag{7.1a}$$

$$X_o = X_i \quad \text{and} \quad Y_o = Y_i, \qquad \text{if } Z_i = 0 \tag{7.1b}$$

$$X_o = Y_i \quad \text{and} \quad Y_o = X_i, \qquad \text{if } Z_i = 1 \tag{7.1c}$$

Since, in principle, the outputs can be used to retrieve the inputs, this gate is also referred to as nondissipative. Equation (7.1) may be rewritten

$$Z_o = Z_i \tag{7.2a}$$

$$X_o = \overline{Z}_i X_i + Z_i Y_i \tag{7.2b}$$

$$Y_o = \overline{Z}_i Y_i + Z_i X_i \tag{7.2c}$$

When $X_i = 0$, for example, X_o corresponds to the AND output of Z_i and Y_i. Accordingly, by setting X_i to zero, a Fredkin gate can be used to accomplish a two-input AND logic operation. In fact, it is relatively easy to arrange a Fredkin gate to also perform NOT and OR logic operations. Figures 7.20a–c show how the Fredkin gate can be used to realize AND, OR, and NOT operations, respectively. Figures 7.20d and 7.20e show circuit configurations for producing fan-out

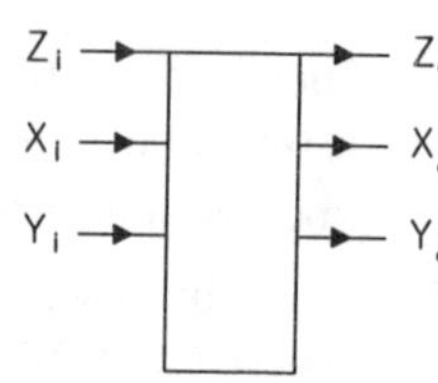

FIGURE 7.19 A Fredkin gate.

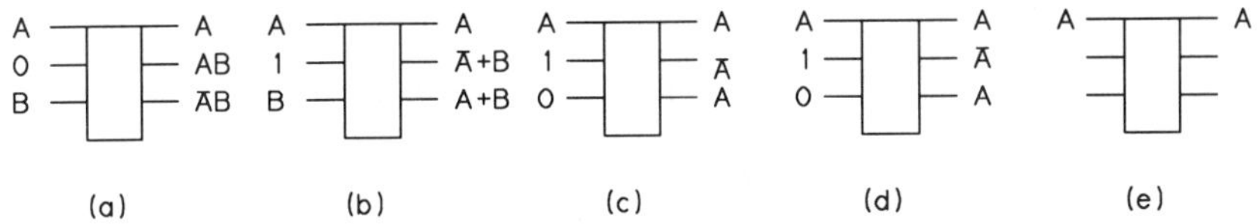

FIGURE 7.20 Fredkin gate realization of (a) AND logic, (b) OR logic, (c) NOT logic, (d) fan-out, and (e) delay.

and delay respectively. Again, since AND, OR, and NOT operations are enough to describe any SOP and POS function, any arbitrary logic circuit can be designed using only Fredkin gates.

7.8 LOGIC FUNCTION USING ONLY MULTIPLEXERS

Combinational circuits may as well be realized using a standardized combinational unit called a multiplexer (MUX). In this integrated device, some of the input variables are used as input selectors, while the remaining variables are entered as data inputs. A multiplexer, known also as a data selector, is a combinational network that has up to 2^n data inputs, n selectors, and an output line. In commercial multiplexers, n is generally limited to values of 1 through 4. An additional control input, referred to as the enable, is often available to turn the MUX on or off. The multiplexer allows the selection of only one of the 2^n data inputs as output. This input selection is determined by the values fed at the control inputs.

The block diagram of an 1-of-8 multiplexer having eight data input lines D_0, D_1, D_2, D_3, D_4, D_5, D_6 and D_7 is shown in Fig. 7.21a. While the MUX output and its complement are available, respectively, at f and $\bar{f}$, the control input E can be manipulated for enabling the device. The internal logic circuit of the corresponding MUX is shown in Fig. 7.21b. By applying appropriate signals to the selector lines, any one of the data lines may be selected. For example, when $(I_2I_1I_0)_2 = (m)_{10}$, the D_m input is routed to the output f provided $E = 0$. Note that whenever $E = 1$, the MUX is completely disabled regardless of the values available at the selectors or data inputs. The enable input E often allows several of these devices to be cascaded together to form a larger sized MUX.

A MUX with 2^n input lines and n selection lines may be used for realizing any arbitrary Boolean function of n variables. This will be illustrated by implementing the function $f(A, B, C, D) = \Sigma m(0, 2, 3, 5, 6, 8, 10, 14)$ using a 1-of-16 MUX. This particular MUX has four selectors just as the function has four variables. Accordingly, the implementation of the given function using a 1-of-16 MUX is relatively straightforward. Since the values available at the selectors I_3, I_2, I_1, and I_0 determine which particular data input values will be selected at the output, the input variables A, B, C, and D of the function may be used as the selectors while binary values (1 or 0) may be used as data inputs in accordance with the minterm representation of the given function. Since $f(A, B, C, D) = \Sigma m(0, 2, 3, 5, 6, 8, 10, 14)$, the binary value of 1 is fed to the data inputs D_0, D_2, D_3, D_5, D_6, D_8,

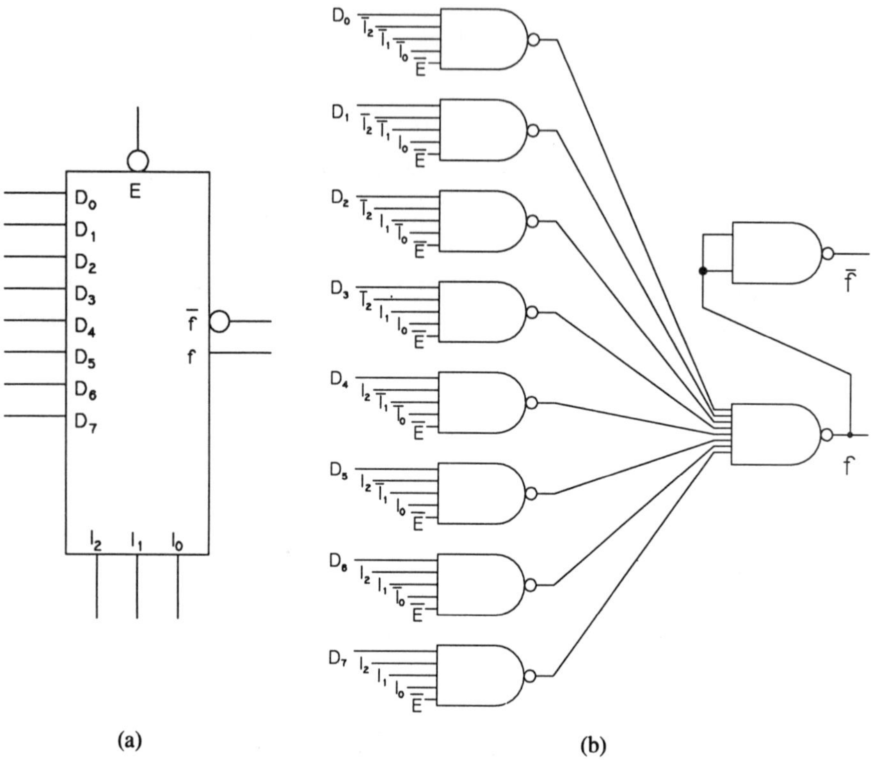

FIGURE 7.21 A 1-of-8 multiplexer: (a) block diagram and (b) logic circuit.

D_{10}, and D_{14} while the binary value of 0 is fed to the remaining eight data inputs. The corresponding MUX circuit is shown in Fig. 7.22a. It must be noted that the order in which the selectors are chosen is very important. The circuit of Fig. 7.22a is going to implement the function only as long as I_3, I_2, I_1, and I_0 are fed, respectively, with A, B, C, and D. As soon as the order in which A, B, C, and D are fed to the selector lines has been changed, the data inputs will have to be changed as well to reflect the corresponding change in the minterm description of the given function.

An alternative approach of using a 1-of-2^n MUX will, however, allow us to implement a function of $n + 1$ variables. We shall illustrate this by implementing the function $f(A, B, C, D) = \Sigma m(0, 2, 3, 5, 6, 8, 10, 14)$ using a 1-of-8 MUX. A 1-of-8 MUX has three selector inputs where all but any one of the four input variables may be entered. For example, we may consider A, B, and C inputs as the three control inputs, I_2, I_1, and I_0, respectively. The technique therefore consists in determining the function output in terms of the fourth input D for every possible combination of the selector inputs. The values so obtained are then entered at the respective data inputs of the multiplexer. The truth table for the said function may then be reorganized as shown in Fig. 7.22b. The entries under the

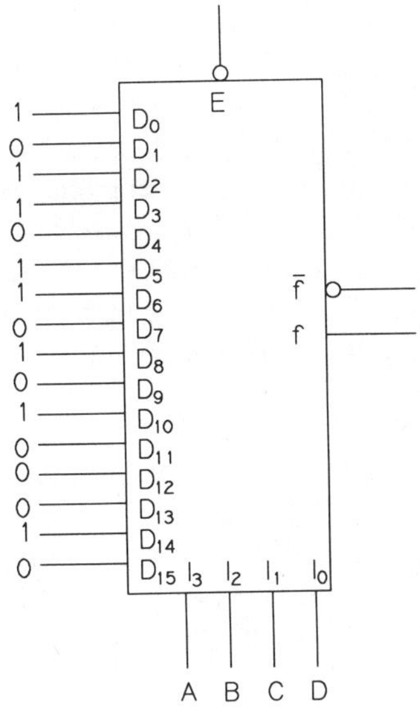

(a)

A	B	C	D	f	$f(D)$
0	0	1	0	1	$\overline{D}$
			1	0	
0	0	1	0	1	1
			1	1	
0	1	0	0	0	D
			1	1	
0	1	1	0	1	$\overline{D}$
			1	0	
1	0	0	0	1	$\overline{D}$
			1	0	
1	0	1	0	1	$\overline{D}$
			1	0	
1	1	0	0	0	0
			1	0	
1	1	0	0	1	$\overline{D}$
			1	0	

(b)

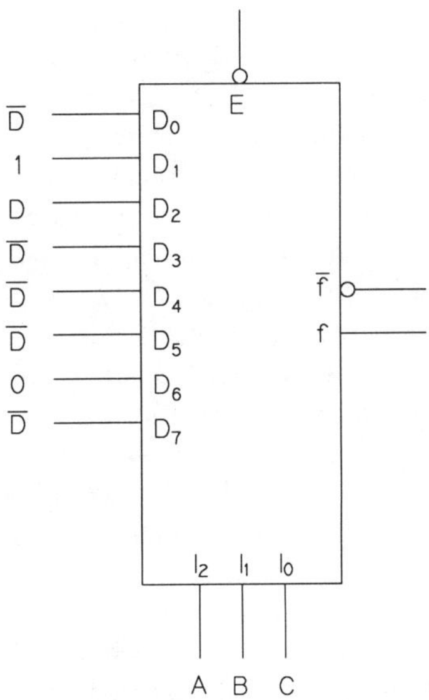

(c)

FIGURE 7.22. The function $f(A, B, C, D) = \Sigma m(0, 2, 3, 5, 6, 8, 10, 14)$ description: (a) using 1-of-16 MUX, (b) using a truth table in terms of $f(D)$, and (c) using a 1-of-8 MUX.

$f(D)$ column are determined by comparing the entries of the column D with those of the column $f(A, B, C, D)$.

For any combination of A, B, and C: (i) if $f(A, B, C, D) = 1$ irrespective of D, then $f(D) = 1$; (ii) if $f(A, B, C, D) = 0$ irrespective of D, the $f(D) = 0$; (iii) if $f(A, B, C, D) = x$ when $D = x$, then $f(D) = D$; and (iv) if $f(A, B, C, D) = \bar{x}$ when $D = x$, then $f(D) = \bar{D}$. The function is therefore implemented as shown in Fig. 7.22c. The eight values of $f(D)$, $\bar{D}$, 1, D, $\bar{D}$, $\bar{D}$, $\bar{D}$, 0, and $\bar{D}$ are fed, respectively, into data inputs, D_0 through D_7. Consequently, for any combination of the three selector inputs, $f = f(D)$ would actually appear at the multiplexer output. The implementation of the same function for the other combinations of the selector inputs can also be done in a similar manner. In fact, by taking different input variables as the selectors as well as by ordering them differently, a total of 24 different circuit configurations may be obtained.

The implementation of an n-variable function using a 1-of-2^n MUX, such as that in the circuit of Fig. 7.22a, can be interpreted as a single-output read-only memory (ROM) device. While the combination of all the values of the selectors serve as addresses, the data inputs are equivalent to values stored at particular address locations. A real ROM, however, may generate multiple outputs. Ideally, a combination of m 1-of-2^n multiplexers in parallel will be equivalent to an n-input m-output ROM device.

7.9 THRESHOLD LOGIC

This section deals with a special type of switching device called threshold elements. A single threshold element can realize a function that otherwise would have required dozens of conventional binary logic gates. Circuits formed of threshold elements generally require less complex interconnections and fewer elements than the equivalent circuits involving conventional logic gates. However, neither the operation nor the manufacturing process of threshold elements are easy. In spite of early interest in such devices, they never became popular. But recent studies have shown that many of the threshold elements mimic the function of neurons. Human neurons are not necessarily fast but they are unbelievably capable of updating outputs as newer experiences (information) are taken into consideration. Obviously, therefore, a surge of activity has now begun involving threshold elements with potential applications to neural networks, adaptive signal processing, adaptive control systems, learning automata, optical computing, and pattern recognition.

A threshold element may have n binary inputs, $X_1, X_2, \ldots, X_n$, and a single binary output f. The workings of a threshold element are, however, determined by a set of n weights so that the weight W_j is associated with the binary input X_j and a threshold value T. The weights as well as the threshold may be real, finite, and either positive or negative. The binary output of the threshold element, shown in

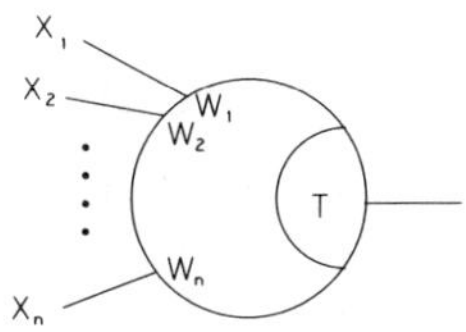

FIGURE 7.23 An n-input threshold element.

Fig. 7.23, is given by

$$f = \begin{cases} 1, & \text{if } \sum W_i X_i \geq T \\ 0, & \text{if } \sum W_i X_i < T \end{cases} \tag{7.3}$$

where $W_i X_i$ may be referred to as the ith inner product. To understand the exact working of such threshold elements it will be better first to consider some examples before resuming our discussion of threshold elements.

Example 7.5 Determine what Boolean function is being realized by the threshold element of Fig. 7.24.

Solution The truth table, as shown in Fig. 7.25, is completed by calculating the sum of inner products and then comparing the sum of inner products with the threshold value. Therefore $f(A, B, C) = \Sigma m(1, 2, 3, 7) = \overline{A}B + \overline{A}C + BC$. The threshold element of Fig. 7.24 is thus realizing a Boolean function that is normally implemented using five conventional logic gates (one NOT, three two-input ANDs, and one three-input OR).

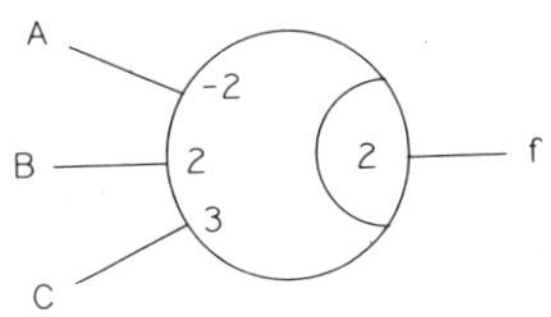

FIGURE 7.24

A	B	C	Sum of Inner Products	f
0	0	0	0	0
0	0	1	3	1
0	1	0	2	1
0	1	1	5	1
1	0	0	−2	0
1	0	1	1	0
1	1	0	0	0
1	1	1	3	1

FIGURE 7.25

A	B	C	f
0	0	0	1
0	0	1	0
0	1	0	1
0	1	1	0
1	0	0	0
1	0	1	0
1	1	0	1
1	1	1	0

FIGURE 7.26

Example 7.6 Obtain the threshold element that realizes the function $f(A, B, C) = \Sigma m(0, 2, 6)$.

Solution The truth table corresponding to the given function is obtained as shown in Fig. 7.26. Therefore the inequalities involving the weights and threshold are determined as follows:

$$0 \geq T$$

$$W_3 < T$$

$$W_2 \geq T$$

$$W_2 + W_3 < T$$

$$W_1 < T$$

$$W_1 + W_3 < T$$

$$W_1 + W_2 \geq T$$

$$W_1 + W_2 + W_3 < T$$

From these inequalities, we may conclude that T must be negative while it has to be larger than both W_1 and W_3 and smaller than W_2. Again, since $W_1 + W_2 \geq T$, W_2 has to be positive. Accordingly, all the aforementioned inequalities may simultaneously be satisfied, for example, by the following values:

$$W_1 = -2, \quad W_2 = 2, \quad W_3 = -4, \quad \text{and} \quad T = -1.$$

A threshold element such as the one described in Fig. 7.24 can be implemented using the resistor–transistor circuit of Fig. 7.27. When the three input resistances are nonidentical, the gate output V_o is determined by the sum of inner products. For example, for the transistor output to be turned on (by going low), V_B has to exceed 0 V (actually, 0.7 V in case the diode is made of silicon). Again, for the transistor output to be turned off, V_B has to be negative. Thus for V_B to be positive,

$$(V_1/R_1) + (V_2/R_2) + (V_3/R_3) \geq V_T/R_T \tag{7.4}$$

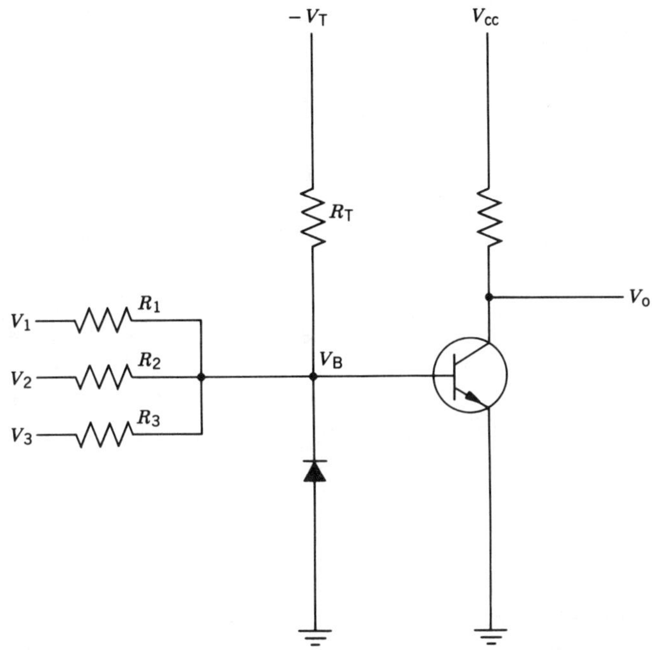

FIGURE 7.27 A three-input threshold element.

Otherwise, V_B is negative. Comparing Eqs. (7.3) and (7.4), we see that for this threshold element, $W_j = 1/R_j$ and $T = V_T/R_T$.

One problem, however, becomes apparent: the resistance values may change significantly from their nominal values. In addition, voltages may vary. Thus the reliability of threshold elements are often questionable. This problem, however, can be overcome by guaranteeing a large difference between the sum of the inner products for which f is 1 and that for which $f = 0$. But, most importantly, threshold elements haven't become popular because of the lack of a simplified procedure with which one can decide if a particular function is realizable using one or more threshold elements. The remaining part of this section addresses some of these complex issues of threshold elements.

A threshold element satisfying the Boolean function $f(x_1, x_2, \ldots, x_n)$ may be specified by a weight-threshold vector

$$V = [W_1, W_2, \ldots, W_n: T] \tag{7.5}$$

and by its input variables $x_1, x_2, \ldots, x_n$. Thus the threshold element of Example 7.6 is specified by the weight-threshold vector $V = [-2, 2, -4: -1]$ and its inputs A, B, and C. The relationship, if there is any, between a Boolean function and its corresponding threshold element is described in terms of the characteristics of the function. A function $f(x_1, x_2, \ldots, x_n)$ is said to be positive in x_i if there exists an SOP or POS expression for f in which x_i appears uncomplemented only. Likewise, f is said to be negative in x_i if there exists a SOP or POS expression for f in which x_i appears as complemented only. Again, if f is either positive or negative in x_i but not both, f is said to be "unate in x_i." And finally, a function that is unate in each one of its variables is "unate." A thorough mathematical investigation (not treated here, for simplicity) confirms that the function corresponding to a valid weight-threshold vector is always unate. This does not, however, imply that all unate functions can be realized using threshold elements.

Consider, for example, the function $f(A, B, C) = \bar{A}B + \bar{A}\bar{B}C + BC$. This function is unate since $\bar{A}B + \bar{A}\bar{B}C + BC = \bar{A}(B + \bar{B}C) + BC = \bar{A}(B + C) + BC = \bar{A}B + \bar{A}C + BC$, which is thus unate in $\bar{A}$, B, and C only. This particular function has already been used in Example 7.5 and realized using the weight-threshold vector $V = [-2, 2, 3{:}\ 2]$.

The sets of input combinations contributing to an output of 1 are called "true vertices" while the remaining input combinations are referred to as "false vertices." For example, the true vertices corresponding to $f = \bar{A}B + \bar{A}C + BC$ are $\{0, 1, 0\}$ and $\{0, 1, 1\}$ obtained from $\bar{A}B$, $\{0, 0, 1\}$ and $\{0, 1, 1\}$ obtained from $\bar{A}C$ and $\{1, 1, 1\}$ and $\{0, 1, 1\}$ from BC. Thus there are only four true vertices corresponding to the function $f = \bar{A}B + \bar{A}C + BC$, namely, $\{0, 1, 0\}$, $\{0, 1, 1\}$, $\{1, 1, 1\}$ and $\{0, 0, 1\}$. The remaining four vertices are false vertices. It is proper to order vertices as true or false. A vertex $\{a_1, a_2, \ldots, a_n\}$ is said to be higher in order than a second vertex $\{b_1, b_2, \ldots, b_n\}$ only if, for all j, $a_j \geq b_j$. For example, the vertex $\{0, 1, 1\}$ has a higher rank when compared to the vertex $\{0, 0, 1\}$ or vertex $\{0, 1, 0\}$. Interestingly, some vertices may not be comparable, for example, $\{0, 0, 1\}$ and $\{0, 1, 0\}$.

In the following, a few of the most important theorems pertaining to threshold elements are listed without proof.

Theorem 1 Provided there exists a weight-threshold vector $V_1 = [W_1, W_2, \ldots, W_n{:}\ T]$ that realizes the function $f(x_1, x_2, \ldots, x_i, \ldots, x_n)$, the function $f(x_1, x_2, \ldots, \bar{x}_i, \ldots, x_n)$ can be realized using the weight-threshold vector $V_2 = [W_1, W_2, \ldots, -W_i, \ldots, W_n{:}\ T - W_i]$.

Theorem 2 Provided there exists a weight-threshold vector $V_1 = [W_1, W_2, \ldots, W_n{:}\ T]$ that realizes the function $f(x_1, x_2, \ldots, x_n)$, the function $\bar{f}(x_1, x_2, \ldots, x_n)$ can be realized using the weight-threshold vector $V_2 = [-W_1, -W_2, \ldots, -W_n{:}\ -T]$.

Theorem 3 The weight-threshold vector, if it exists, corresponding to a Boolean function $f(x_1, x_2, \ldots, x_n)$ that is unate and also positive in each one of its variables can be determined by considering only minimal true vertices and maximal false vertices.

The general step-by-step rules for determining the threshold element corresponding to a Boolean function are listed below.

Step 1 Test the function for unateness. If the test fails, the function is not realizable using a threshold element.

Step 2 The unate function is made positive in each one of its variables by relabeling operations, such as $\overline{X}_i$ being replaced by Y_i.

Step 3 Determine all minimal true vertices (of the lowest rank) and maximal false vertices (of the highest two ranks).

Step 4 For each pair of a minimal true vertex $\{a_1, a_2, \ldots, a_n\}$ and a maximal false vertex $\{b_1, b_2, \ldots, b_n\}$, set up the inequality $a_1 W_1 + a_2 W_2 + \cdots + a_n W_n > b_1 W_1 + b_2 W_2 + \cdots b_n W_n$. Solve these inequalities simultaneously for the weights.

Step 5 Determine threshold and therefore the weight-threshold vector.

Step 6 Convert the weight-threshold vector (using Theorem 1) to find the true weight-threshold vector by making the $Y_i \rightarrow \overline{X}_i$ transformation.

However, if the function is not unate, one can split the function into two or more functions (using the OR operation) such that each one of the decomposed functions is unate. A separate threshold element is then determined for each of these decomposed functions.

Example 7.7 Obtain the threshold element, if there is any, for realizing $f(A, B, C, D) = AB + A\overline{B}C\overline{D}$.

Solution We see that $AB + A\overline{B}C\overline{D} = A(B + \overline{B}C\overline{D}) = A(B + C\overline{D}) = AB + AC\overline{D}$. Therefore the function is unate. Next, for the corresponding unate function that is completely positive, we choose

$$f(A, B, C, D) = AB + ACD$$

The minimal true vertices are $\{1, 1, 0, 0\}$ and $\{1, 0, 1, 1\}$ and the maximal false vertices are $\{0, 1, 1, 1\}$, $\{1, 0, 1, 0\}$, and $\{1, 0, 0, 1\}$. The total number of inequalities is thus $2 \times 3 = 6$. They are obtained as follows:

$$\{1, 1, 0, 0\} \text{ and } \{0, 1, 1, 1\} \longrightarrow W_1 + W_2 > W_2 + W_3 + W_4$$

$$\{1, 1, 0, 0\} \text{ and } \{1, 0, 1, 0\} \longrightarrow W_1 + W_2 > W_1 + W_3$$

$$\{1, 1, 0, 0\} \text{ and } \{1, 0, 0, 1\} \longrightarrow W_1 + W_2 > W_1 + W_4$$

$$\{1, 0, 1, 1\} \text{ and } \{0, 1, 1, 1\} \longrightarrow W_1 + W_3 + W_4 > W_2 + W_3 + W_4$$

$$\{1, 0, 1, 1\} \text{ and } \{1, 0, 1, 0\} \longrightarrow W_1 + W_3 + W_4 > W_1 + W_3$$

$$\{1, 0, 1, 1\} \text{ and } \{1, 0, 0, 1\} \longrightarrow W_1 + W_3 + W_4 > W_1 + W_4$$

Therefore the constraints are

$$W_1 > W_3 + W_4$$

$$W_2 > W_3$$

$$W_2 > W_4$$

$$W_1 > W_2$$

$$W_4 > 0$$

$$W_3 > 0$$

which can be summarized as

$$W_1 > W_3 + W_4 > 0$$

$$W_1 > W_2 > W_3 > 0$$

$$W_1 > W_2 > W_4 > 0$$

One of the solutions can be, for example, $W_1 = 4$, $W_2 = 3$, $W_3 = 2$, and $W_4 = 1$. Considering a pair of minimal true and maximal false vertices, $\{1, 1, 0, 0\}$ and $\{0, 1, 1, 1\}$, for example, the sum of inner products corresponding to the first one is $4 + 3 = 7$ and that corresponding to the second is $3 + 2 + 1 = 6$. Now since $7 \geq T \geq 6$, we may choose $T = \frac{13}{2}$. The weight-threshold vector V_1 corresponding to $AB + ACD$ is $[4, 3, 2, 1: \frac{13}{2}]$. Therefore the weight-threshold vector V_2 corresponding to $AB + AC\overline{D}$ is given by $[4, 3, 2, -1: \frac{13}{2} - 1] = [4, 3, 2, -1: \frac{11}{2}]$.

This chapter covers what are considered to be the fundamentals of logic gates, logic simplification, and how such simplification relates to the design of combinational logic circuits. This was done by introducing Karnaugh's map scheme, which was shown suitable for minimizing functions of up to six variables. The Q-M tabular reduction method was shown to be a useful technique for handling multiple-variable logic functions. In the final few sections of this chapter, we used the concepts of universal logic gates, Fredkin gates, multiplexers, and threshold gates for implementing arbitrary logic circuits.

7.10 PROBLEMS

1. Verify the Boolean theorems as well as their duals (wherever appropriate) using Huntington's postulates.
2. Minimize the following Boolean functions:
   (a) $\overline{(A + B)(A + \overline{B}\overline{C})D + \overline{A}B + ABC + A(\overline{B} + \overline{C})}$
   (b) $AB + \overline{A}B + AB\overline{C}D$
   (c) $ABC + BCD + A\overline{B}\overline{D} + \overline{A}BC\overline{D} + \overline{A}\overline{B}\overline{D}$

3. Complement the Boolean expressions of Problem 2 and then reduce the complemented functions.
4. Obtain the optimal logic circuit for converting 4-bit binary numbers to their equivalent gray code numbers.
5. Obtain the optimal logic circuit for converting a 4-bit gray code into their equivalent 4-bit binary number.
6. Obtain the minimal and hazard-free SOP expressions for the following functions:
   (a) $f(A, B, C, D) = \sum m(0, 1, 3, 5, 8, 9, 11, 15)$
   (b) $f(A, B, C, D) = \sum m(1, 3, 5, 8, 11, 15) + d(0, 2, 9)$
   (c) $f(A, B, C, D) = \sum m(0, 4\text{–}13, 15)$
   (d) $f(A, B, C, D, E) = \sum m(1, 7\text{–}12, 21, 22, 29)$
   (e) $f(A, B, C, D, E) = \sum m(1, 8\text{–}11, 22, 26, 27) + d(3, 5, 23, 31)$
7. Repeat Problem 6 but obtain POS expressions.
8. Repeat Problem 6 without neglecting hazard conditions.
9. Obtain the minimal equivalent of $f = (\overline{X} + Y)(X + \overline{Y} + Z)(X + \overline{Z})$ using the K-map.
10. Obtain the minimal SOP forms of the following functions using the Q-M tabular reduction method.
   (a) $f(A, B, C, D) = \sum m(0, 1, 4\text{–}10, 12, 14)$
   (b) $f(A, B, C, D, E, F) = \sum m(0, 2, 4, 5, 7, 8, 16, 18, 24, 32, 36, 40, 48, 56)$
   (c) $f(A, B, C, D, E) = \sum m(0, 2, 3, 7, 10, 11, 12, 17, 18, 24) + d(13\text{–}15)$
   (d) $f(A, B, C, D) = \sum m(2, 4, 6, 8, 11, 15)$
   (e) $f(A, B, C, D, E) = \sum m(0, 2, 3, 7, 8, 10, 11, 12, 24) + d(5, 16, 18)$
11. Obtain the minimized POS form for the functions of Problem 10.
12. Design a combinational circuit that turns on a light emitting diode for each even BCD input.
13. Using Fredkin gates, design circuits for two-input (a) NAND, (b) NOR, (c) X-OR, and (d) X-NOR logic operations.
14. Repeat Problem 13 but for three-input logic operations.
15. Obtain the circuit for the following functions using (i) only NAND gates, (ii) NOR gates, and (iii) Fredkin gates.
   (a) $f(A, B, C, D) = \sum m(0, 1, 2, 4, 10, 11, 13, 15)$
   (b) $f(A, B, C, D) = \sum m(1, 3, 4, 9, 10\text{–}13)$
   (c) $f(A, B, C, D) = \sum m(1, 8\text{–}10, 15)$
   (d) $f(A, B, C, D, E) = \sum m(1, 3\text{–}8, 11, 14, 17, 22, 24\text{–}27, 30)$
   (e) $f(A, B, C, D, E) = \sum m(1\text{–}5, 8\text{–}10, 13\text{–}17, 21, 25\text{–}27, 30, 31)$
16. (a) Use a single level of a 1-of-8 MUX and a few assorted gates (if needed) for obtaining a combinational circuit for each of the functions of Problem 15. (b) For each of your solutions, complement the variable B, and rearrange the inputs so that the function is still correct.

17. Using 1-of-4 MUX modules obtain a two-level MUX circuit for each of the functions of Problem 15.

18. Implement the function $f(A, B, C, D) = \Sigma m(3, 4, 8\text{–}10, 13\text{–}15)$ by (a) using a 1-of-4 MUX and a few assorted gates and (b) using 1-of-2 and 1-of-4 MUXs in two levels.

19. Obtain the circuit for the function $f(X, Y, Z, U, V) = \Sigma m(0, 1, 6, 7, 9, 12, 13, 15, 18, 20, 22, 24\text{–}26, 28)$ using two levels of a 1-of-4 MUX and a few assorted gates.

20. Design (a) a 1-of-8 MUX and (b) a 1-of-16 MUX using 1-of-4 MUX units.

21. Without using a multiplexer, design a six-input combinational circuit whose output, y_0, is given by

$$y_0 = \begin{cases} x_{-1}, & \text{if } d = 1,\ s = 1, \text{ and } E = 1 \\ x_1, & \text{if } d = 0,\ s = 1, \text{ and } E = 1 \\ x_0, & \text{if } d = -,\ s = 0, \text{ and } E = 1 \\ 0, & \text{if } d = -,\ s = -, \text{ and } E = 0 \end{cases}$$

where d, s, E, x_0, x_1, and x_{-1} are the inputs. Describe your reasoning as to why this circuit is called a shifter.

22. Using multiplexers, design a multi-bit shifter that has an $(n + 2)$-bit input (x), and n-bit output (y), and three control inputs $(s, d,$ and $E)$.

23. Determine which of the following functions are realizable using a threshold element: (a) $A + \overline{B}\overline{C}$, (b) $AB + CD$, (c) $AB + BC$, and (d) $BC + ABD + CD$.

24. Determine the threshold element for realizing (a) $\overline{A}B + \overline{A}D + B\overline{C} + \overline{A}\overline{C}$, (b) $\Sigma m(2, 3, 6, 7, 15)$, (c) $\Sigma m(3, 5, 7, 15)$, and (d) $\Sigma m(10, 12, 14, 15)$.

25. Replace the circuit of Fig. 7.28 using a single threshold element.

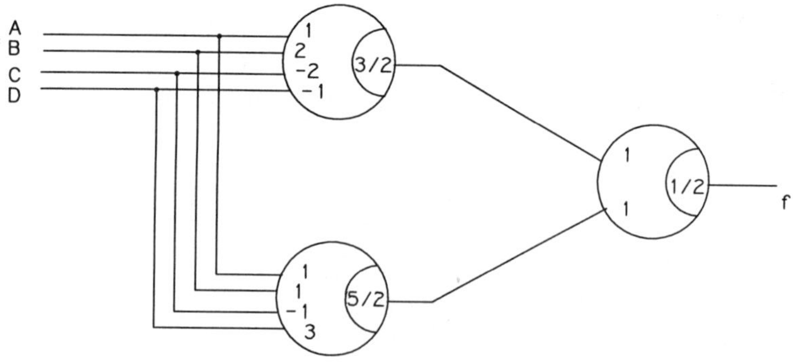

FIGURE 7.28

26. Obtain the truth table for the following weight-threshold vectors corresponding to the inputs A_2, A_1, and A_0:

(a) [2, 2, 2:5]

(b) [6, 6, 6, 6:5] when its inputs are, respectively, [−2, 3, 3:5], [6, 2, −2:5], [3, 2, 3:5], and [3, 3, −2:5]

(c) [6, 6, 6, 6:5] when its inputs are, respectively, [−2, −2, 6:5], [−2, 6, −2:5], [3, −2, 3:5], and [3, 3, −2:5].

Obtain a composite truth table for the three inputs having three outputs $f_2 f_1 f_0$, where f_2, f_1, and f_0 are the outputs obtained using threshold circuits of (a), (b), and (c), respectively. Show that the composite truth table describes the binary sum of $(A_2A_1)_2$ and $(A_0)_2$.

7.11 SUGGESTED READINGS

Culliney, J. N., M. H. Young, T. Nakagawa, and S. Muroga, Results of the Synthesis of Optimal Networks of AND and OR Gates for Four-Variable Switching Functions, *IEEE Trans. Comput.*, **C-27,** 76 (1979).

Curtis, H. A., Short-cut Method of Deriving Nearly Optimal Arrays of NAND Trees, *IEEE Trans. Comput.*, **C-28,** 521 (1979).

Cuykendall, R., and D. McMillin, Control-Specific Optical Fredkin Circuits, *Appl. Opt.*, **26,** 1959 (1987).

Davio, M., Read-Only Memory Implementation of Discrete Functions, *IEEE Trans. Comput.*, **C-29,** 931 (1980).

Ektare, A. B., and D. P. Mital, Multiplexer Logic Circuit Design Using Cubical Complexes, *Electron. Lett.*, **16,** 495 (1980).

Ektare, A. B., and D. P. Mital, Probabilistic Approach to Multiplexer Logic Circuit Design, *Electron. Lett.*, **16,** 686 (1980).

Ellis, D. T., A Synthesis of Combinational Logic and NAND or NOR Elements, *IEEE Trans. Electron. Comput.*, **EC-14,** 701 (1965).

Fredkin, E., and T. Toffoli, Conservative Logic, *Int. J. Theor. Phys.*, **21,** 219 (1982).

Halder, A. K., Karnaugh Map Extended to Six or More Variables, *Electron. Lett.*, **18,** 868 (1982).

Hong, S. J., R. G. Cain, and D. L. Ostapko, MINI: A Heuristic Approach for Logic Minimization, *IBM J. Res. Dev.*, **18,** 443 (1974).

Johnson, E. L., and M. A. Karim, *Digital Design: A Pragmatic Approach*, PWS–Kent Publishing, Boston, 1987.

Jullien, G. A., Residue Number Scaling and Other Operations Using ROM Arrays, *IEEE Trans. Comput.*, **C-27,** 325 (1978).

Kambayashi, Y., Logic Design of Programmable Logic Arrays, *IEEE Trans. Comput.*, **C-28,** 609 (1979).

Li, H. F., Variable Selection in Logic Synthesis Using Multiplexers, *Int. J. Electron.*, **49,** 185 (1980).

Lotfi, Z. M., and A. J. Tosser, Systematic Search for Minimum Synthesis of Logical Functions with Multiplexers, *Int. J. Electron.*, **47,** 569 (1980).

Mirsalehi, M. M., J. Shamir, and H. J. Caulfield, Residue Arithmetic Processing Utilizing Optical Fredkin Gates, *Appl. Opt.*, **26,** 3940 (1987).

Muroga, S., *Threshold Logic and Applications*, Wiley, New York, 1971.

Papachristou, C. A., An Algorithm for Optimal NAND Cascade Logic Synthesis, *IEEE Trans. Comput.*, **C-27,** 1099 (1978).

Rhyne, V. T., P. S. Noe, M. H. McKinney, and U. W. Pooch, A New Technique for the Fast Minimization of Switching Functions, *IEEE Trans. Comput.*, **C-26,** 757 (1977).

Sasao, T., An Algorithm to Derive the Complement of a Binary Function with Multiple-Valued Inputs, *IEEE Trans. Comput.*, **C-34,** 131 (1985).

Shamir, J., H. J. Caulfield, W. Micelli, and R. J. Seymour, Optical Computing and the Fredkin Gates, *Appl. Opt.*, **25,** 1604 (1986).

Su, S. Y. H., and P. T. Cheung, Computer Minimization of Multi-valued Switching Functions, *IEEE Trans. Comput.*, **C-21,** 995 (1972).

Taconet, B., New Prime Implicant Algorithm, Based on Monotonicity Properties, *Electron. Lett.*, **20,** 89 (1984).

Weinberger, A., High-Speed Programmable Logic Array Adders, *IBM J. Res. Dev.*, **23,** 163 (1979).

CHAPTER EIGHT

Combinational Logic

8.1 INTRODUCTION

A combinational logic circuit is characterized by one or more logic outputs whose behavior depends only on the present values of the input and not on the past values. In the previous two chapters, we introduced the tools necessary to design combinational logic circuits. The design algorithm leading to the realization of a complex combinational circuit consists of several essential steps. First, the logic problem is intuitively analyzed and then decomposed into a set of smaller but nontrivial functional units. Next, truth tables for each of these functional units are identified and then the minimized logic functions for each of the functional units are identified. Finally, logic circuits corresponding to each of the functional units are assembled and then cascaded to form the complex logic circuit.

There are many different combinational logic circuits available, each for a specific application. In this chapter, however, some of the more significant combinational logic circuits used in computing machines such as adders, subtracters, decoders, and encoders are presented. These devices have played important roles in the development of more advanced digital systems. The application of these medium-scale electronic integrated circuits is not governed by any well defined set of design procedures, such as those for individual logic gates. Interestingly, heuristic conclusions often lead to more sensible designs. We must emphasize that a complete understanding of the various logic devices to be covered in this chapter is important since many of these ideas can finally be translated to the optical domain.

8.2 BINARY ADDERS

A binary adder is perhaps the most basic arithmetic circuit since it can be used repeatedly to accomplish more complex operations, such as multiplication, division, the trigonometric function, and the square root. When the addition circuit has no carry-in, as in the least significant bit of the operands, it involves only two inputs, addend and augend. The addition of the remaining bits would require a carry-in coming in from the less significant bit position.

Multi-bit adders can be designed in a number of different ways. Each such design may have different speed and gate-count characteristics. A two-level network derived directly from its truth table would obviously work as a fast logic circuit. However, this otherwise regular design approach would lead to circuits that require a large number of gates as well as a large number of gate inputs per gate. For example, a NAND-NAND addition circuit capable of adding two n-bit numbers may consist of 2^{2n} NAND gates each with $2n + 1$ inputs and one 2^{2n}-input NAND gate. Even for a moderate value of n, this circuit thus becomes reasonably complex. Of course, this problem could be solved partially by using a ROM, for example. But even in that case, the ROM will have to have $2n$ lines and at least $n + 1$ output lines. The memory size requirement of the ROM is thus nontrivial.

An alternative to the aforementioned expensive designs is rather straightforward. It follows directly from the observation of the mechanics of the n-bit addition operation. The process of adding augend and addend is identical at each of the columns except at the least significant position. An n-bit addition circuit can therefore be realized by designing a total of n single-bit addition circuits and then cascading them together to form an n-bit adder. All but one of these single-bit adder units need to be full adders (i.e., adders with three inputs—addend, augend, and carry-in—and two outputs—sum and carry-out). The least significant single-bit adder can be either a half adder (with only two inputs—addend and augend—and two outputs—sum and carry-out) or a full adder whose carry-in is tied to a 0. While cascading the adder units, the carry-out output of each of these units is tied to the carry-in input of the next significant unit. Thus an n-bit addition circuit may consist of either n single-bit full adders or $n - 1$ single-bit full adders and one half adder.

8.2.1 Half Adder

The half adder unit, as already described, is a two-input/two-output combinational circuit. It uses the least significant bits of any two n-bit operands as the inputs. The truth table corresponding to its two inputs, A_0 and B_0, and the outputs, sum S_0 and carry-out C_0, is shown in Fig. 8.1a. From the truth table, the equations for

A_0	B_0	C_0	S_0
0	0	0	0
0	1	0	1
1	0	0	1
1	1	1	0

(a)

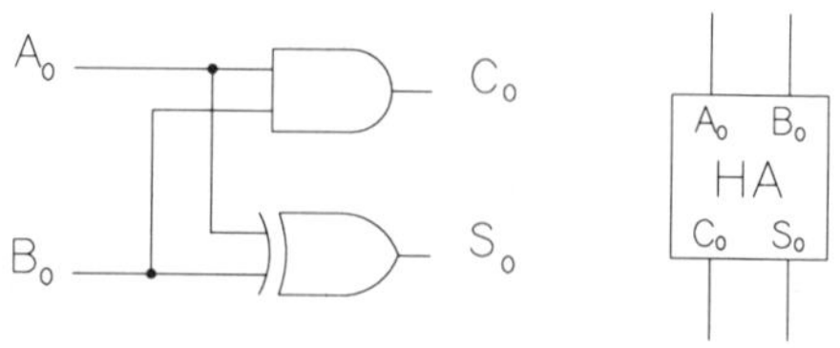

(b) (c)

FIGURE 8.1 The half adder: (a) truth table, (b) logic circuit, and (c) block diagram.

the sum S_0 and carry-out C_0 are found to be

$$S_0 = \overline{A}_0 B_0 + A_0 \overline{B}_0 = A_0 \oplus B_0 \tag{8.1}$$

and

$$C_0 = A_0 B_0 \tag{8.2}$$

These logic functions may be realized in several ways. Figure 8.1b illustrates one version of this circuit while Fig. 8.1c shows its block diagram.

8.2.2 Full Adder

A full adder is a three-input/two-output combinational logic circuit that when placed at the ith bit location adds two binary digits, A_i and B_i, and a carry-in from the $(i - 1)$-bit position, C_{i-1}. The full adder truth table for the outputs C_i and S_i is shown in Fig. 8.2a. The logic equations for the sum and carry-out are

$$S_i(A_i B_i C_{i-1}) = \overline{A}_i \overline{B}_i C_{i-1} + \overline{A}_i B_i \overline{C}_{i-1} + A_i \overline{B_i C_{i-1}} + A_i B_i C_{i-1} \tag{8.3}$$

and

$$C_i(A_i, B_i, C_{i-1}) = A_i B_i + A_i C_{i-1} + B_i C_{i-1} \tag{8.4}$$

Again, as in the case of the adder, there are several ways to implement the full adder equations. Figure 8.2b shows the NAND-only two-level version of the full adder circuit.

Equations (8.3) and (8.4) can be rewritten, respectively, as

$$S_i = A_i \oplus B_i \oplus C_{i-1} \tag{8.5}$$

and

$$C_i = A_i B_i + C_{i-1}(A_i \oplus B_i). \tag{8.6}$$

It is interesting to note that both of these equations have $A_i \oplus B_i$ common. Also recall that $A_i \oplus B_i$ is the sum output and $A_i B_i$ is the carry-out output of a half adder whose inputs are, respectively, A_i and B_i. This points to a newer design for the full adder. The full adder equations can be expressed in terms of the half adder outputs as

$$S_i = S_{HA} \oplus C_{i-1} \tag{8.7}$$

and

$$C_i = C_{HA} + C_{i-1} S_{HA} \tag{8.8}$$

where S_{HA} and C_{HA} are the sum and carry-out of the half adder. Equation (8.7) also suggests that if both the carry-in and the half adder sum were to be fed into

A_i	B_i	C_{i-1}	C_i	S_i
0	0	0	0	0
0	0	1	0	1
0	1	0	0	1
0	1	1	1	0
1	0	0	0	1
1	0	1	1	0
1	1	0	1	0
1	1	1	1	1

(a)

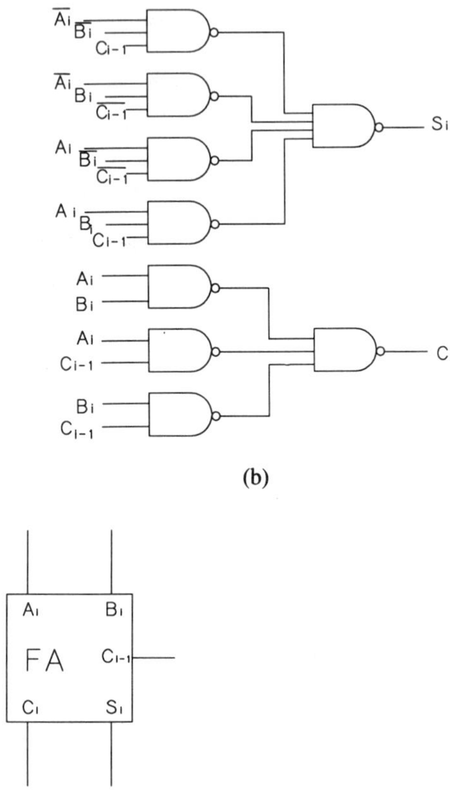

(b)

(c)

FIGURE 8.2 Full adder (a) truth table, (b) circuit, and (c) block diagram.

another half adder, it would generate the final sum. Again, if the carry-outs from both of these half adders were to be ORed together, it would correspond to the full adder carry-out C_i as per Eq. (8.8). Figure 8.3a shows this alternative full adder circuit using two half adders and an OR gate, while Fig. 8.3b shows its equivalent form using only NAND logic gates. This NAND-only circuit requires nine NAND gates and a total of 18 gate inputs. In comparison, the circuit of Fig. 8.2b uses 12 NAND gates and 31 gate inputs. It must, however, be noted that while the circuit of Fig. 8.3b uses three fewer gates and 13 fewer gate inputs, this circuit is relatively slower because the signals have to go through six levels of NAND logic.

Figure 8.4a shows the connection of n full adders, while Fig. 8.4b shows $n - 1$ full adders and one half adder to produce an n-bit adder. The total propagation delay of this n-bit ripple adder is nD, where D represents the propagation delay of a single full adder. This delay accounts for the fact that the carry in a multi-bit adder has to propagate through all the full adders. This delay thus becomes more and more significant with increasing n. The common digital electronic adders available in the market are generally of only 4-bit size.

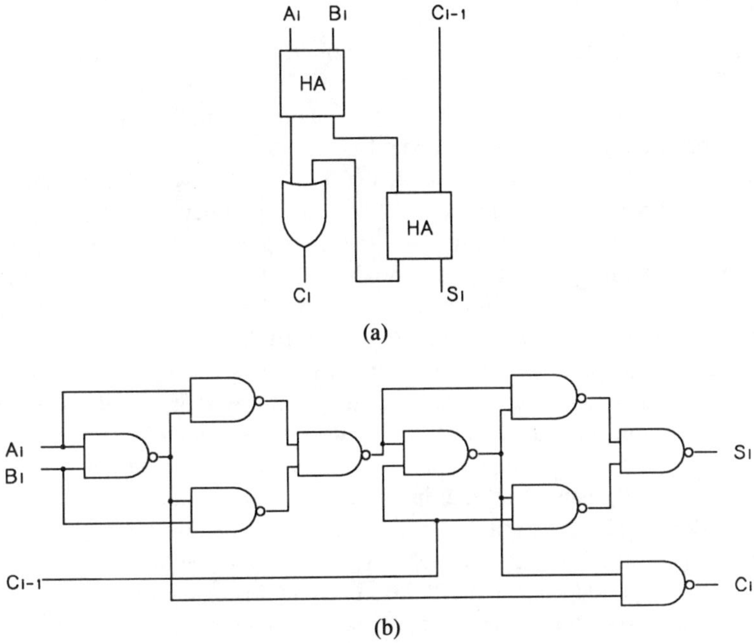

FIGURE 8.3 Full adder circuits: (a) using half adders and (b) using only NAND gates.

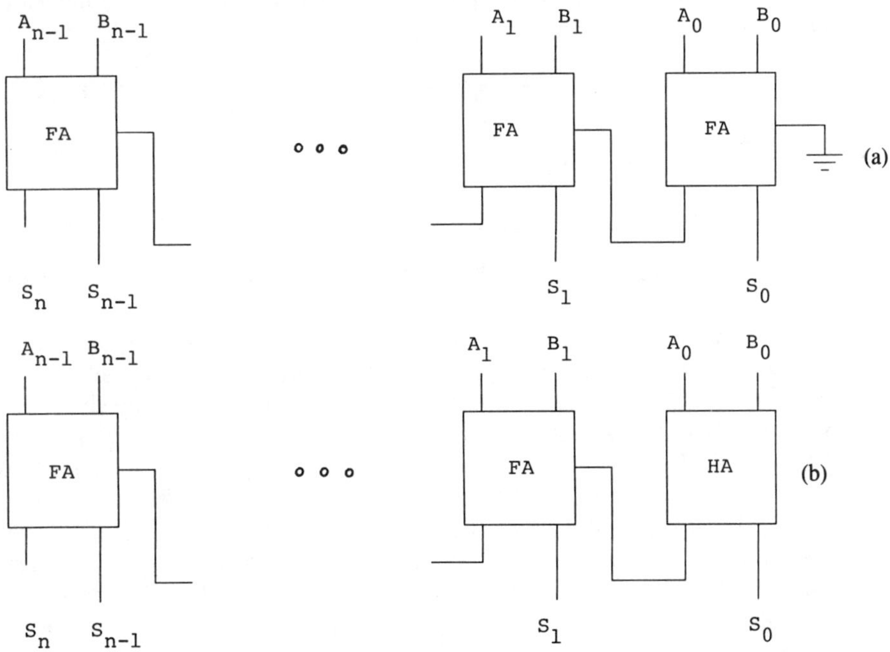

FIGURE 8.4 An n-bit parallel adder circuit: (a) using n full adders and (b) using $n - 1$ full adders and one half adder.

8.2.3 Applications of Binary Adders

It is important to note that we can combine multiple full adders not only to obtain a multi-bit binary adder but also to obtain a multiplication circuit. Multiplication of binary numbers consists of repeated addition operations. Consider, for example, the multiplication of two 4-bit number $A = A_3A_2A_1A_0$ and $B = B_3B_2B_1B_0$. Fortunately, in the binary number system, the product p_{xy} of two binary numbers a_x and b_y is obtained by ANDing the input bits. The product of the multi-bit binary numbers can therefore be obtained by first finding the partial products and then adding the relatively displaced partial products as shown in Fig. 8.5a. Figure 8.5b shows the corresponding circuit resulting in an 8-bit product output. In the next chapter, sequential design techniques are presented that allow one to design a multiplier that uses a repetitive algorithm for multiplication.

In general, a computer designed to handle n-bit operands can also perform operations on $2n$-bit quantities. Such an application is commonly referred to as a multiple precision operation. Consider, for example, multiplication of two 8-bit numbers, A and B, using 4-bit multipliers and 4-bit binary adders. While the 4-bit binary adders can be the ripple adders that we have just discussed, the 4-bit multipliers may be in the form of a ROM. The two 4-bit quantities to be multiplied make up an 8-bit address for the ROM. Each ROM location may store an 8-bit product of the two 4-bit quantities that make up its address.

In the multiplication circuit design being considered here, it is wise to break the given task into several simpler tasks in order to make the problem fit the devices that are available. Each 8-bit operand can be expressed in terms of the least significant four bits, L, and most significant four bits, M, as

$$A = 2^4(M_A) + L_A. \tag{8.9}$$

Accordingly, the product, P, may be expressed as

$$\begin{aligned} P &= [2^4(M_A) + L_A][2^4(M_B) + L_B] \\ &= [2^8(M_AM_B) + 2^4(M_AM_B + M_BL_A) + (L_AL_B) \end{aligned} \tag{8.10}$$

These partial products may be realized using four 4-bit multipliers of inputs; (1) M_A and M_B (2) M_A and L_B, (3) M_B and L_A, and (4) L_A and L_B. These four 4-bit multipliers will result in four 8-bit outputs, X, Y, Z, and W, respectively. Note also that both Y and Z have the same order of significance. Accordingly, Y and Z may be added using two 4-bit full adders. The 8-bit quantity W is next added to the sum of Y and Z, which has been shifted to the left four places, that in turn must be added to X, which has been further shifted to the left four places. The product of two 8-bit quantities finally results in a 16-bit product, which may therefore be generated by the circuit of Fig. 8.6, which uses five 4-bit full adders.

A single ROM, if it were to be designed for multiplying two 8-bit numbers, would have required 16 address lines and 16 output lines. This would have amounted to a total memory size of $2^{16} \times 16 = 1{,}048{,}576$ bits. Each ROM that

				A_3	A_2	A_1	A_0
				B_3	B_2	B_1	B_0
				A_3B_0	A_2B_0	A_1B_0	A_0B_0
			A_3B_1	A_2B_1	A_1B_1	A_0B_1	
		A_3B_2	A_2B_2	A_1B_2	A_0B_2		
	A_3B_3	A_2B_3	A_1B_3	A_0B_3			
P_7	P_6	P_5	P_4	P_3	P_2	P_1	P_0

(a)

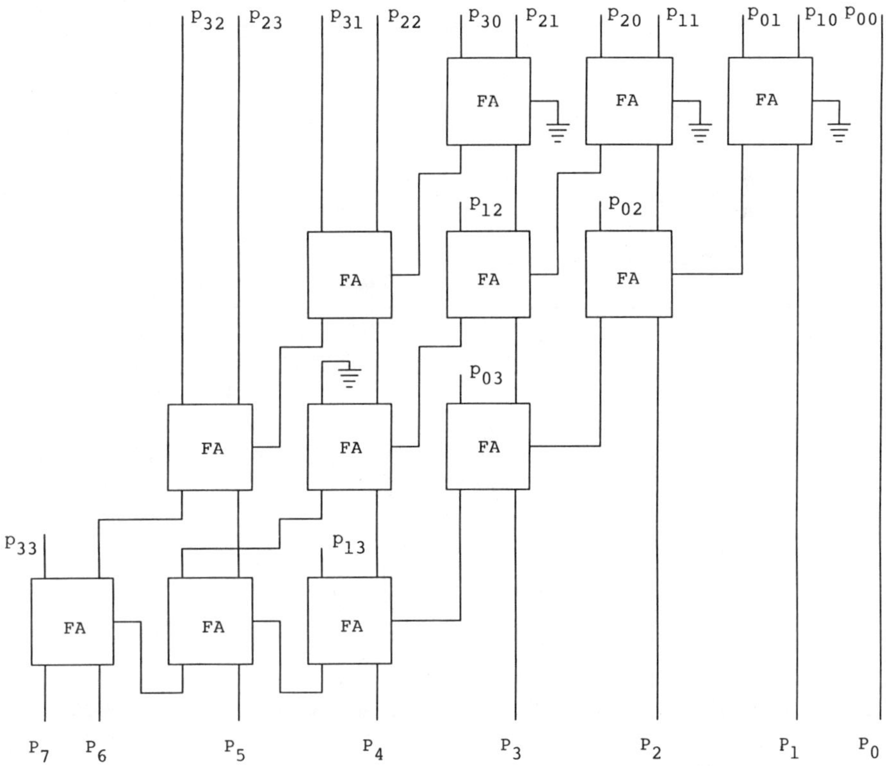

(b)

FIGURE 8.5 Multiplication of two 4-bit numbers: (a) operation and (b) logic circuit.

we used in the circuit of Fig. 8.6 has a memory size of only $2^8 \times 8 = 2048$ bits, and therefore the total memory size of the circuit is only 8192 bits.

Another application of the binary adders is in the design of binary subtracters. An arithmetic subtracter circuit, however, can also be designed from scratch. It is possible to design a multi-bit subtracter from a single-bit full subtracter and a half subtracter. The design scheme is very much similar to that of a multi-bit adder circuit. In practice, however, stand-alone subtracters are neither used nor designed.

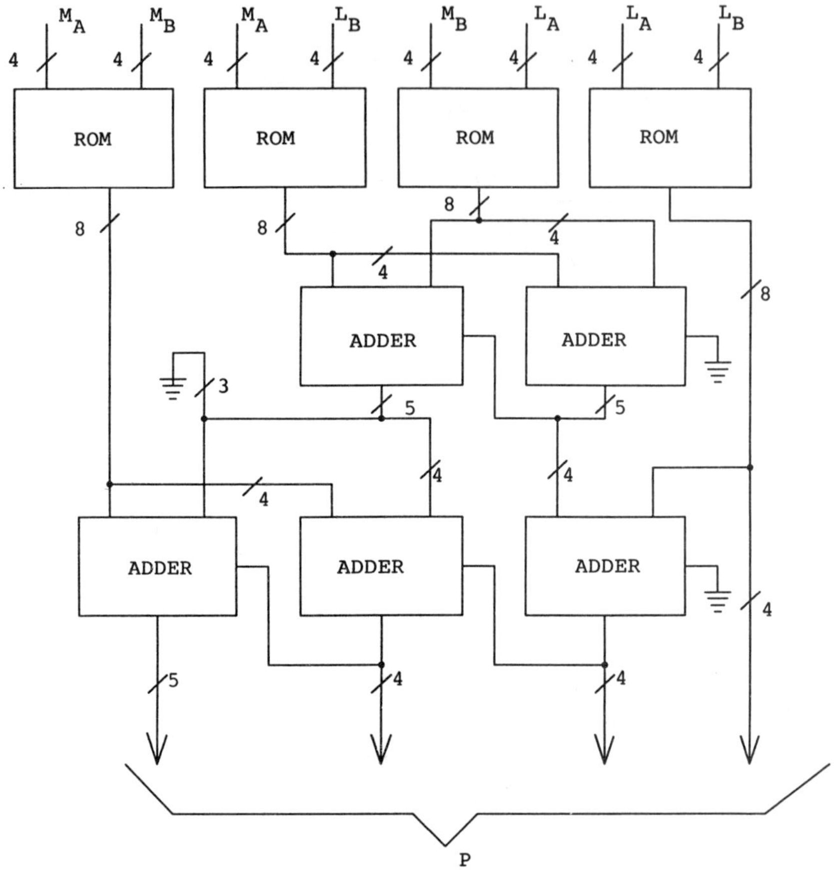

FIGURE 8.6 Multiplication circuit for multiplying two 8-bit numbers.

Since the subtraction operation is nothing but the addition of minuend and the complement of subtrahend, rather than designing a straightforward subtracter, multi-bit subtracters are designed from a multi-bit adder in accordance with complement arithmetic. The multi-bit adder circuit may be fed directly with minuend and the complemented subtrahend. We have already pointed out that the 2's complement is preferred over the 1's complement scheme. If each bit of the subtrahend is individually complemented using a NOT gate, for example, the corresponding 1's complement would be obtained. However, a 2's complement could be obtained by adding a 1 to the least significant bit of the corresponding 1's complement. Fortunately, this should not be a problem if we use a ripple adder whose least significant adder is a single-bit full adder. The carry-in input of this least significant bit full adder can be fed with a 1 as shown in the circuit of Fig. 8.7. It should be noted that X-OR gates, instead of NOTs, are used in the circuit for performing the complement operation. In this circuit, the two-input X-OR gates are serving as

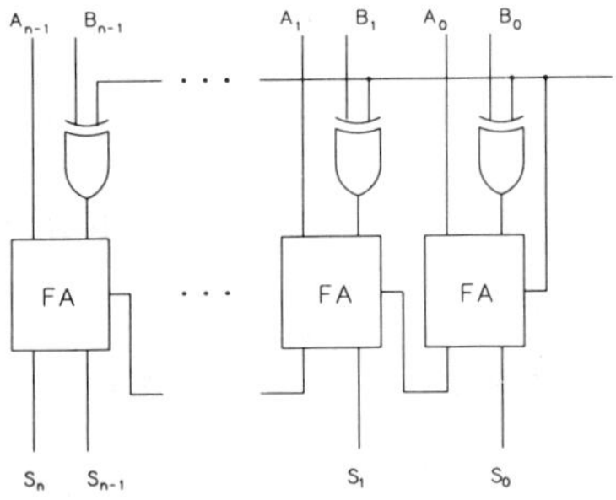

FIGURE 8.7 A mulit-bit adder/subtracter circuit.

programmable inverters with one input tied to E. Whenever E is a 0, the X-OR serves as a delay element, but when E is a 1 it works as an inverter since $0 \oplus X = X$ and $1 \oplus X = \overline{X}$. Thus, when $E = 1$, the X-OR gates are generating the 1's complement of the subtrahend, but when $E = 0$, the X-OR gates are leaving the inputs unchanged.

The use of X-OR logic gates is offering the possibility of having both an adder and a subtracter. The enable input E allows the circuit to add or to subtract by taking 1's complement and adding a 1 to the least significant bit. The circuit adds with a carry-in of 0 when $E = 0$; otherwise the circuit complements the subtrahend and adds that to the minuend with a carry-in of 1.

8.3 CARRY LOOK-AHEAD ADDER

The multi-bit adder circuit designed in the last section is referred to as a ripple adder since in it the carriers would have to actually ripple through several units before yielding the sum. The performance of a ripple adder or adder/subtracter is thus limited by its propagation delay, which has to account for the rippling of carriers/borrows. The worst-case propagation delay thus becomes directly proportional to the number of individual units, which in turn is dependent on the bit size of the operands used.

A speed-up scheme is sought for both the adder and adder/subtracter circuits since almost all computation operations directly involve them. One rather effective method of speeding up the addition process is known as the carry look-ahead (CLA) scheme. The CLA scheme is based on the concept that the circuit delay can drastically be reduced if the full adders at the higher-order bit locations can be made independent of the carry-in inputs. To design a CLA, we therefore must have a good understanding of both the generation and propagation of carries.

There are two particular properties pertaining to the carries. First, the carry-out is the same as the carry-in as long as only one of the other two inputs is a 1. Second, the carry-out is always a 1 independent of the carry-in when both of the other inputs are 1's. Consequently, two useful functions referred to as carry-propagate, P_i, and carry-generate, G_i, can be defined for the ith bit by

$$P_i = A_i \oplus B_i \tag{8.11}$$

$$G_i = A_i B_i \tag{8.12}$$

where A_i and B_i are, respectively, the ith addend and augend bits.

Equations (8.5) and (8.6) can now be rewritten in terms of the carry-propagate and carry-generate functions as

$$S_i = P_i \oplus C_{i-1} \tag{8.13}$$

$$C_i = G_i + P_i C_{i-1} \tag{8.14}$$

Both of these equations for the ith bit still depend on the carry coming in from the $(i - 1)$th bit. It may be noted that we know of only one carry-in, C_{-1}, that is 0 by default. C_0 can be determined using Eq. (8.14) from a knowledge of C_{-1}, which in turn can be used again in Eq. (8.14) to generate C_1 and so on. Using this idea, the carries for a 4-bit CLA adder, for example, can be generated as shown below:

$$C_0 = G_0 + P_0 C_{-1} \tag{8.15}$$

$$\begin{aligned} C_1 &= G_1 + P_1 C_0 \\ &= G_1 + P_1 G_0 + P_1 P_0 C_{-1} \end{aligned} \tag{8.16}$$

$$\begin{aligned} C_2 &= G_2 + P_2 C_1 \\ &= G_2 + P_2 G_1 + P_2 P_1 G_0 + P_2 P_1 P_0 C_{-1} \end{aligned} \tag{8.17}$$

$$\begin{aligned} C_3 &= G_3 + P_3 C_2 \\ &= G_3 + P_3 G_2 + P_3 P_2 G_1 + P_3 P_2 P_1 G_0 \\ &\quad + P_3 P_2 P_1 P_0 C_{-1} \end{aligned} \tag{8.18}$$

However, once the carries have been obtained, the corresponding four sum bits can be generated readily using Eq. (8.13).

Each of the generate and propagate functions can be realized from the data bits using a maximum of three levels of NAND gates. Note that an X-OR function is realized using four NAND gates in three layers. Again, using Eqs. (8.15) through (8.18), the carry-out for each of the bit locations can be implemented with a two-level NAND circuit. And, finally, the sum bits can be generated using Eq. (8.13) via a three-level NAND circuit. The circuit diagram of a 4-bit CLA adder, for example, is shown in Fig. 8.8. The circuit consists of three sections, each having four subunits. The propagate-generate (PG) section consisting of identical subunits generates the carry-propagate and carry-generate functions. The carry section intakes the outputs of the previous section as well as C_{-1} and generates the carries.

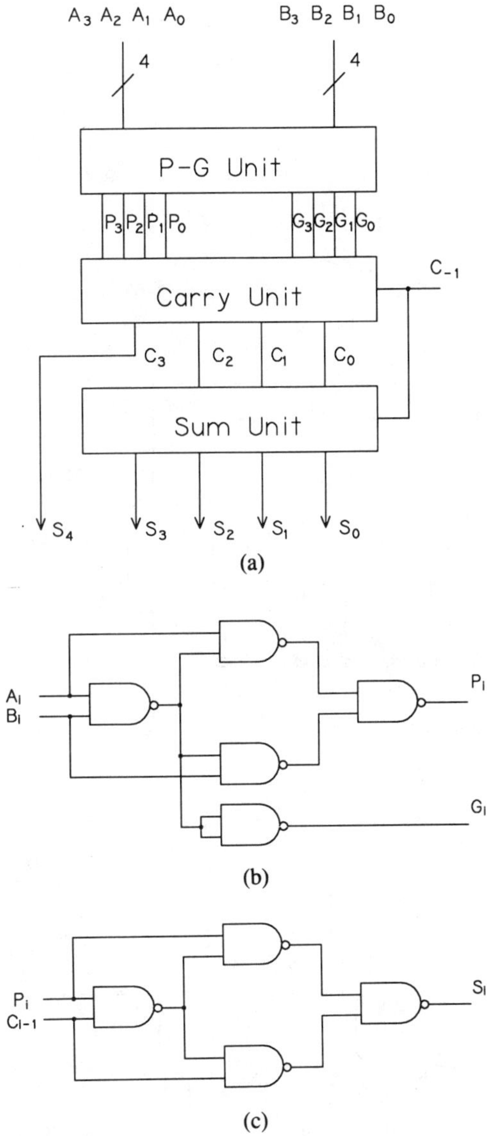

FIGURE 8.8 A 4-bit CLA adder: (a) block diagram, (b) PG unit, (c) carry unit, and (d) sum unit.

Each subunit is different. The complexity of each of the carry subunits increases as one moves further away from the least significant bit. Finally, the sum section consisting of identical subunits generates the sum bits.

Each propagation-generation subunit requires five NAND gates, while each sum subunit requires four NAND gates as shown in Fig. 8.8. The n-bit carry unit requires a total of $(n^2 + 3n)/2$ NAND gates. An n-bit CLA adder therefore re-

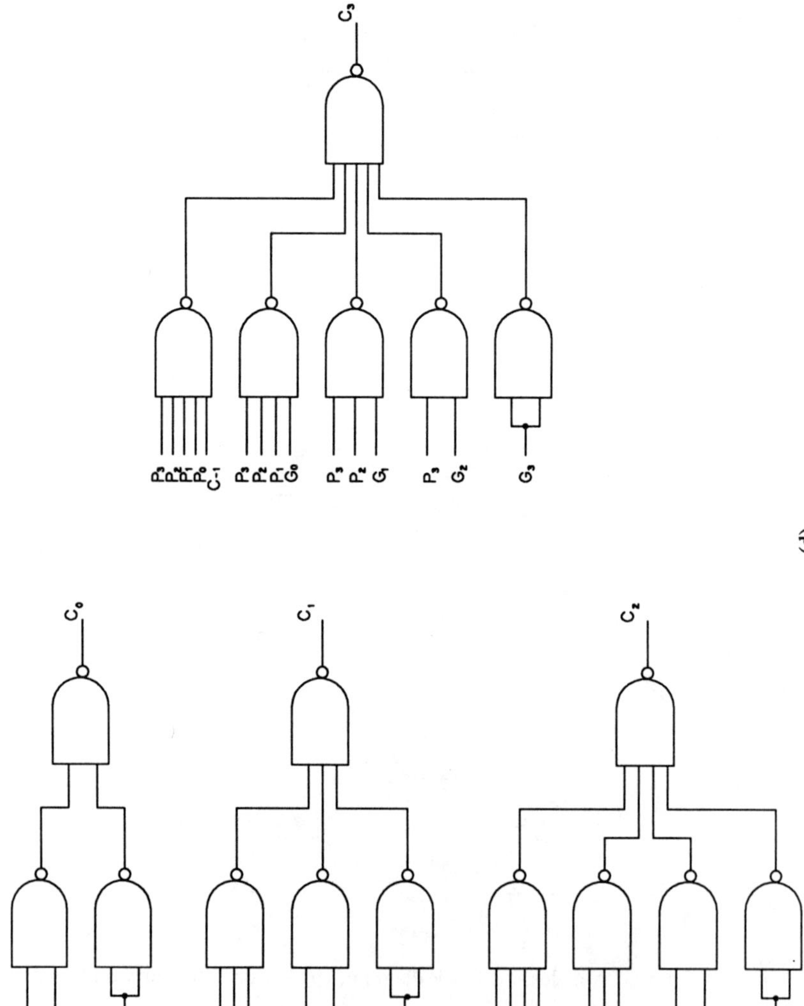

FIGURE 8.8 *(Continued)*

quires a total of $(n^2 + 21n)/2$ NAND gates and involves a total of eight units of NAND gate delay. As n increases, the computation time saved increases with the use of CLAs. For example, while a 4-bit CLA cuts down the time factor by about one-third, a 64-bit CLA adder reduces the computation time by a factor of 17. However, two practical aspects become obvious. As n increases, the number of gates needed also increases. More importantly, with larger values of n, the number of NAND gate inputs increases. First, NAND gates with too many inputs are hard to find commercially. This is compounded by the fact that the carry-in must drive a large number of NAND gates. In addition, the propagate functions will be subjected to a high fan-out requirement, resulting in a serious fan-in problem for large n. At times, however, acute fan-in and fan-out problems are solved by having combined multiple CLA adders in a ripple format.

8.4 BCD ARITHMETIC CIRCUITS

Electronic computers are equipped to handle binary numbers. However, most users of computers are more familiar with decimal digits. In this section, we explore the design of arithmetic circuits that can take in decimal numbers in the form of BCD. Obviously, the BCD based systems would require more memory since this particular coding is less efficient.

BCD arithmetic is reasonably complicated because when two BCD numbers are added, for example, using a binary addition circuit, some of the sums turn out to be invalid. When two 4-bit binary numbers are added on a binary adder, it is possible to obtain 16 different sums as output. Unfortunately, six of these 16 numbers are undefined in BCD. In addition, carries present in a binary addition may or may not correspond to carries present in decimal addition. Thus the logic circuit designer has two choices. The first option involves designing a BCD adder circuit from scratch while the second option may involve modifying the already-designed binary adder circuit to suit the needs of BCD numbers.

Designing a BCD adder from scratch involves identifying the corresponding truth table. Such an adder circuit will have a total of nine inputs and five outputs. Four of the nine inputs correspond to the augend, another four to the addend, and the ninth input is the carry-in. One of the five outputs is the carry-out while the remaining four produce the BCD sum. The truth table therefore consists of $2^9 = 512$ different input combinations. This circuit may be designed using either a ROM or PLA-like integrated circuit or by minimizing the function using the Q-M technique and then generating a reduced circuit using logic gates. In either case, the design process is rather involved.

The simplest approach to a BCD adder design is to incorporate 4-bit binary adder modules. A table relating the to-be-corrected and the corresponding corrected BCD sums and carry is shown in Fig. 8.9. Note that the largest to-be-corrected sum could be 10011, which is obtained by adding two BCD nines and a carry-in of 1. The sums that are greater than 1001 are all undefined and therefore, need correction. Whenever the output of the corresponding nine inputs exceeds

To-be-Corrected					Corrected				
C_3'	S_3'	S_2'	S_1'	S_0'	C_3	S_3	S_2	S_1	S_0
0	0	0	0	0	0	0	0	0	0
		⋮					⋮		
0	1	0	0	1	0	1	0	0	1
0	1	0	1	0	1	0	0	0	0
0	1	0	1	1	1	0	0	0	1
0	1	1	0	0	1	0	0	1	0
0	1	1	0	1	1	0	0	1	1
0	1	1	1	0	1	0	1	0	0
0	1	1	1	1	1	0	1	0	1
1	0	0	0	0	1	0	1	1	0
1	0	0	0	1	1	0	1	1	1
1	0	0	1	0	1	1	0	0	0
1	0	0	1	1	1	1	0	0	1

FIGURE 8.9 Table for the to-be-corrected and corrected BCD sums.

nine (1001), a carry is generated and the sum needs to be corrected by means of a correction circuit.

A close examination of the table in Fig. 8.9 reveals that the corrected sum may be obtained by adding 0110 to the corresponding to-be-corrected sum. Furthermore, whenever $C_3'S_3'S_2'S_1S_0'$ is in need of correction, C_3 is a 1. The condition for the generation of C_3 can be determined by means of K-map minimization, according to which

$$C_3 = C_3' + S_3'S_1' + S_3'S_2' \tag{8.19}$$

The output of a logic circuit corresponding to Eq. (8.19) can be used to determine whether or not 0110 should be added to $S_3'S_2'S_1'S_0'$ for correction. Accordingly, the design of a BCD adder from a 4-bit binary adder will involve a carry decoder circuit and a second 4-bit full adder circuit for completing the correction.

Figure 8.10 shows the logic circuit for BCD addition. The sum bits of the first binary adder are introduced at the augend inputs of the second binary adder. When the corrected carry-out C_3 is a 0, the to-be-corrected sum is added to 0000 and thus remains unchanged. However, when C_3 is a 1, the to-be-corrected sum is added to 0110 to give the corrected sum.

Another alternative to designing BCD adders from scratch may involve first converting the BCD numbers into their binary equivalent, then adding the binary numbers using binary adders and then transforming the binary sum into its equivalent BCD form. This latter design requires both postcorrection and preconversion circuitries and is therefore more expensive than the one already designed.

We next consider the design of BCD subtracters. This is more complex than that for BCD addition, since in it there is a possibility of having a negative difference. However, as we have already dealt with binary subtraction, BCD subtraction

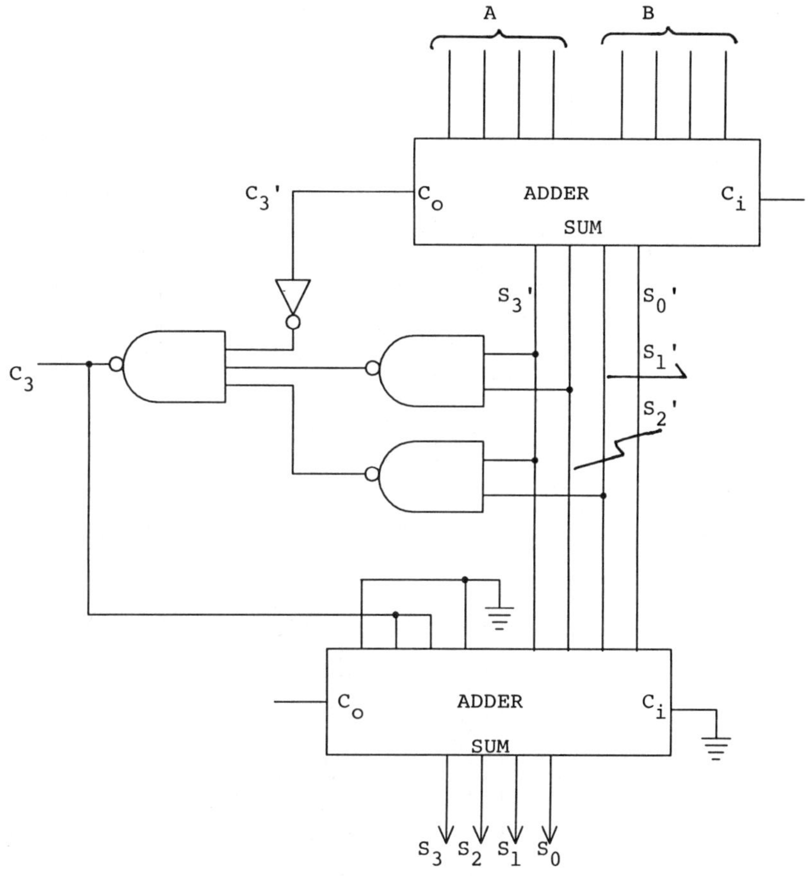

FIGURE 8.10 The BCD adder circuit.

can also be handled using a complement operation. The technique using the 10's complement would result in a circuit analogous to that of Fig. 8.7. On the other hand, scheme using the 9's complement would require an end-around-carry into the system. The BCD subtracter using 9's complement can be designed using a BCD adder logic circuit such as that of Fig. 8.10 and a 9's complementer logic circuit. For example, the 9's complement of the BCD number 7 is 1001 − 0111 = 0010. Interestingly, however, the 9's complement of any BCD digit is found by adding its 1's complement to 1010. For example, the BCD digit 7 can be processed to give 1000 + 1010 = 0010.

Figure 8.11a shows a BCD 9's complementer logic circuit that uses inverters for performing 1's complement. These inverters can be replaced by X-OR gates as in the circuit of Fig. 8.11b for designing a BCD adder/subtracter unit. For each BCD digit, there are three 4-bit binary adders. The first adder serves as the 9's complementer, the second serves as the adder, while the third serves as the post-processor. The unit performs simple BCD addition, when E is 0, since the com-

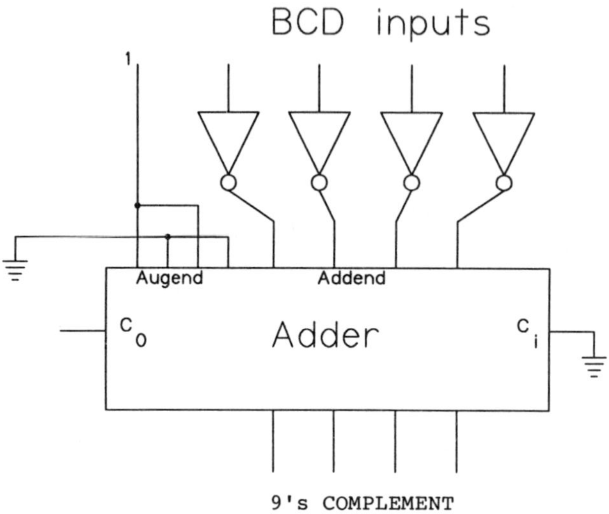

(a)

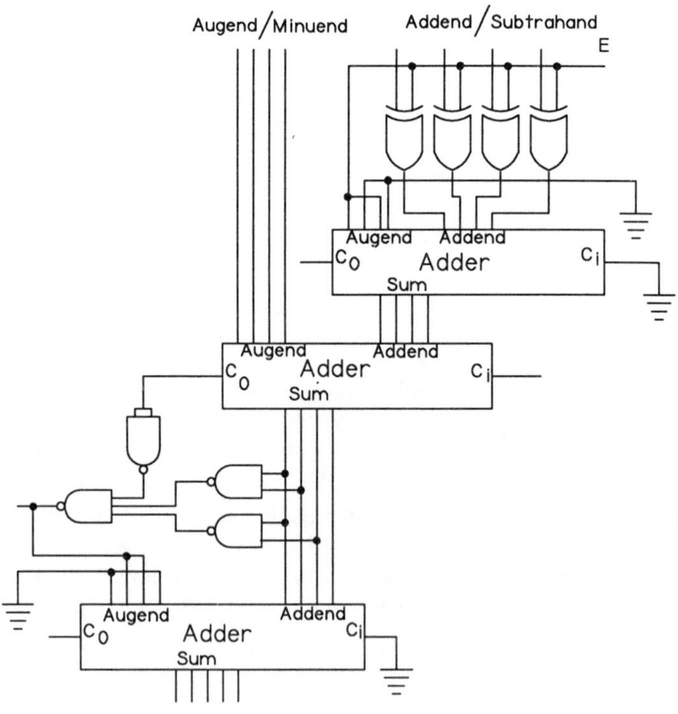

(b)

FIGURE 8.11 The logic circuit for (a) 9's complementation and (b) BCD adder/subtracter.

plementer leaves the augend input unchanged. But when E is a 1, the subtrahend is 9's complemented and is then added to the minuend. Consequently, when E is 0 the BCD sum appears at the output and when E is a 1 the BCD difference is obtained. When multiple units are cascaded to make a multi-decade BCD adder/subtracter, the carry-out from the most significant BCD adder unit is fed as input into the carry-in of the least significant BCD adder unit. Such end-around-carry completes the requirement of 9's complement subtraction.

It is important to realize that if one were to use XS3 numbers instead of BCD, finding the 9's complement of a number becomes too trivial. The 9's complement of an XS3 number is obtained simply by taking its 1's complement. However, XS3 addition is not necessarily the same as BCD addition. While adding XS3 numbers, as a general rule, if a carry is produced the correct sum is obtained by adding 0011 to the sum, otherwise 1101 (2's complement of 0011) is added to the sum. Accordingly, the XS3 adder/subtracter is obtained as shown in Fig. 8.12.

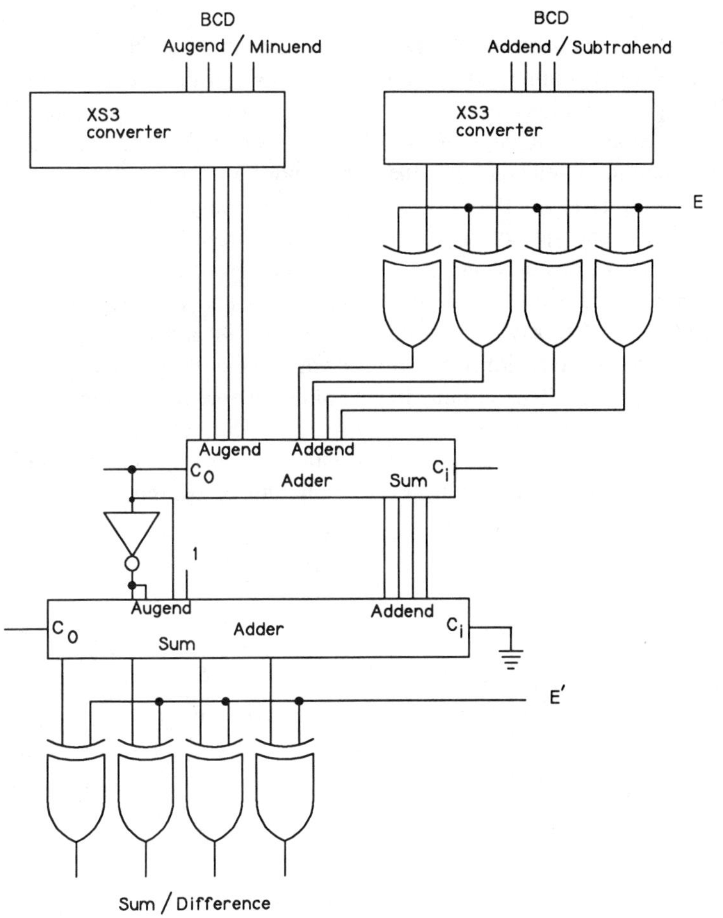

FIGURE 8.12 An XS3 adder/subtracter unit.

When the enable input E is a 0, either the augend or 1's complement of the subtrahend arrives at the first binary adder along with the addend. The output of this binary adder is added either to 0011 or 1101, depending on whether or not a carry is generated. The other enable input E' is activated to complement the output when there is no end-around-carry.

8.5 ARITHMETIC LOGIC UNIT

We have already encountered several examples of multi-bit arithmetic operations. Often, however, it is necessary to also perform different types of bit-by-bit logic operations between two multi-bit operands. A multi-function circuit operating on groups of bits serves such a purpose and is of importance in digital electronic circuits. Such multi-function circuits are commonly referred to as arithmetic logic units (ALUs). The different operations of an ALU are usually selected by means of control inputs.

The ALU logic operations are realized by directing first the input bits to appropriate circuits performing various logic functions and then selecting one of these possible logic operations as output using a multiplexer. The commercially available ALUs generally operate on two 4-bit values and allow for 32 arithmetic and 16 logic operations selectable by the combination of four control inputs, a mode selector, and a carry-in. Between any two input binary variables, there can be a total of 16 different types of logic operations, all of which are achievable by means of logic gates. In this section, however, the design of a relatively less complex ALU is considered just to demonstrate the process and some typical functions.

Consider an ALU having two multi-bit inputs, A and B. The circuit of Fig. 8.13a shows a logic circuit for the ith bit position consisting of only a 1-or-4 MUX and four logic gates. This circuit, for example, is able to perform up to four different logic operations between its inputs, A_i and B_i. Any specific operation can be selected by means of the two control inputs, S_1 and S_0. The enable input, S_2, is set to a 0 to enable the multiplexer in all four cases. The truth table in Fig.

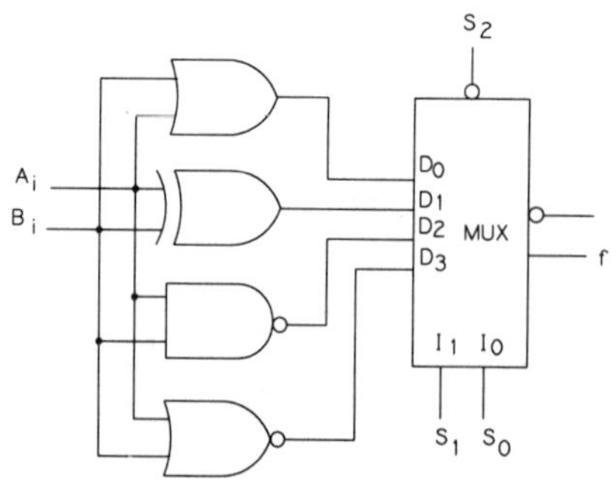

(a)

S_2	S_1	S_0	Function
0	0	0	$A_i + B_i$
0	0	1	$A_i \oplus B_i$
0	1	0	$\overline{A_i B_i}$
0	1	1	$\overline{A_i + B_i}$

(b)

FIGURE 8.13 An ALU logic section: (a) circuit and (b) menu table.

8.13b lists the possible logic operations and the corresponding control conditions. One could have replaced this 1-of-4 MUX with a 1-of-16 MUX for realizing 16 different logic operations. In that case, however, two additional control inputs would be necessary.

The ALU arithmetic section may be designed around a multi-bit adder circuit. If the carry-in is utilized, it is possible to obtain twice the number of arithmetic operations for the same number of control inputs. For every control condition the arithmetic output when $C_{-1} = 1$ is always 1 greater than the corresponding output when $C_{-1} = 0$. Figure 8.14a shows only the *i*th bit of an ALU arithmetic section. This section, which also uses a 1-of-4 MUX, can perform a total of eight different arithmetic operations. The MUX output has been fed into one of the adder inputs. In addition, this circuit consists of a full adder and two logic gates. The control input S_2 is used to differentiate between the arithmetic and the logic operations. Figure 8.14b lists all the arithmetic operations and the corresponding control inputs

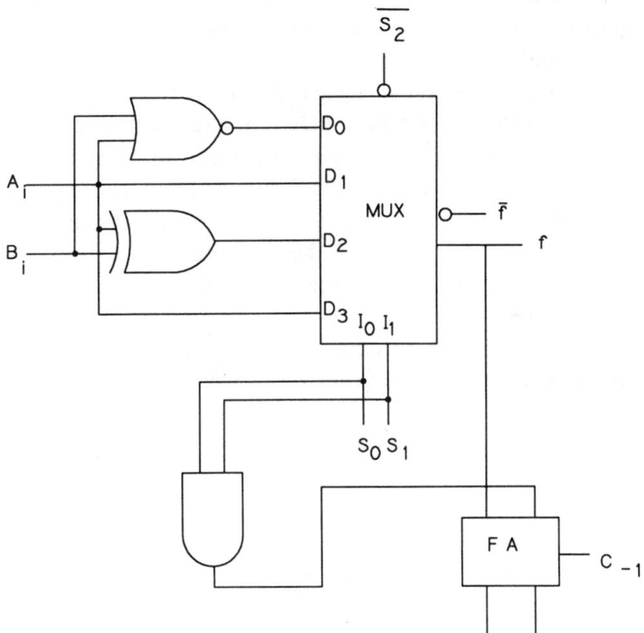

(a)

S_2	S_1	S_0	$C_{-1} = 0$	$C_{-1} = 1$
1	0	0	$\overline{A_i + B_i}$	$\overline{A_i + B_i}$ plus 1
1	0	1	A_i	A_i plus 1
1	1	0	$A_i \oplus B_i$	$A_i \oplus B_i$ plus 1
1	1	1	A_i minus 1	A_i

(b)

FIGURE 8.14 The ALU arithmetic section: (a) circuit and (b) menu table.

needed for operating this arithmetic unit. Note that when n-bit A is added to a string of n 1's, it is equivalent to adding A to the 2's complement of 1.

Both of these units, arithmetic and logic, may be combined together to make an ALU. Accordingly, the outputs are internally ORed together to produce the ALU output. Whenever $S_2 = 0$, the ALU processes one of the four logic outputs; otherwise, it generates one of the eight arithmetic outputs.

8.6 DECODERS AND ENCODERS

In digital systems, electronic or optical, it is necessary to convert one code to another. A decoder is a specially designed logic device that translates a code into a more useful standard form while an encoder is a logic device that converts a standard digit or information into a coded version. A decoding function, for example, is encountered in the BCD to seven-segment decoders. This particular decoder, for example, accepts as inputs binary coded decimal and provides outputs to drive a standard seven-segment light emitting diode display device in order to decode bits into readable digits.

In this section, we consider the design of standard decoders whose primary function is to identify the exact binary value of a set of inputs. These decoders accept n bits of binary input and convert them to up to 2^n unique outputs. For example, a decoder of this type receiving 2 bits of input data will have at least four output lines since binary values at the input may appear in four different ways. Depending on the actual binary value of the inputs, only one of its many outputs are turned on or off. This particular output is usually not enabled by any other binary value of the inputs.

Figure 8.15 shows one such decoder, commonly referred to as a 2–4 line decoder. When the enable input E is a 0, the decoder is enabled, and when E is a 1,

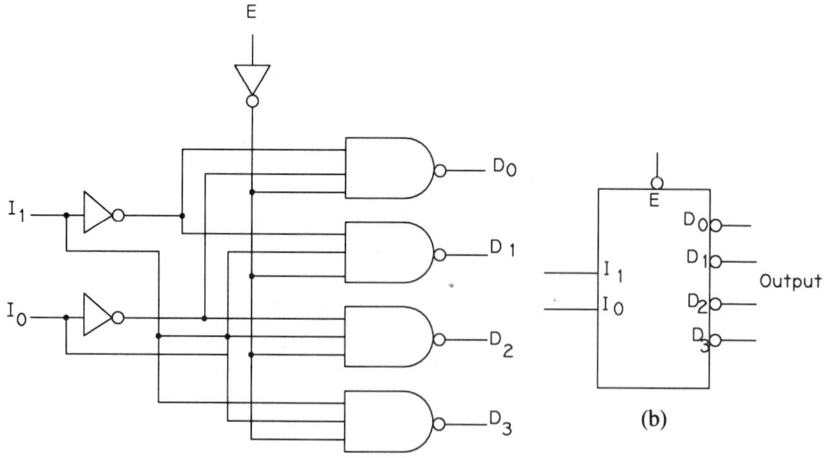

FIGURE 8.15 A 2-4 line decoder: (a) logic circuit and (b) block diagram.

the decoder is disabled. All outputs are 1 except for the line labeled D_n when n is the decimal equivalent of the binary values $I_1 I_0$. For example, when $AB = 10$ only the D_2 output is turned off. Such a decoder scheme allows us, for example, to activate four different operations by means of the four decoder outputs. By means of only these two inputs A and B, the four operations can be controlled directly.

The standard decoder design layout can be reorganized to design a demultiplexer (DMUX) logic circuit. A demultiplexer is a logic device that functions like a multiplexer except in it the data flow in the reverse direction. A DMUX receives data on a single entry line and outputs these data on one of its many output lines. Figure 8.16 shows a 1–4 line demultiplexer circuit where the address loaded at $I_1 I_0$ determines which particular output will take on the DMUX input value. The function of the various inputs are quite similar between a decoder and a DMUX. In a decoder, the outputs are just turned on or off, depending on the values of $I_1 I_0$. In comparison, in a DMUX, the selected output line (determined by the values at $I_1 I_0$) takes on the value of the DMUX input.

Smaller sized decoder/demultiplexer circuits are often cascaded together to design a larger decoder/demultiplexer circuit. Figure 8.17 gives an example where two 2–4 line decoders are cascaded together by means of their enable inputs to form one 3–8 line decoder. When $I_2 = 0$, only the top 2–4 line decoder is enabled and the lower one is disabled. Again, when $I_2 = 1$, the top decoder is disabled and the bottom decoder is enabled. The combined circuit functions as a 3–8 line decoder, where I_2, I_1, and I_0 function as the selectors. For example, when $I_2 I_1 I_0 = 010$ and D_2 output of the top 2–4 line decoder is enabled, but when $I_2 I_1 I_0 =$ 110 the D_2 output of the bottom 2–4 line decoder is enabled. Accordingly, the D_1

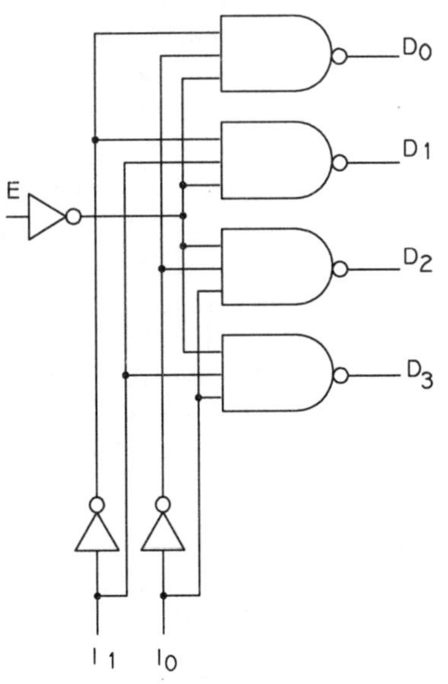

(a)

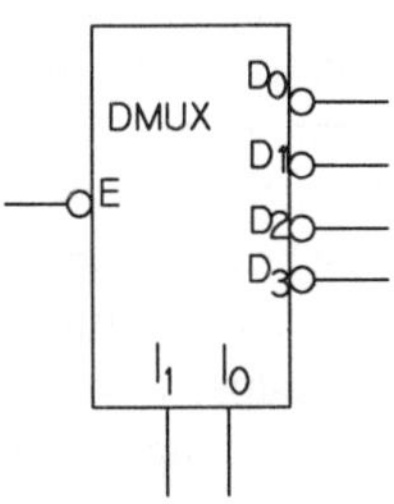

(b)

FIGURE 8.16 A 1-4 line demultiplexer: (a) logic circuit and (b) block diagram.

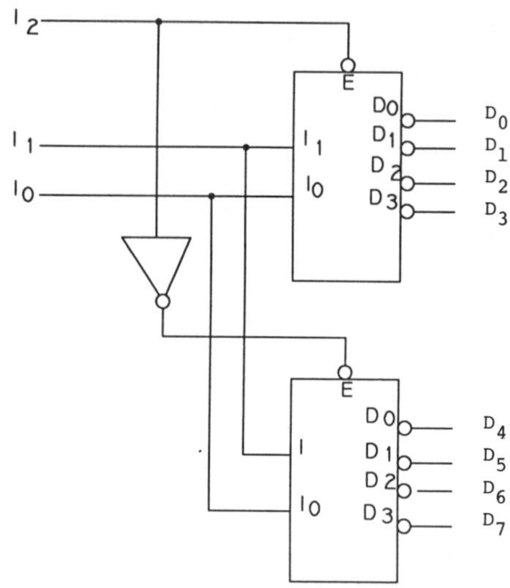

FIGURE 8.17 A 3-8 line decoder using two 2-4 line decoders.

through D_3 output lines of the bottom 2–4 line decoder functions, respectively, as D_4 through D_7 output lines of the corresponding 3–8 line decoder.

Following the design of the 2–4 line decoder, as shown in Fig. 8.15, one can extrapolate the design of any $n - 2^n$ line decoder. In the first level, there will be $n + 1$ NAND gates in the form of inverters (for the enable input and complemented inputs) and the second level will be formed of 2^n NAND gates each with $n + 1$ inputs. Figure 8.18 shows an alternative form of a 4–16 line decoder, for example, that also uses NAND gates but each with a reduced number of gate inputs. Such a decoding structure is commonly referred to as "tree-type" decoding. Usually, the number of transistors used in an electronic decoder is approximately equal to the number of inputs to each gate and the number of gates used. The circuit designers prefer to deal with a decoder scheme that provides for a smaller sum of gates and gate inputs. A tree-type decoder would require a smaller number of transistors than a regular decoder of the type shown in Fig. 8.15.

Figure 8.19 shows the logic circuit of a 4–16 line decoder using yet another interesting scheme referred to as "balanced decoding." In this scheme, the inputs are first lumped into groups and decoding is performed first within each of the individual groups, with the final level performing intergroup decodings. The significance of this improvement becomes obvious as larger sized decoders are considered. The regular decoder scheme is still the fastest since it involves only two levels of NAND gates.

A device that accepts an input digit and then converts into a coded output is referred to as an encoder. They are used not only in electronics but have potential to be used in optical systems. In fact, in a subsequent chapter, we shall see the need for such or similar encoders for encoding optical inputs. In reality, an encoder is the inverse of a decoder. Again, like a decoder, each encoder is different and serves only a special purpose. Consider, for example, the design of an encoder

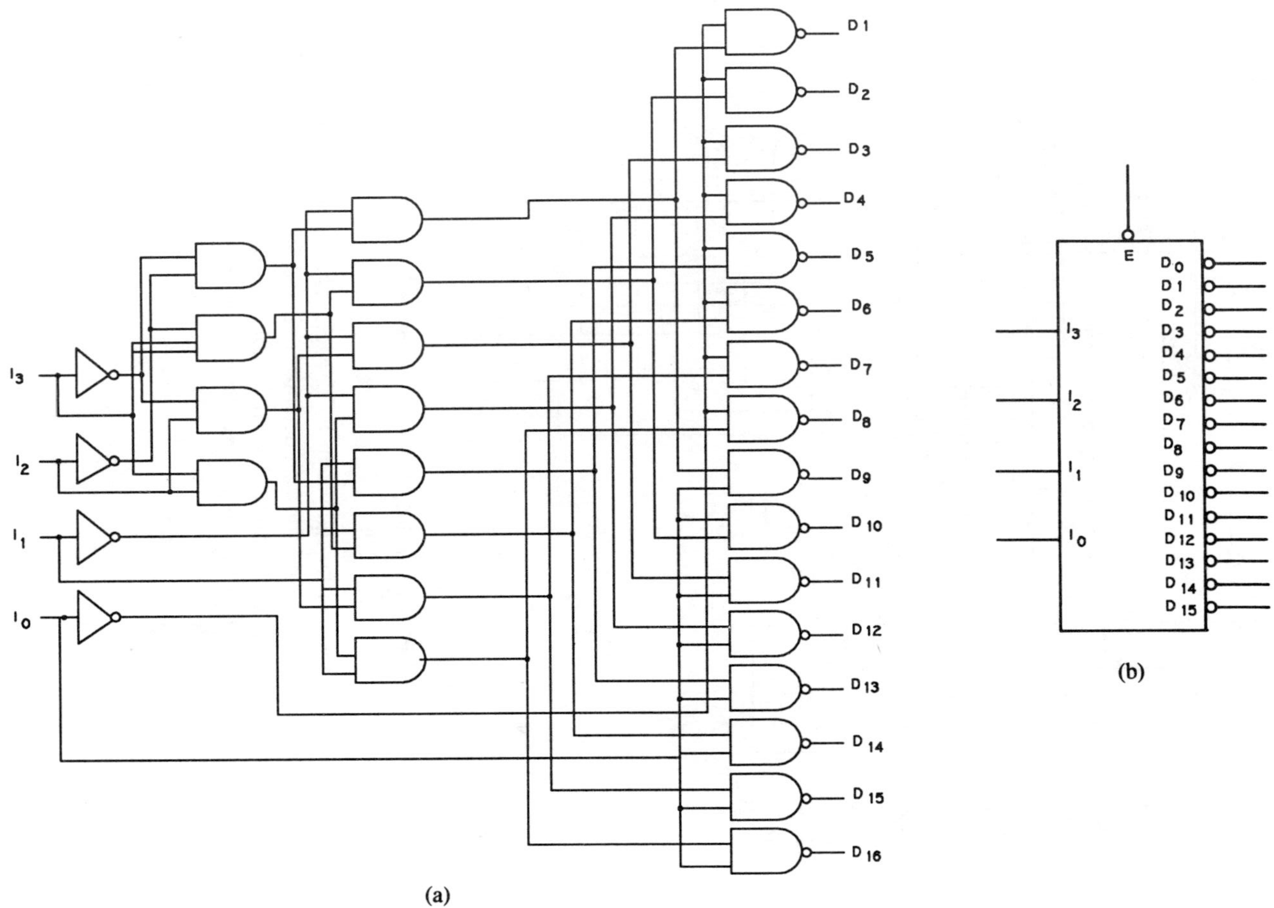

FIGURE 8.18 A tree-type 4-16 line decoder: (a) logic circuit and (b) block diagram.

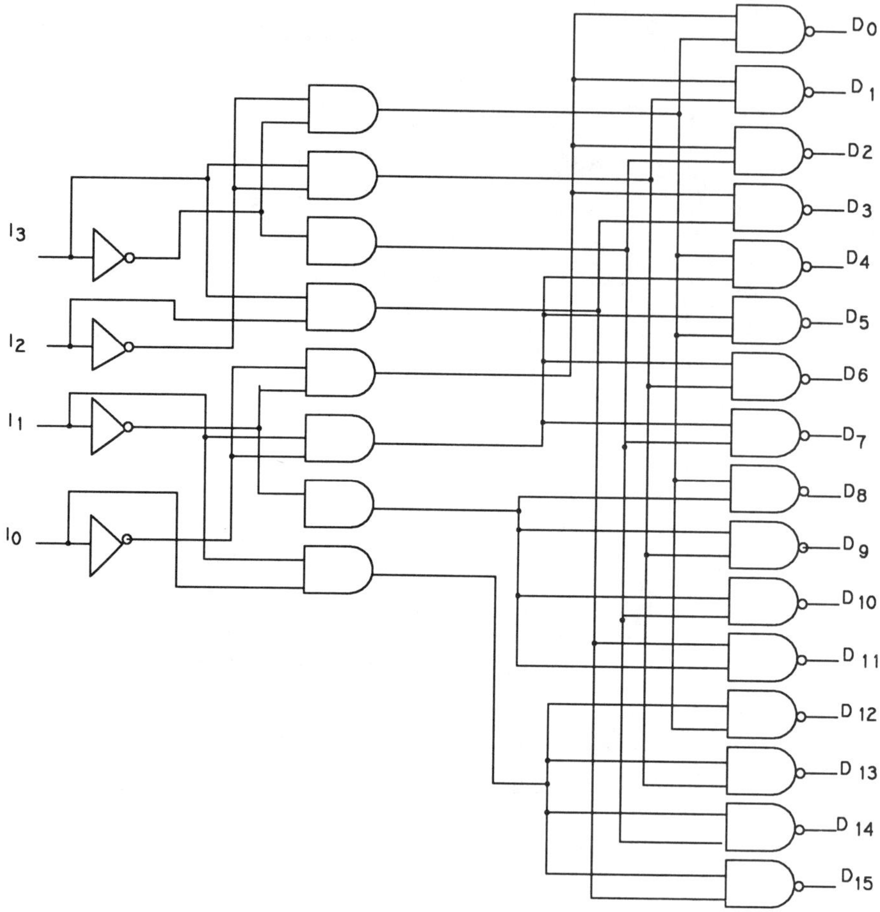

FIGURE 8.19 A balanced 4-16 line decoder.

that encodes decimal inputs to its equivalent BCD code. This particular encoder will have a total of ten inputs—one for each decimal digit and four output lines to represent the corresponding BCD codes. The truth table for the encoder is listed in Fig. 8.20a, from which the BCD outputs can be obtained as follows:

$$f_3 = 8 + 9$$

$$f_2 = 4 + 5 + 6 + 7$$

$$f_1 = 2 + 3 + 6 + 7$$

$$f_0 = 1 + 3 + 5 + 7 + 9$$

The resulting circuit is very straightforward and is obtained using only four OR gates, as shown in Fig. 8.20b.

Decimal Input	$f_3f_2f_1f_0$
0	0000
1	0001
2	0010
3	0011
4	0100
5	0101
6	0110
7	0111
8	1000
9	1001

(a)

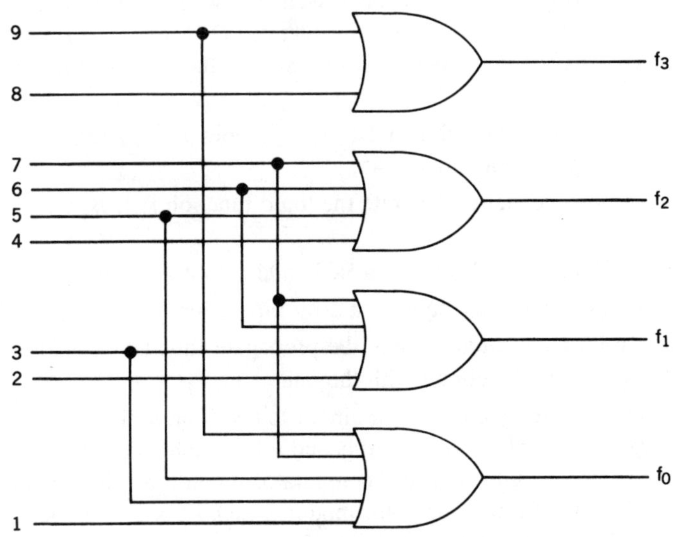

(b)

FIGURE 8.20 The decimal to BCD encoder: (a) truth table and (b) logic circuit.

8.7 PROBLEMS

1. Design a half adder using (a) NOR gates and (b) multiplexers.
2. Design a full adder using (a) NOR gates and (b) multiplexers.
3. Design a full adder circuit using logic gates (without cascading single-bit full adders) that can be used to add two bits of addend, two bits of augend, and a carry-in.
4. Obtain a combinational circuit for realizing the 2's complement of a 5-bit binary number.

5. Design a 32-bit fast adder where the fan-in and fan-out problems are avoided by allowing the carries to ripple through after the addition of every 8 bits.
6. Design a logic circuit for dividing an 8-bit number by a 4-bit number.
7. Design a single-bit full subtracter using half subtracter modules.
8. Design decimal code converters for the following conversions: (a) 8-4-2-1 to Excess-3, and (b) Excess-3 to BCD.
9. Design a circuit capable of comparing two unsigned 8-bit binary integers X and Y.
10. Design a dual-purpose 4-bit converter that converts binary to gray and also gray to binary.
11. Use single-bit full adders to add (a) 6 single-bit numbers, (b) 12 single-bit numbers, and (c) 15 single-bit numbers.
12. Design a four-input priority encoder such that when two inputs, D_i and D_j, are high simultaneously D_i has priority over D_j when $i > j$. The encoder produces a binary output code corresponding to the input that has the highest priority.
13. Organize an ALU to (a) subtract 1 from a number and (b) add 1 to a number. Show the setup if the number is 45_{10}.
14. Use a 3–8 line decoder to generate the logic function $f(A, B, C) = \Sigma m(0, 1, 4, 5)$.
15. Use Excess-3 code for designing a BCD adder/subtracter unit.
16. Design a logic circuit that squares a 2-bit binary input.
17. Design a n-bit CLA circuit where the propagate function is defined as $P_i = A_i + B_i$. Compare the circuit with the one using $P_i = A_i \oplus B_i$.
18. Design an n-bit binary comparator circuit to test if an n-bit number A is equal to, or larger than, or smaller than a second n-bit number B. The problem could be broken into one unit of a half comparator module for the 0th bit position and $n - 1$ units of full comparator modules each of one of the remaining bit positions. The ith bit module is generating two outputs, G_i and L_i, such that (a) $G_j = 1$ and $L_j = 0$ if $A_j > B_j$; (b) $G_j = 0$ and $L_j = 1$ if $A_j < B_j$; and (c) when $A_j = B_j$, then $G_j = G_{j-1}$ and $L_j = L_{j-1}$ for a full comparator and $G_j = L_j = 0$ for a half comparator. Describe the working principles of the design and show how you could improve on this design.
19. Design an n-bit comparator module where the most significant bits are compared first. The comparisons are made at the less significant bits only if the comparisons(s) made at higher significant bit(s) was inconclusive.

8.8 SUGGESTED READINGS

Agrawal, D. P., Negabinary Carry Look Ahead Adder and Fast Multiplier, *Electron. Lett.*, **10,** 312 (1974).

Arazi, B., An Electrooptical Adder, *Proc. IEEE*, **73,** 162 (1985).

Daws, D. C., and E. V. Jones, Hardware-Efficient Bit Sequential Adders and Multipliers Using Mode-Controlled Logic, *Electron. Lett.*, **16,** 434 (1980).

Johnson, E. L., and M. A. Karim, *Digital Design: A Pragmatic Approach*, PWS–Kent Publishing, Boston, 1987.

Ling, H., High-Speed Binary Adder, *IBM J. Res. Dev.*, **25,** 156 (1981).

Liu, T. K., K. R. Hohulin, L. E. Shiau, and S. Muroga, S, Optimal One-Bit Full Adders with Different Types of Gates, *IEEE Trans. Comput.*, **C-23,** (1974).

Swartzlander, E. E., Digital Optical Arithmetic, *Appl. Opt.*, **25,** 3021–3032 (1986).

CHAPTER NINE

Sequential Logic

9.1 INTRODUCTION

While the previous chapter dealt with combinational circuits, in this chapter, we examine sequential circuits. In these types of circuit, the present value of the outputs are dependent on both the present values of the inputs and the past values of the inputs. In comparison, the steady-state output(s) of a combinational circuit depends only on its current inputs. The most common example of these circuits is the flip-flop (FF), which has two stable states. While of the two states one is assigned the logic 0 value, the other is assigned the logic 1 value. There are a number of FFs in common usage and they differ from one another in the number of inputs they have and in the manner in which they behave.

A complex sequential circuit, such as those of counters and registers, generally employs both combinational circuits and flip-flops. All circuit action takes place under the control of a periodic sequence of pulses called a clock. Each clock pulse permits the circuit to either remain in its present state (present set of FF values) or move to another state (a new set of FF values). In a counter, for example, a change in the FF states follows a particular pattern, such as counting up or counting down. The advantage of these clocked sequential circuits is that glitches that occur due to the imperfect nature of the logic devices have no effect.

In this chapter, the to-be-covered topics will include a description of the existing FFs, design of clocked sequential circuits, design and applications of counters and registers, and memory devices. Finally, examples of several arithmetic algorithms that can be implemented using the clocked sequential circuit are presented. It should be noted that sequential circuits can be synchronous (clocked) as well as asynchronous. In this chapter, however, we limit ourselves to synchronous sequential circuits because of their obvious and immediate application in optical computing.

9.2 FLIP-FLOPS

In general, FFs are two-output devices, such that the outputs are complements of each other. The functional characteristics of these flip-flops are described in terms of their inputs and time. After the set of inputs has been introduced at the FF, the FF outputs are generated a short time later. The output values often depend on the previous values of the outputs, which, however, depend in turn on the past values of the inputs. FFs form the basic building blocks that, when cascaded to some carefully selected combinational units, result in complex sequential circuits. There are four types of basic FFs. They are the Set-Reset (S-R) FF, J-K FF, Toggle (T) FF, and Delay (D) FF. We shall see later, however, that both T and D FFs are generated using J-K FFs.

9.2.1 S-R Flip-Flop

Figure 9.1a shows a primitive S-R FF circuit that has two inputs, S and R, two outputs, Q and $\overline{Q}$, and is made up of two two-input NOR gates, for example. The two outputs, Q and $\overline{Q}$, are both fed back as input to one of these two NOR gates. It is important to realize that the presence of these feedback connections makes a sequential device different from a combinational device. The table in Fig. 9.1b lists all the characteristics of this S-R flip-flop. All the inputs are listed along with variable time t to indicate their time dependence, while the output is obtained at a time T later. The propagation delay T is the time duration between a change in the

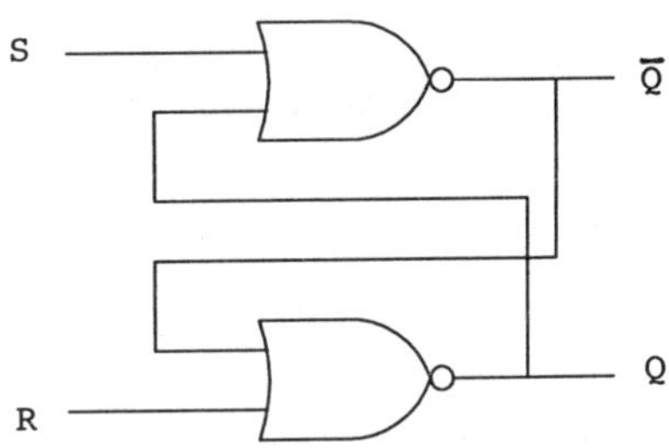

(a)

$S(t)$	$R(t)$	$Q(t)$	$Q(t+T)$
0	0	0	0
0	0	1	1
0	1	0	0
0	1	1	0
1	0	0	1
1	0	1	1
1	1	0	—
1	1	1	—

(b)

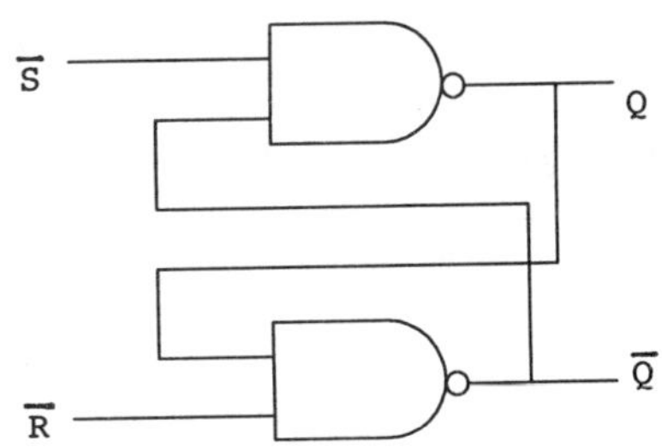

(c)

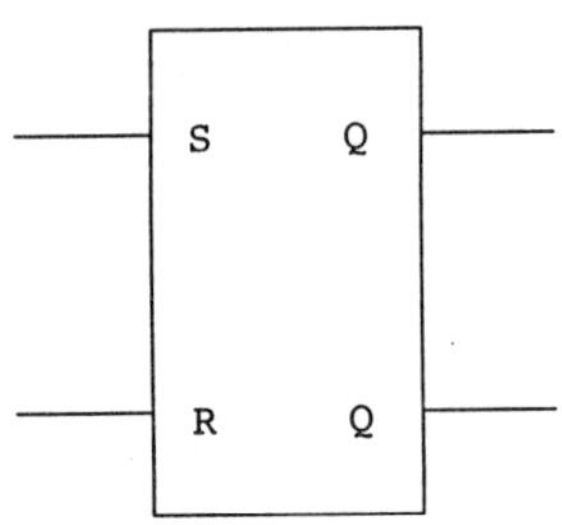

(d)

FIGURE 9.1 S-R FF: (a) logic circuit using NOR gates, (b) characteristic table, (c) logic circuit using NAND gates, and (d) logic symbol.

input and possible change in the output and is equivalent to the lumped delay of the circuit. We find that when $S = R = 0$ the Q and $\overline{Q}$ values remain unchanged, while when $S = 1$ and $R = 0$, Q takes the value of 1 and when $S = 0$ and $R = 1$, Q takes the value of 0. These three conditions of the FF are referred to, respectively, as "hold," "set," and "reset" conditions. However, when $S = R = 1$, the FF outputs are no longer complements of each other and therefore this state of the FF is forbidden.

The FF equation for $Q(t + T)$ may be expressed as

$$Q(t + T) = \overline{\overline{S(T) + Q(t)} + R(t)} \tag{9.1}$$

which, however, can also be rewritten in the form of

$$Q(t + T) = \overline{\overline{S(T)}\;\overline{\overline{R(t)}\,Q(t)}} \tag{9.2}$$

The logic circuit for the NAND-based S-R FF follows from Eq. (9.2) and is shown in Fig. 9.1c.

It is advantageous to control the exact time as to when the S-R FF output may change. This is accomplished by adding an additional input CK, referred to as the clock, to the S-R FF circuit, as shown in Fig. 9.2a. This CK input can be made periodic so that the FF outputs change only periodically. We see that when CK input is a 0, the FF outputs remain unchanged as in the hold mode. But when CK = 1, the FF is free to have any one of the three permissible operation modes provided that both S and R inputs are not 1. However, in order that the FF functions as expected, the CK input should remain turned on for a duration long enough so that the FF outputs are able to reach steady states.

9.2.2 J-K Flip-Flop

The S-R FFs are somewhat limited in their applications since they involve the possibility of encountering a forbidden state. This limitation is overcome by a particular FF that is referred to as the J-K FF. Like the S-R FF, it is also a two-

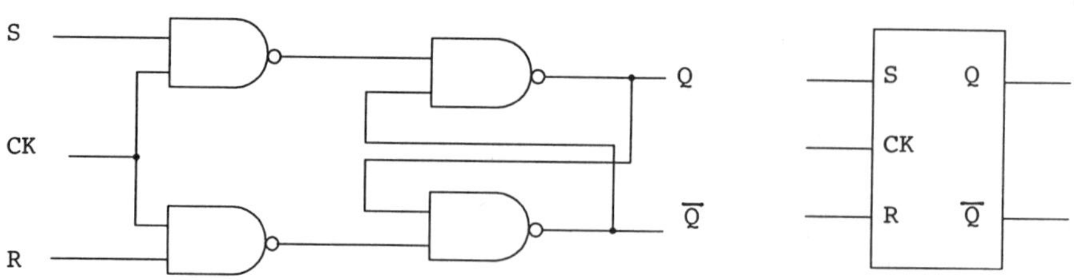

(a) (b)

FIGURE 9.2 A clocked S-R flip-flop: (a) logic circuit and (b) logic symbol.

input sequential device. Three of its modes of operation are exactly similar to those of the three permissible S-R modes of operation. In addition, however, the J-K FF is equipped with a fourth mode operation.

Figure 9.3a shows the logic circuit of a J-K FF that is slightly different from that of Fig. 9.2a. The J-K FF includes two additional feedback connections whereby Q is fed as input along with K, and $\overline{Q}$ is fed as input along with J. The next state value of Q is now expressed as

$$Q(t + T) = J(t)\overline{Q(t)} + \overline{K(t)}Q(t). \tag{9.3}$$

The J and K inputs of the J-K FF are identical in behavior to the S and R inputs, respectively, of the S-R FF except for its fourth mode of operation. When $J = K = 1$, the Q output takes the value of $\overline{Q}$, while the $\overline{Q}$ output takes the value of Q. This mode of operation is referred to as the toggle mode.

The J-K FF of Fig. 9.3a is also not without problems. When the clock duration is too long, a J-K FF output may undergo more than one change during the same clock pulse. As a consequence, Q may be indeterminate at the termination of the clock input. This problem can be avoided by making sure that the clock pulses are relatively narrower. Such problems are also avoided by making use of two J-K FFs in a master–slave configuration, as shown in Fig. 9.4. While the clock input is fed directly to the master J-K FF, the complement of the clock is fed to the slave input. Consequently, the master FF changes on the positive edge of the clock but the corresponding change in the slave FF doesn't take place until the tail-end of the clock input.

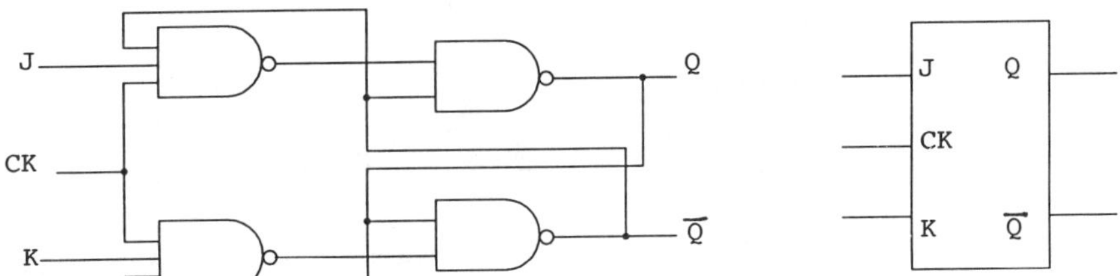

(a) (b)

$J(t)$	$K(t)$	$Q(t)$	$Q(T + T)$
0	0	Q	Q
0	1	—	0
1	0	—	1
1	1	Q	$\overline{Q}$

(c)

FIGURE 9.3 The J-K flip-flop: (a) logic diagram, (b) logic symbol, and (c) function table.

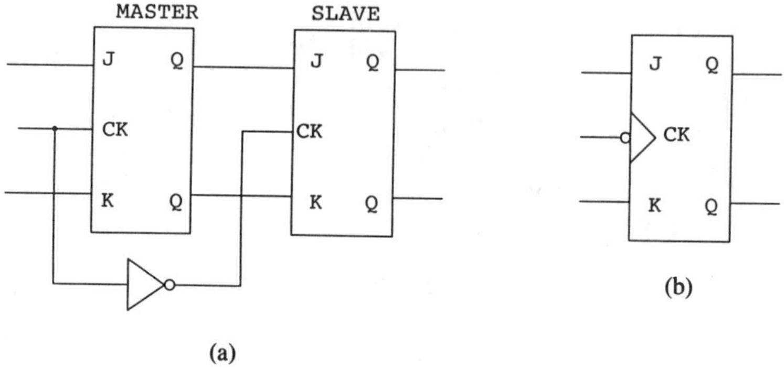

FIGURE 9.4 Master–slave J-K flip-flop: (a) logic diagram and (b) logic symbol.

9.2.3 T and D Flip-Flops

Toggle (T) and Delay (D) flip-flops are in effect special cases of the J-K FF. They are obtained by externally manipulating the inputs of a J-K FF. Figure 9.5 shows these two flip-flops. The T FF is obtained by tying both J and K inputs together. In the case of the D FF, however, the J input is renamed D and $\overline{D}$ is fed directly into its K input. In the T FF, when $T = 0$, the FF operates in the hold mode, but when $T = 1$, the FF operates in the toggle mode. In comparison, the D FF output takes on the value of the input. In reality, the D FF serves as a delay element with input appearing at the output.

The next-state equation of the D FF is given by

$$Q(t + T) = D(t) \tag{9.4}$$

while that of the T FF is given by

$$Q(t + T) = Q(t) \oplus T(t) \tag{9.5}$$

In general, a FF involves only four types of state transition. The values of Q may undergo any one of the four transitions, $0 \to 0$, $0 \to 1$, $1 \to 0$, and $1 \to 1$. The four FFs accomplish these transitions in different ways. The design of a se-

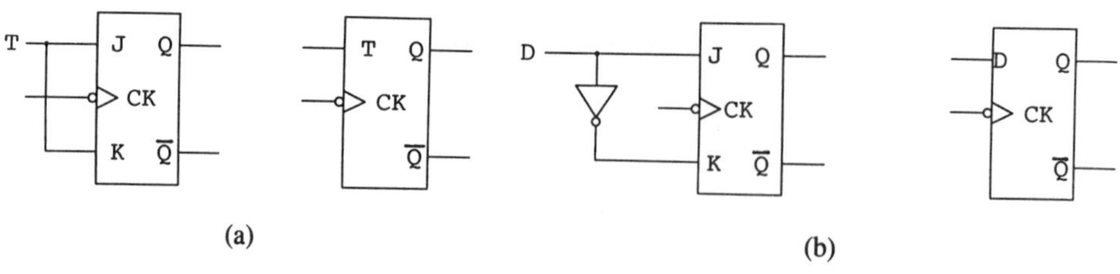

FIGURE 9.5 (a) T FF logic circuit and its logic symbol; and (b) D FF logic circuit and its logic symbol.

TABLE 9.1 Contol Conditions

$Q(t)$	$\longrightarrow$	$Q(t+T)$	S-R	J-K	T	D
0		0	0–	0–	0	0
0		1	10	1–	1	1
1		0	01	–1	1	0
1		1	–0	–0	0	1

quential circuit using a particular type of FF is therefore different from that using a different type of FF. The behavior of the four flip-flops is summarized in Table 9.1.

9.3 SYNCHRONOUS SEQUENTIAL CIRCUITS

A synchronous sequential circuit consists of both a combinational component and sequential devices designed so that their combination generates outputs that are functions of inputs and previous states of the circuit. In particular, the circuit is referred to as synchronous whenever the states are expected to change only periodically in accordance to a clock input. Sequential circuits can also be asynchronous, but a discussion of these is avoided in this book for simplicity.

The input, the output, the present state, and the next state of a sequential circuit are customarily related to one another by means of either a state diagram or a state table. The state diagram of a specific sequential circuit is only a graphical representation of its behavior. Customarily, the states are represented by circles while arrows between these states represent the input-induced transitions. For example, the transition from state A to state B caused by an input x can be represented by drawing a line between A and B and by labeling the said transition line as x/z, where z may represent the corresponding circuit output. For n circuit inputs, a total of 2^n transition lines originate from each of the states. The tabular version of the state diagram is referred to as the state table.

In each state it is necessary for the circuit to be able to determine which state it is in, what the current value of x is, and then set up the flip-flop inputs such that the correct state is entered when the clock input occurs. The implementation of an m-state sequential circuit requires p FFs, where p is the smallest integer for which $2^p \geq m$. The outputs of these FFs are called the state variables and are used to describe the state of the circuit.

The design of a synchronous sequential circuit is a relatively involved process. The design process consists of the following steps.

1. Obtain the state diagram.
2. Obtain the state table for the state diagram.
3. Eliminate redundant states.
4. Make state assignments.

5. Determine the types of FFs.
6. Obtain the corresponding excitation maps.
7. Determine the output and FF equations.
8. Construct the sequential logic circuit.

In constructing a state diagram from the word statement of a design problem, a state that is identical to another state may inadvertently be included. In general, the redundant state or states increase the number of total states, and therefore it may lead to a circuit that uses an extra flip-flop, thus making the circuit more complicated and expensive. Furthermore, fault diagnostic techniques are generally based on the assumption that no redundant states exist. One of the design objectives is therefore to eliminate all redundant states. It must be noted, however, that step 3 of the aforementioned algorithm is optional. Sometimes redundant states are obvious. Any two states will be considered equivalent if they have identical outputs and make transitions to the same states for each of the circuit inputs.

The total design process is rather lengthy and varies from problem to problem. The following examples illustrate the implementation of the sequential design algorithm.

Example 9.1 Design a 3-bit sequential counter that counts clock pulses.

Solution We can envision a counter as a device that receives clock pulses as input. Each time a clock pulse is received, the counter should count-up. No control variable is needed, only the occurrence of the clock pulse is necessary for a state change. We may assume, for example, that the states change at the trailing edge of the clock input. The 3-bit counter could be designed such that the outputs go through the sequence 000 → 001 → 010 → 011 → 100 → 101 → 110 → 111 and repeat. Since the counter is only a 3-bit device, it would have to reset at the eighth clock. The state diagram and the corresponding state table for the counter are obtained as shown in Fig. 9.6, where PS and NS, respectively, stand for present state and next state.

The eight states, A through H, are all different since they represent completely different events. Consequently, we may conclude that none of these are redundant states. Again, since there are eight unique states, the corresponding circuit requires a minimum of three FFs.

Since the sequential circuit can be designed with three FFs, each of the eight states can have any one of the eight assignments, 000 through 111. We may choose, however, the states to have the same assignments as the outputs. In that event, the outputs can be obtained directly from the FF outputs. Accordingly, the assignments may be chosen as $A = 000$, $B = 001$, $C = 010$, $D = 011$, $E = 100$, $F = 101$, $G = 110$, and $H = 111$. For example, an assignment of 110 for state G implies that correspondingly the three FFs will have 1, 1, and 0 as their respective Q values. The transition table may now be obtained as shown in Fig. 9.7, where Q_1, Q_2, and Q_3 represent the three FF outputs, respectively.

The number of FFs is three. We may choose any type of FF for our purpose.

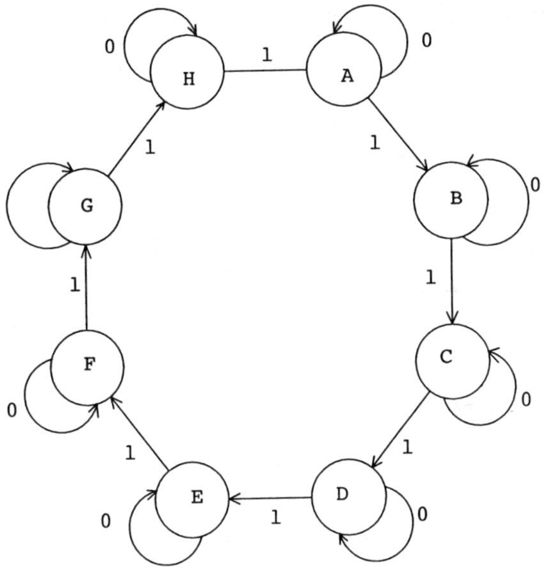

PS	NS CK = 0	NS CK = 1	Output
A	*A*	*B*	000
B	*B*	*C*	001
C	*C*	*D*	010
D	*D*	*E*	011
E	*E*	*F*	100
F	*F*	*G*	101
G	*G*	*H*	110
H	*H*	*A*	111

FIGURE 9.6

PS $Q_1Q_2Q_3$	NS CK = 0	NS CK = 1	Output
000	000	001	000
001	001	010	001
010	010	011	010
011	011	100	011
100	100	101	100
101	101	110	101
110	110	111	110
111	111	000	111

FIGURE 9.7

$Q_1Q_2Q_3$	CK = 0	CK = 1	CK = 0	CK = 1	CK = 0	CK = 1	$Z_1Z_2Z_3$
000	0−	0−	0−	0−	0−	1−	000
001	0−	0−	0−	1−	0−	−1	001
010	0−	0−	−0	−0	0−	1−	010
011	0−	1−	−0	−1	−0	−1	011
100	−0	−0	0−	0−	0−	1−	100
101	−0	−0	0−	1−	−0	−1	101
110	−0	−0	−0	−0	0−	1−	110
111	−0	−1	−0	−1	−0	−1	111

FIGURE 9.8

For our illustration, we may choose J-K FFs, for example. Consequently, the excitation table for the J-K inputs may be obtained as shown in Fig. 9.8.

The excitation equations for each of the FF inputs, J_1, K_1, J_2, K_2, J_3, and K_3, may be obtained using four-variable K-maps to give

$$J_1 = K_1 = Q_2Q_3$$

$$J_2 = K_2 = Q_3$$

$$J_3 = K_3 = 1$$

while the outputs are obtained directly from the three Q's. The resulting logic circuit for the 3-bit counter is shown in Fig. 9.9.

State assignment allows the circuit to remember which state it is in. Each bit in the code represents the output of a FF and is called a state variable. In Example, 9.1, state assignments were made rather arbitrarily. It can be shown that the combinational circuit complexity is different for different sets of state assignments.

The number of possible state assignments for any given nontrivial sequential circuit is stunning. For n present states and p FFs, there may be a total of $[(2^p - 1)!]/[(2^p - n)!p!]$ unique assignments. For example, the number of unique assignments for a ten-state sequential circuit can be as large as 75,675,600. The optimum state assignment is one that reduces the combinational logic complexity

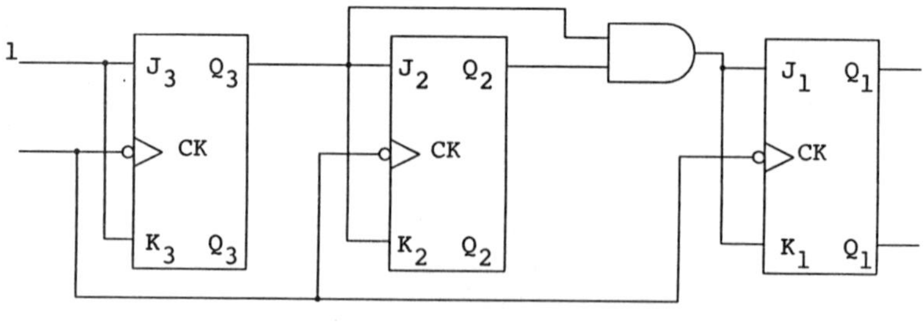

FIGURE 9.9

of a sequential circuit to a minimum. Any unique assignment is valid; however, it is always better if an attempt is made to assign codes in a way such that the number of cases where more than one bit in a code changes, when states change, is kept to a minimum.

9.4 COUNTERS

With the introduction of flip-flops and synchronous sequential circuits behind us, the study of counters will be a natural extension. The application of counters are many. For example, they may be used to keep track of the sequence of instructions in a program, to introduce the sequence of timing signals, for frequency division, for counting, and for many other operations. The basic operation characteristic of a counter is sequential such that for every present state there is a well-defined next state. The design of a counter like any other sequential circuit involves designing combinational logic that decodes the present state and enables entry into the next state of the sequence.

Counters may be classified into two types: synchronous and asynchronous. In a synchronous counter all FFs change state synchronously with the clock input. In comparison, in an asynchronous counter FF outputs do not change simultaneously with the clock input. However, our discussion in this section is limited to synchronous counters. The simplest possible counter is a single-bit counter that toggles between two states, 0 and 1. A J-K FF in toggle mode will function as a single-bit counter alternating between the two states with the occurrence of each clock. Accordingly, the FF output of a single-bit counter has a frequency that is one-half the input clock frequency.

A 3-bit binary up-counter, having eight states, was designed in Example 9.1. This counter consists of three J-K flip-flops whose states $Q_2Q_1Q_0$ move in sequence 000 through 111. We can repeat the same procedure to design any n-bit binary up-counter with states being represented by $Q_{n-1}, Q_{n-2}, \cdots, Q_1$, and Q_0. It turns out that the design of all binary up-counters follows a degree of modularity. The output Q_0 changes with every clock pulse and, in general, Q_i changes state if all less significant bits Q_{i-1} through Q_0 are 1. Accordingly, J and K equations for the ith bit of an n-bit binary up-counter are given by

$$J_i = K_i = Q_{i-1}Q_{i-2} \cdots Q_1Q_0 \tag{9.6}$$

Alternatively, we can also express the excitation equation as

$$J_i = K_i = J_{i-1}Q_{i-1} \tag{9.7}$$

Depending on the form of excitation equations used, there can be two types of binary up-counter logic circuits. Figure 9.10a shows a configuration where the FF outputs are ANDed in parallel. The propagation delay for each of the flip-flops is the same. However, the fan-in to the AND gate and the fan-out of each FF increase

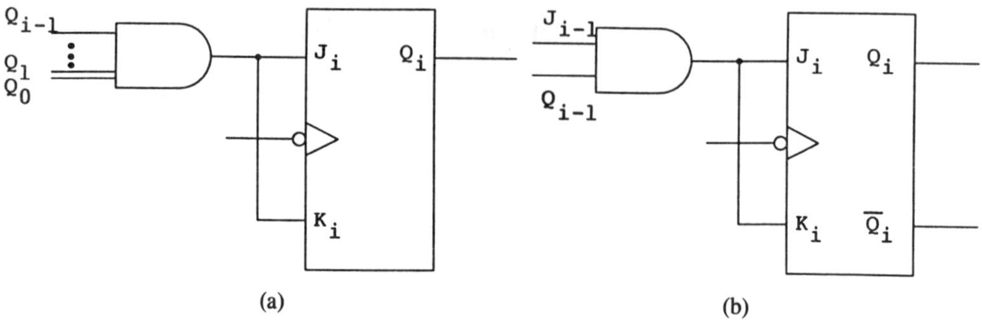

FIGURE 9.10 The *i*th bit of a synchronous counter using (a) multiple-input AND gates and (b) two-input AND gates.

with increasing i. Again, Fig. 9.10b shows an arrangement where the fan-in of the AND gates is always two but the propagation delay increases with increasing i. Note that the first method allows for faster clocking but is not without fan-in and fan-out problems.

In a down-counter, a binary number is set into the counter, which then counts toward zero as the clock pulses begin to come in. The down-counters may as well be designed in the same way as up-counters. Like that in the case of up-counters, the down-counter excitation equation is also seen to have a periodicity. The J and K equations for an n-bit binary down-counter are found to be

$$J_i = K_i = \overline{Q_{i-1}}\,\overline{Q_{i-2}} \cdots \overline{Q_1}\,\overline{Q_0} \tag{9.8}$$

By comparing Eq. (9.7) with Eq. (9.6), it is obvious that a binary down-counter logic circuit may also have two forms.

Equations (9.6) and (9.8) can be combined next to design the logic circuit for a binary up–down counter. A control input may be introduced in the circuit almost as it was done in the case of the binary adder–subtractor. The control input E can be used to select either the up or down mode of the counter. The corresponding J and K equations for the up–down counter are given by

$$J_i = K_i = EQ_{i-1} \cdots Q_0 + \overline{EQ_{i-1}} \cdots \overline{Q_0} \tag{9.9}$$

Consequently, when E is a 1, the binary counter counts up, and when E is a 0, the counter counts down.

A 4-bit gray code counter, for example, may count 0000, 0001, 0011, 0010, 0110, 0111, 0101, 0100, 1100, 1101, 1111, 1110, 1010, 1011, 1001, and 1000 in that order and then resets to 0000 before resuming the count-up operation again. Consider, for example, the design of a BCD decade counter. The synchronous circuit design algorithm reveals that the corresponding excitation equations are

$$
\begin{aligned}
J_3 &= Q_2 Q_1 Q_0 \\
K_3 &= Q_0 \\
J_2 &= K_2 = Q_1 Q_0 \\
J_1 &= \overline{Q_3} Q_0 \\
K_1 &= Q_0 \\
J_0 &= K_0 = 1
\end{aligned} \tag{9.10}
$$

Similarly, those for the BCD down-counter are found to be

$$
\begin{aligned}
J_3 &= K_3\, \overline{Q_2}\, \overline{Q_1}\, \overline{Q_0} \\
J_2 &= Q_3 \overline{Q_0} \\
K_2 &= \overline{Q_1}\, \overline{Q_0} \\
J_1 &= \overline{Q_0}(Q_3 + Q_2) \\
K_1 &= \overline{Q_0} \\
J_0 &= K_0 = 1
\end{aligned} \tag{9.11}
$$

It must be noted that it is not necessary for a designer to go through the prescribed but cumbersome design steps outlined in the previous section. An existing circuit can sometimes be modified to suit the requirement of a newer circuit. For example, a BCD decade counter can be designed also from a 4-bit binary up-counter. A synchronous BCD counter works exactly as a 4-bit binary counter until count 9 ($Q_3Q_2Q_1Q_0 = 1001$) has been reached. The 4-bit binary counter advances to 1010, while the BCD counter starts all over at 0000. Accordingly, it is necessary to reset FFs when they are about ready to go to 1010. This modification process involves the following considerations:

1. The bits Q_3 and Q_0 should be reset.
2. Q_1 should be prevented from becoming a 1.

We note that the least significant FF of the binary up-counter is always in a toggle mode, which implies that Q_0 will reset all by itself. The Q_3 bit of the 4-bit counter changes only when the count reaches either 0111 and 1111. This occurs because the inputs J_3 and K_3 are both 0 during all other counts. For example, if the K_3 input were a 1 instead of a 0, when the count reaches 1001 Q_3 would be reset as desired in a BCD up-counter. This can be accomplished, however, if K_3 is fed

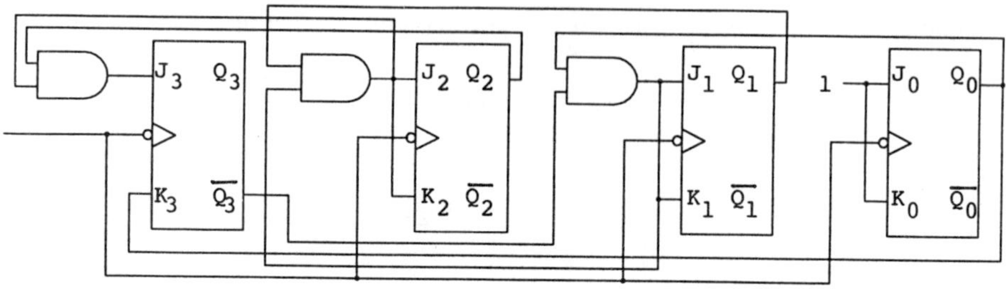

FIGURE 9.11 A synchronous BCD up-counter.

directly with Q_0 (instead of J_3). The next consideration is to prevent Q_1 from switching back to 1 once the count has reached 1001. This objective is realized by supplying both J_1 and K_1 with $\overline{Q_3}Q_0$. Fortunately, this modification does not cause any problem since $\overline{Q_3}$ is a 1 until the count reaches 1000. Finally, once the count had reached 1000, Q_1 need not be turned on. Accordingly, the BCD counter circuit may be obtained as shown in Fig. 9.11.

9.5 REGISTERS

Registers are formed of multiple FFs and used primarily in storing and manipulating data. They are classified according to the way information bits are stored and retrieved. If data are stored and removed at either end of a multi-bit register, one bit at a time, the register is referred to as a *serial* or *shift* register. However, if all bits of the word are stored or retrieved simultaneously, the register is referred to as a *parallel* register.

The flip-flops in a shift register are connected so that each flip-flop transfers its bit of information to the adjacent flip-flop coincident with each clock pulse. It is reasonably easy to design registers using either D FFs or J-K FFs configured in the form of D FFs. Figure 9.12a shows the ith bit of a shift-right serial register while Fig. 9.12b shows that of the shift-left serial register. With each clock input

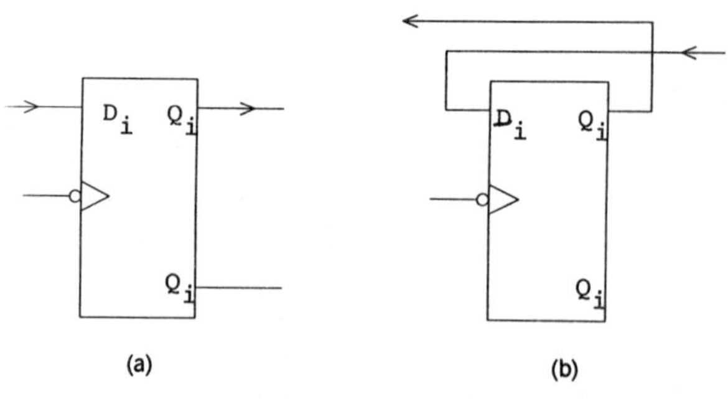

FIGURE 9.12 The serial register: (a) shift-right and (b) shift-left.

the bits move one position to the right or to the left depending on the shift register used.

A useful application of shift-right registers, for example, requires a connection of the rightmost FF output to the input of the leftmost FF. In that case the least significant bit is not lost but appears at the most significant bit. Such a shift-right register is known as a *circulate right* register. Similarly, by feeding the output of the most significant FF to the input of the least significant FF, a left-shift register can be converted to a *circulate left* register.

Shift registers play an important role in arithmetic operations. A binary number can be multiplied by 2 by shifting the stored number 1 bit to the left and divided by 2 by shifting the register content 1 bit to the right. The bits shifted in at one end and out at the other end are not necessarily wasted. They are also used in arithmetic operations in many instances.

Both shift-right and shift-left capabilities may be combined as in the circuit of Fig. 9.13 to obtain a bidirectional serial-in serial-out shift register. This is similar in design to that of an up–down counter obtained by combining an up-counter and a down-counter. The direction of shift is controlled by the input SLE. When SLE = 1, the bidirectional register operates in left-shift mode provided HOLD = 0, and when SLE = 0 the bidirectional register works as a right-shift register.

An n-bit serial-in serial-out shift register requires n clock pulses to load an n-bit word. In comparison, a parallel-in shift register loads all information bits simultaneously. Both serial-in and parallel-in shift registers have specific applications in digital systems. A parallel-in serial-out shift register using D FFs is shown in Fig. 9.14. When LOAD is low, the data are loaded into the register in parallel. Again, if LOAD is high, the Q output of the FF of every stage is serially shifted to the right. Parallel-in serial-out shift registers allow the system to accept data n bits at a time on n lines and then to send them one bit after another on one line. This is a standard mode of communication between digital systems.

A universal shift register is a versatile shift register that has capabilities for parallel loading, parallel outputs, bidirectional shifting, and bidirectional serial input and output. In other words, it is capable of operating in all the register modes described previously. An n-bit universal shift register can be designed by means

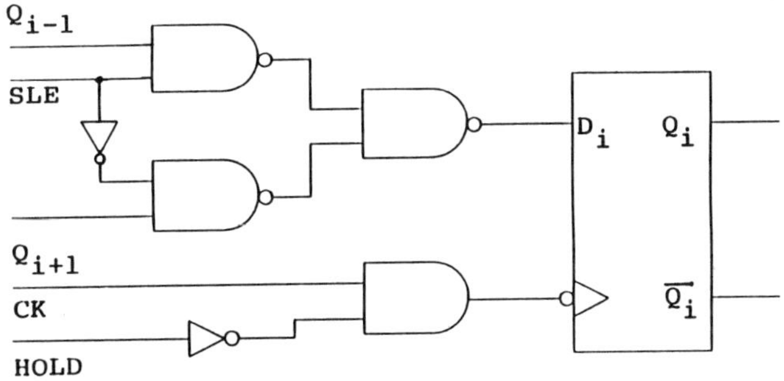

FIGURE 9.13 The ith bit of a bidirectional shift register.

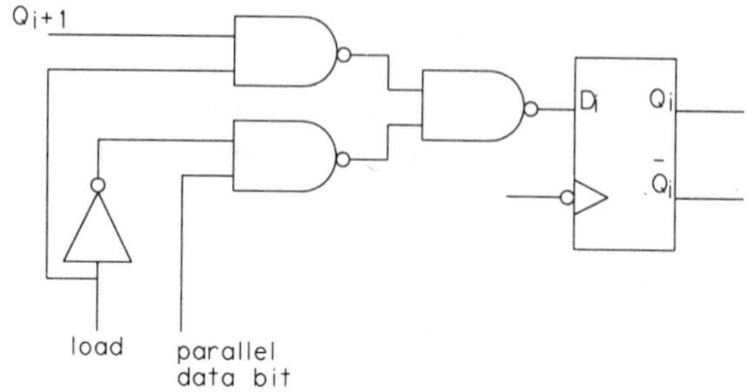

FIGURE 9.14 The ith bit of a parallel-in serial-out shift register.

of n D FFs and n 1-of-4 multiplexers. The ith bit of such a universal register is shown in Fig. 9.15. When $I_1 I_0 = 00$, the register performs the parallel-in operation; when $I_1 I_0 = 11$, the register retains its old value; when $I_1 I_0 = 01$, the register performs a right-shift operation, and when $I_1 I_0 = 10$, the register performs a left-shift operation.

Registers are often assembled to form storage arrays, referred to as scratch pad memory. Figure 9.16a shows an arrangement of registers that can store up to m 2^n-bit words. The device consists of m registers each made of 2^n FFs. When the write enable WE is low, the m data inputs, I_0 through I_{m-1}, are routed to a particular location of each register as specified by the values entered in the n write select (WS) lines. For example, a value of 101 (say, when $n = 3$) at the WS lines of a scratch pad memory will cause the m data inputs to be loaded at the corresponding bit-5 FFs. Again, stored data from the scratch-pad memory may be retrieved through the m output lines by applying a low to the read enable (RE) and necessary address bits to the read select (RS) lines. Figure 9.16b shows the corresponding logic diagram of the memory, for example, when $n = 2$. These memories, although very fast, are not used extensively since the corresponding LSI format is required to have too many pin-outs.

The availability of many different register types and their potential for varied applications have resulted in a concise language for describing the flow of infor-

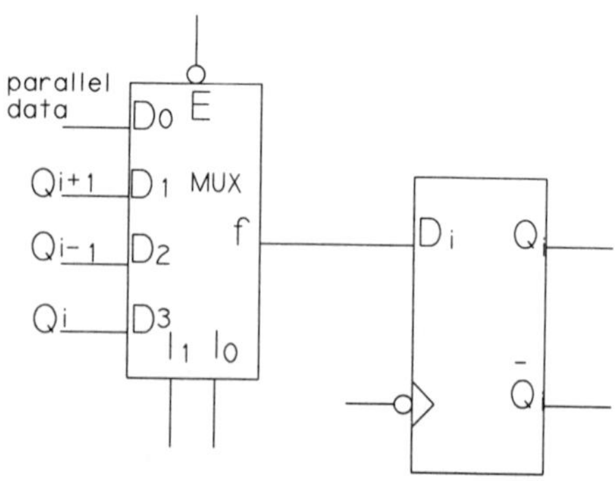

FIGURE 9.15 The ith bit of a universal shift register.

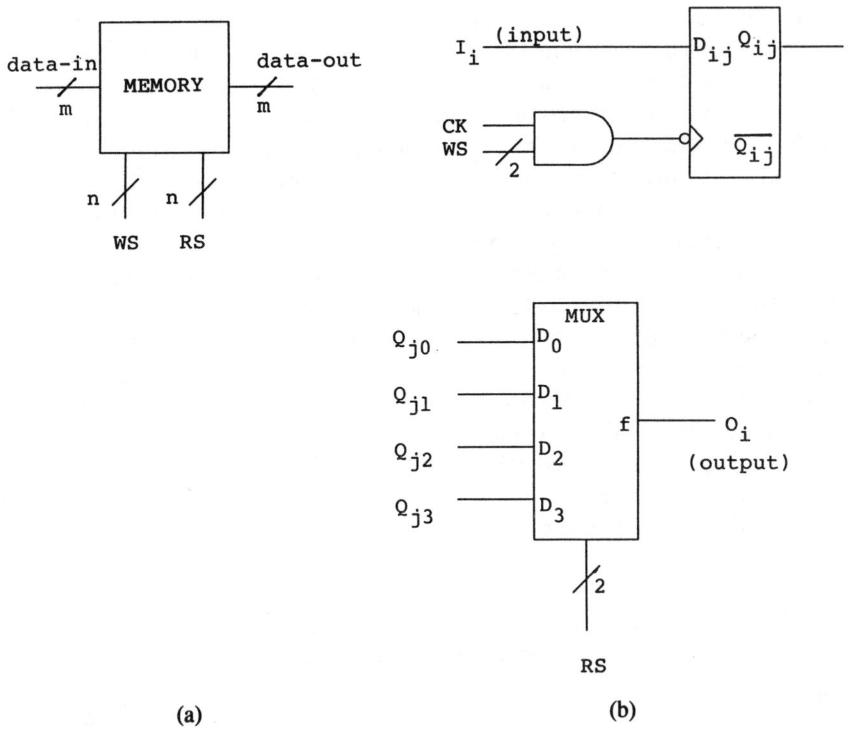

FIGURE 9.16 Scratch-pad memory: (a) block diagram and (b) logic diagram when $n = 2$.

mation between registers. *Register transfer language* (RTL) is just a way to describe and then design registers used in a data processor. A complete register transfer description involves two important units, data and control units. The data unit involves registers, data paths, and the logic necessary to implement a prescribed set of register transfers. On the other hand, the control unit generates appropriate signals in a certain specific sequence to orchestrate the register transfers between the data units.

The RTL operation represented by $A \leftarrow B$, implies that the data in register B are being loaded in register A. It is implicit that both of these registers have same number of FFs. Such a RTL operation can be executed in a single clock period, in which case the RTL description will correspond to a single state transition of the sequential machine. Certain RTL standardization, however, may be considered essential. For example, the contents of a register are usually denoted by a single-letter acronym of the name of the register. A concatenated register formed by joining two or more registers may therefore be represented by the corresponding register acronyms separated only by commas. For uniformity, the bits of each register could be numbered from the right to left. For example, A_0 and A_{n-1}, respectively, represent the least and the most significant bits of an n-bit register named A.

Some of the RTL arithmetic, bit-by-bit logic, shift, rotate, scale, and conditional operations are shown in Table 9.2. Note that a scale-right operation is very similar to a shift-right operation except that the MSB retains its old value. The shift, rotate, and scale operations are generally represented by two lowercase letters. For shift and rotate, the first letter indicates the type (r for rotate and s for shift) and the second letter indicates the particular direction (r for right and l for left). In particular, scl may be used to indicate a scale-left operation and scr may be used to indicate a scale-right operation.

Control conditions are generally included to distinguish one set of RTL executions from another. The RTL operations describe the internal operations of a digital system and accordingly it may be used in the design of both data and control units. In general, the control condition is separated from the corresponding RTL operation by a colon (:) sign while a semicolon (;) sign is used to indicate the end of the RTL operation. For example, "C: $A \leftarrow B$; $B \leftarrow \overline{B}$ plus 1;" corresponds to two simultaneously performed RTL operations. It implies that the register A is loaded with data from register B and the 2's complement of the data available in register B is loaded in register B when the control condition C is a 1.

TABLE 9.2 RTL Examples

Initial values: A = 10100, B = 11010, and C = 00011

Operation	Meaning	Consequence
(a) ARITHMETIC		
$A \leftarrow 0$	Clear A	A = 00000
$A \leftarrow B$ plus C	Sum of B and C to A	A = 11101
$B \leftarrow B$ plus 1	Increment B by 1	B = 11011
(b) LOGIC		
$A \leftarrow BC$	B AND C (bit-by-bit) to A	A = 00010
$A \leftarrow A + B$	A OR B (bit-by-bit) to A	A = 11110
$B \leftarrow \overline{B}$	Complement B	B = 00101
$B \leftarrow \overline{B}$ plus 1	2's Complement of B	B = 00110
(c) SERIAL		
$B \leftarrow$ sr B	Shift right B one bit	B = 01101
$B \leftarrow$ sl B	Shift left B one bit	B = 10100
$B \leftarrow$ sr2 B	Shift right B two bits	B = 00110
$B \leftarrow$ rr B	Rotate right B one bit	B = 01101
$B \leftarrow$ scr B	Scale right B one bit	B = 11101
$B,C \leftarrow$ sr2 B,C	Shift right concatenated B and C two bits	B,C = 0011010000
(d) CONDITIONAL		
If (B_0 = 1) $C \leftarrow 0$	If bit 0 of B is 1, then C is cleared	C = 00011
If ($B > C$) $A \leftarrow 0$, $C \leftarrow 1$	If B is greater than C then clear A and C is set to 1	A = 00000 C = 00001

A scratch-pad memory is a primitive example of a location-addressable memory (LAM), where each datum stored in a cell is retrieved via its location (i.e., address). In comparison, there is a class of storage device that is accessed or loaded on the basis of their contents. This latter type of memory, referred to as content-addressable memory (CAM), allows for sophisticated data transfer and is finding use not only in electronic computers but in optical computing systems. In the next section, we shall investigate the characteristics of such LAM and CAM devices.

9.6 MEMORY DEVICES

9.6.1 Location-Addressable Memory

There are several different types of location-addressable memory. They can be used either only in read mode or in both read and write modes. While read-only memory (ROM) is of the first type, a random access memory (RAM) is of the latter type. Again, memories can be either semiconductor, or magnetic, or optical in nature. A general memory device implies units where relatively large quantities of information are stored, perhaps for a relatively long period. For example, a single register would not be considered a standard memory unit unless it could hold a large number of bits.

In particular, the read-only memory (ROM) is combinational in nature and is used only in its read mode. The data to be read are initially written (or programmed) into the ROMs by the manufacturer. Once programmed, a ROM cannot usually be deprogrammed. There is, however, a modified version of the ROM available, which is user-programmable, and is commonly referred to as programmable ROM (PROM). ROMs are often used for permanent storage of programs and data.

ROMs consist of an internal $n - 2^n$ line decoder, which is capable of generating any one of 2^n minterms available through its outputs, as shown in Fig. 9.17. These decoder outputs form a matrix structure along with m orthogonally placed ROM output lines. At each of the $m \times 2^n$ intersections, an electronic element, such as a diode or a transistor, is housed. When two or more electronic elements along a given ROM output line are properly enabled, an OR operation is performed between the corresponding minterms. Accordingly, by using such a ROM, a total of m different Boolean functions can be generated.

The internal decoder decodes the n address inputs and selects one of the 2^n memory words. For example, 10 address lines may select up to $2^{10} = 1024$ words. Memory size is usually given as integer values of 1K words, the other name for 1024 words. Likewise, 16 address lines would allow selection of 64K words ($64 \times 1024 = 65{,}536$ active locations).

LAM devices can also be optical besides the semiconductor type. Optical memories employ either thermal or high-field effects of laser power in certain materials to store information. They offer, in principle, a potential improvement in storage density and access time. We have already covered holographic storage schemes in an earlier chapter. This technique has reached a certain maturity in the sense of

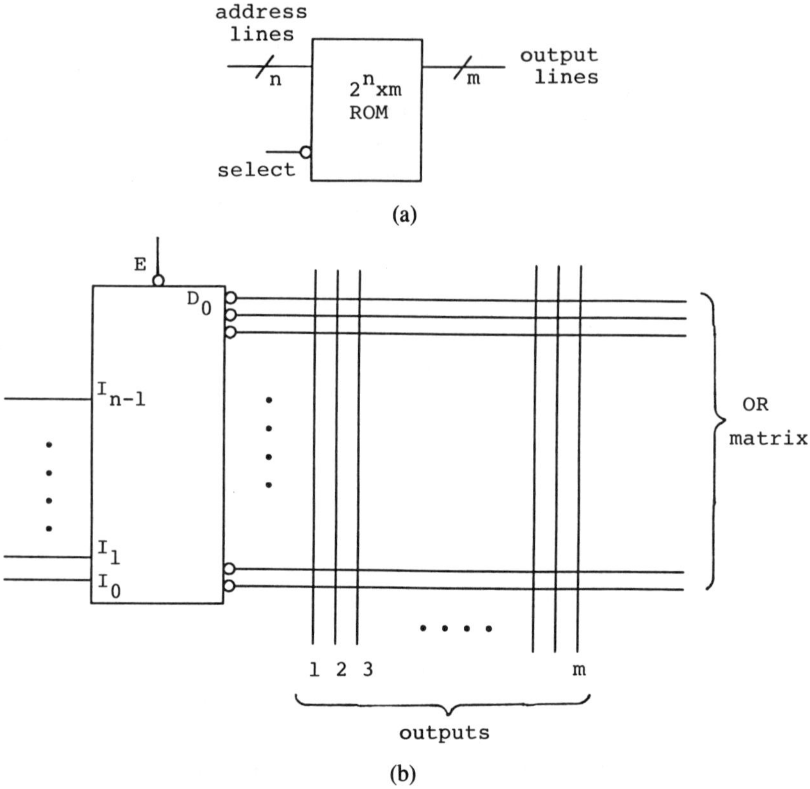

FIGURE 9.17 A read-only memory: (a) block diagram and (b) internal logic diagram.

being a readily useful method of processing data. Holography provides an enormous potential for data-handling capacity. The key to high storage density is the submicron wavelength of light, since the theoretical limit of closeness of stored bits to each other is set by the wavelength of exposing light. The practical storage limit is set by particle size in the recording medium.

9.6.2 Content-Addressable Memory

The conventional von Neumann type of computers, which bank on LAM devices, have frequently been regarded as unsatisfactory, especially when dealing with large problems. The conventional computing schemes involve operands that must be read and written serially. Content-addressable memory (CAM), in comparison, provides an alternative since the operands can be called out by their data contents and therefore perhaps many operands can be called out simultaneously. While CAMs are already finding some use in electronic computers, they have their greatest potential in an optical computing system. Even though discussions in this section are limited to semiconductor-based CAMs, optical holographic schemes can also be used to produce efficient CAMs.

Since the accessing of data, in a CAM, is on the basis of content, it will involve comparison of an external search argument with part or all of the information stored. It is desirable that such bit-by-bit comparison operations are performed in parallel. Figure 9.18 shows the logic circuit of a single word location. The search argument is carried via a set of parallel lines to the respective positions of all word locations. Each of the X-NOR gates tests for the logical equivalences while the output of the many-input AND gate indicates whether or not the search argument is the same as the content stored. This type of circuit is to be expected at every CAM word location. Alternatively, mismatch of the bits could be identified using X-OR logic gates at each of the bit locations, but then the final AND gate will have to be replaced by a NOR gate.

Alternatively, multi-bit search argument $a_{n-1}a_{n-2} \cdots a_2a_1a_0$ can be compared sequentially with the jth memory word $b_{j,n-1}b_{j,n-2} \cdots b_{j,2}b_{j,1}b_{j,0}$ by starting a comparison at the most significant bit. Each set of a_k and $b_{j,k}$ bits may be processed to generate two comparator outputs, $g_{j,k}$, referred to as the greater than output, and $l_{j,k}$, referred to as the less than output. The kth comparator outputs are then given by

$$g_{j,k} = g_{j,k+1} + \overline{a_k} b_{j,k} \overline{l_{j,k+1}} \tag{9.12a}$$

$$l_{j,k} = l_{j,k+1} + a_k \overline{b_{j,k}} \overline{g_{j,k+1}} \tag{9.12b}$$

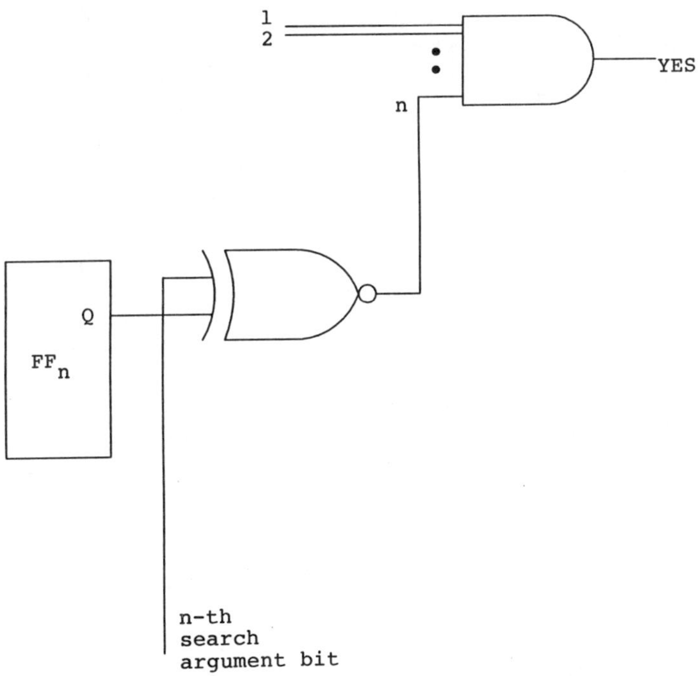

FIGURE 9.18 Comparison circuit for a CAM word location.

such that $g_{j,n} = l_{j,n} = 0$. If $a_{n-1} = 0$ and $b_{j,n-1} = 1$, then surely the search argument is greater than the jth memory word, in which case $g_{j,k} = 1$ and $l_{j,k} = 0$ for all $k < n - 1$. But if instead $a_{n-1} = 1$ while $b_{j,n-1} = 0$, then $g_{j,k} = 0$ and $l_{j,k} = 1$ for all $k < n - 1$. However, if a_{n-1} and $b_{j,n-1}$ are equal, then both of the comparator outputs $g_{j,n-1}$ and $l_{j,n-1}$ are zero. In this latter case, the same process of comparison will have to be carried out in the $n - 2$ bit position, and so on, until the comparison is finalized. Finally, if only $g_{j,0}$ is a 1, then the jth memory word is larger than the search argument; if only $l_{j,0}$ is a 1, then the search argument is greater than the jth memory word; and if $g_{j,0} = l_{j,0} = 0$, the search argument is the same as the jth memory word. Note that both $g_{j,0}$ and $l_{j,0}$ cannot both be 1. Figure 9.19 shows the iterative logic circuit for determining the status of the jth memory word.

While the comparison scheme as summarized in Eq. (9.12) is relatively slower, the speed of the sequential search can be expedited by beginning at the least significant bit. The search can be stopped in case the numbers contain leading zeros. For this alternative sequential search, the comparator outputs will now be given by

$$g_{j,k+1} = \overline{a_k}b_{j,k} + \overline{a_k\overline{b_{j,k}}}g_{j,k} \tag{9.13a}$$

$$l_{j,k+1} = a_k\overline{b_{j,k}} + \overline{\overline{a_k}b_{j,k}}l_{j,k} \tag{9.13b}$$

with $g_{j,0} = l_{j,0} = 0$. The search argument is determined to be greater than the jth memory word if $g_{j,n}$ is a 1 and less than the jth memory word if $l_{j,n}$ is a 1. Whenever $g_{j,n} = l_{j,n} = 0$, the search argument is the same as the jth memory word.

In masked search, only a subset of the bits of a search argument is compared with the respective bits of the jth memory word. Customarily in a CAM, those memory words that agree in the specified unmasked bit positions with the search argument are accessed. The bits of the mask word $c_{n-1}c_{n-2} \cdots c_2c_1c_0$ has a value of 1 in all masked bits and 0 otherwise. Accordingly, the masked match at

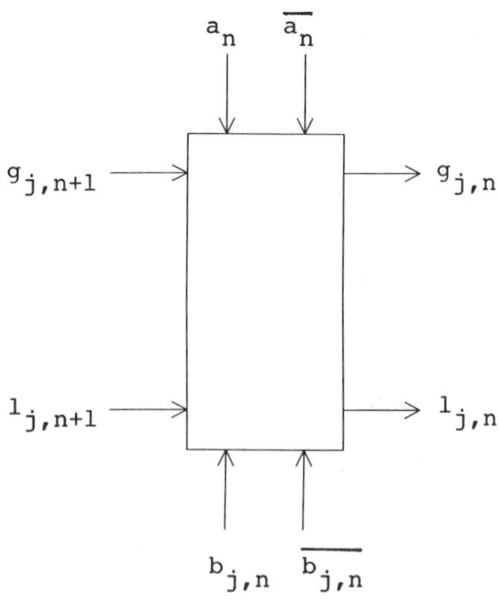

FIGURE 9.19 Sequential logic diagram for comparison of two n-bit numbers.

bit k is given by

$$m_{j,k} = a_k b_{j,k} + \overline{a_k} \overline{b_{j,k}} + c_k \tag{9.14}$$

Alternatively, the masked mismatch at bit k may be obtained as

$$\overline{m_{j,k}} = (\overline{a_k} b_{j,k} + a_k \overline{b_{j,k}}) \overline{c_k} \tag{9.15}$$

The matching of the search argument with the jth memory word is then obtained as

$$m_j = m_{j,n-1} m_{j,n-2} \cdots m_{j,2} m_{j,1} m_{j,0} \tag{9.16a}$$

$$m_j = \overline{(\overline{m_{j,n-1}} + \cdots + \overline{m_{j,2}} + \overline{m_{j,1}} + \overline{m_{j,0}})} \tag{9.16b}$$

The results of such bit comparisons in a word have to be processed using logic gates having very high fan-in. This problem is thus circumvented using sequential processing. A serious drawback to this solution is a delay that drastically increases the CAM access time. Sometimes such drawbacks are partially eliminated by making use of a wired-AND function, which produces an output equivalent to a switch connected between the output line and the ground potential.

9.7 SEQUENTIAL ARITHMETIC

Whenever an arithmetic or logic operation is carried out on a digital computer, it generally involves multiple transfer of multi-bit data from register to register. In general, the arithmetic operations are expected to be executed in parallel since this is a faster mode of operation. In comparison, however, sequential arithmetic schemes are slower but require less complicated and less expensive circuits. In the previous chapter, for example, we have already seen the design and complexities involved with a carry look-ahead adder. The CLA is undoubtedly faster than a ripple adder but it is associated with difficult fan-in and fan-out issues.

The register transfers of a sequential arithmetic processor involve a versatile control unit that coordinates the transfers and operations in a certain specific order. The RTL algorithm describing the corresponding sequential operations provides the necessary clues for designing both data and the corresponding control units. In this section, however, we refrain from discussing designs of such control units. Instead, we limit ourselves to the key ideas and algorithms used for such arithmetic and logic operations.

9.7.1 Sequential Addition/Subtraction

Figure 9.20 shows a logic block diagram suitable for sequential addition and subtraction operations. First, consider the addition operation. One bit of addend and one bit of augend starting with the least significant bit are shifted right into a single-bit full adder. The resulting sum bit of the full adder is stored at the most significant bit of a sum register. When the sum bits are being stored, the carry-out bit may

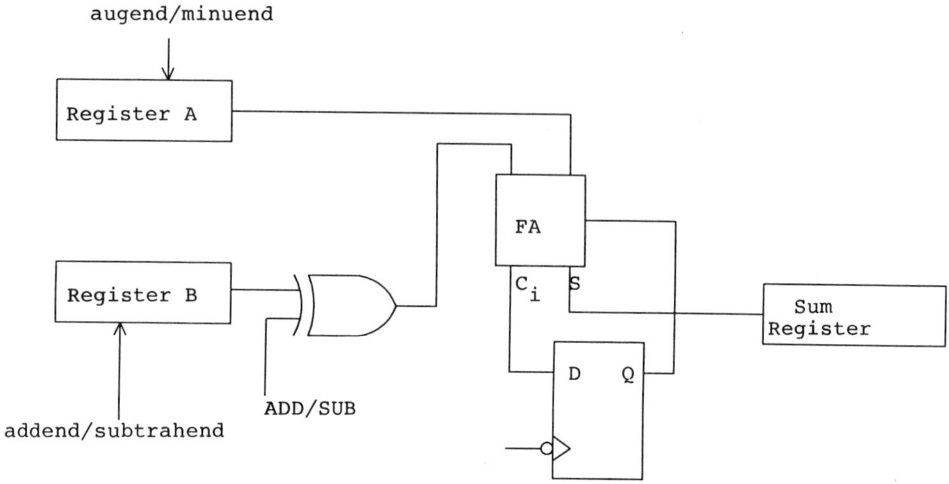

FIGURE 9.20 The logic block diagram of a sequential adder/subtracter.

be fed back as a carry-in bit to the same full adder along with the next pair of shifted addend and augend bits. This shift-add scheme is continued until all bits of the addend and augend have been processed. Finally, for subtraction, the 2's complement capability may be incorporated into the circuit.

The sequential arithmetic unit is fed with a number of control inputs to begin and run the complete operation. As is obvious from the circuit of Fig. 9.20, at least one control input to the unit is used to indicate whether the intended operation is addition or subtraction. More importantly, a system clock pulse must be provided so that sequencing of the different operations may be synchronized and controlled. While two *n*-bit shift registers may be enough to store the addend and augend, the single-bit carry-borrow flip-flop is used for storing the carry-out temporarily. The X-OR logic gate is necessary to complement the subtrahend. Furthermore, a combinational unit may be designed and incorporated for indicating an overflow condition if it were to occur.

The subtraction operation involves taking the 1's complement of each bit of register *Y* as it passes through the X-OR logic gate while the 2's complement operation is completed by setting the initial content of the carry/borrow storage flip-flop, $CBFF_0$, to a 1. The control variable, *E*, determines whether or not the circuit should perform addition. For example, when *E* is a 0, the addition operation is selected, but when *E* is a 1, the subtraction is selected. Initially, the two operands are to be loaded into the registers in parallel. However, when $E = 0$, the initial value of the $CBFF_0$ should be reset to 0. The RTL algorithm of the *n*-bit adder/subtracter unit can be expressed as follows:

T_1: X ← (augend/minuend); Y ← (addend/subtrahend); $CBFF_0$ ← E; OI ← 0; N ← 0;
T_2: X_{n-1} ← X_0 plus [Y_0 ⊕ E] plus $CBFF_0$; Y ← rr Y; $CBFF_0$ ← (carry-out); $X_{0\ldots(n-2)}$ ← $X_{1\ldots(n-1)}$; N ← N plus 1;

T_3: IF (N < n − 1) T_2 ← 1 ELSE IF (N = n minus 1)
 T_4 ← 1 ELSE T_5 ← 1;
T_4: IF ($X_0(Y_0 \oplus E)\overline{CBFF_0} + \overline{X_0}\,(\overline{Y_0 \oplus E})\,CBFF_0$ = 1) OI
 ← 1; T_2 ← 1;
T_5: HALT;

where T_x is the control condition, which must be set to a 1 for the execution of the corresponding RTL statement. Note that even after the detection of an overflow, the algorithm goes on to determine a sum that is wrong.

The sequential addition/subtraction begins at T_1 when the registers are loaded with the n-bit operands and the $CBFF_0$ is loaded with E. When $E = 0$, the initial carry-in is 0, that is, $CBFF_0 \leftarrow 0$. When $E = 1$, a 1 is stored at the $CBFF_0$ for completing the 2's complement operation. The variable N keeps track of the add operations. A comparator could determine if the current count is less than, equal to, or greater than $n - 1$. The overflow indicator, OI, indicates whether or not there is an overflow. An overflow cannot be determined until there is only one more addition.

Whenever two numbers are added, there is always the possibility of an overflow if the result's word-length is constrained to be the same as that of the operands. In order for the result to be valid, the carry into and out of the sign bit must be the same. At T_4, the two sign bits are in X_0 and Y_0. An overflow results whenever (1) $X_0 = Y_0 = 1$, $E = 0$, and $CBFF_0 = 0$; (2) $X_0 = Y_0 = 0$, $E = 0$, and $CBFF_0 = 1$; (3) $X_0 = 1$, $Y_0 = 0$, $E = 1$, and $CBFF_0 = 0$; and (4) $X_0 = 0$, $Y_0 = 1$, $E = 1$, and $CBFF_0 = 1$. Note that $CBFF_0$ holds the carry-in into the sign bit.

Upon reexamination of the RTL statements it should be apparent that at T_3 and T_4 there are no actions that cannot be performed at T_2. Accordingly, the RTL steps T_2 through T_5 can be reduced to give a two-step RTL sequence:

T_2: X_{n-1} ← X_0 plus (Y ⊕ E) plus $CBFF_0$; Y ← rr Y;
 $CBFF_0$ ← (carry-out); X ← sr X; N ← N plus 1;
 IF (N = n minus 1) OI ← $X_0(Y_0 \oplus E)\overline{CBFF_0} + \overline{X_0}$
 $(\overline{Y_0 \oplus E})\,CBFF_0$; IF (N = n) T_3 ← 1 ELSE T_2 ← 1;
T_3: HALT;

Example 9.2 Add 000010 and 001101 using a sequential arithmetic scheme.

Solution We set up the following table:

T	X	Y	$CBFF_0$	OI	N
T_1	000010	001101	0	0	0
T_2	100001	100110	0	0	1
T_2	110000	010011	0	0	2
T_2	111000	101001	0	0	3
T_2	111100	110100	0	0	4
T_2	011110	011010	0	0	5
T_2	001111	001101	0	0	6
T_3	001111	001101	0	0	6

The sum is 001111 without the presence of an overflow.

9.7.2 Sequential Multiplication

The algorithm for the multiplication of two multi-bit binary numbers follows from the classical rules. Consider, for example, the multiplication of 14 and 6 in both decimal and in binary:

```
   14        1110
 × 06      × 0110
 ----      ------
   84        0000
  00        1110
 ----      1110
  084     0000
          -------
          1010100
```

We see that 84_{10} is actually 1010100_2. In both decimal and binary multiplications, the product size can be twice as large as the multiplier size. Accordingly, when multiplying two n-bit words, a register of size $2n$ bits must be used for storing the product.

There are several means by which multiplication can be realized. Consider that both multiplicand and multiplier are positive, for simplicity. Accordingly, the corresponding multiplier circuit would require two shift registers, X and Y, for storing the two numbers, a $2n$-bit shift register P for storing the product, and a $2n$-bit CLA. The RTL program for the sequential multiplication scheme can be summarized as follows:

```
T1: P ← 0; N ← 1; X ← (multiplicand); Y ←
    (multiplier);
T2: IF (Y0 = 1) P ← X plus P; N ← N plus 1;
T3: Y ← sr Y; X ← sl X; If (N ≠ n plus 1) T2 ← 1;
T4: HALT;
```

where N is the count of partial products. The register that stores the multiplicand should be of $2n$-bit size since the multiplicand needs to be shifted left repetitively. Figure 9.21 shows the corresponding logic block diagram for the multiplication algorithm.

The multiplicand is loaded into the right-half of register X while the n-bit register Y is loaded with the multiplier. Since the CLA carry-out is not expected to be a 1, it is not even stored. The test performed during T_2 guarantees that the value of X will be added to the partial product only if the least significant bit of Y is a 1. Such conditional addition is performed by ANDing each bit of the multiplicand, X, with Y_0.

We may note that in the circuit of Fig. 9.21, while the multiplier is being shifted to the right, the register Y has unused FFs on the extreme left. Furthermore, the register P holds leading zeros in the beginning. In fact, as register P utilizes an

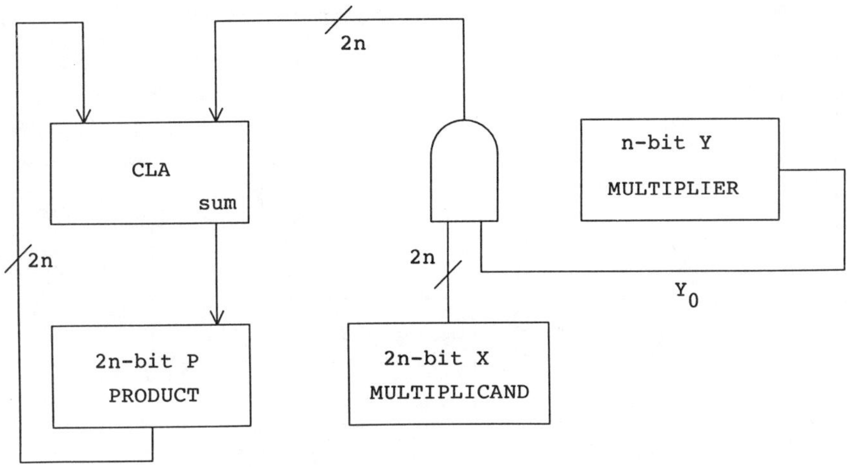

FIGURE 9.21 Logic block diagram for sequential multiplication.

additional FF on its left, register Y vacates one FF. Accordingly, an improved version of the RTL program may be developed as follows:

```
T1: P ← 0; N ← 0; X ← (multiplicand); Y ←
    (multiplier); C ← 0;
T2: IF (Y0 = 1) C,P ← X plus P; IF (N = n minus 1)
    T4 ← 1;
T3: C,P,Y ← sr C,P,Y; N ← N plus 1; T2 ← 1;
T4: HALT;
```

Note that C is a single-bit register that is tied to the left-end of register P such that it is able to store the CLA carry-out. Note that the registers C, P, and Y are all physically linked so that the shift-right operation at T_3 involves all of them. The data from C shift into the most significant bit of P and the least significant bit of P occupies the most significant bit of Y. Figure 9.22 shows the logic block diagram for this latter multiplication scheme.

The bits in register X are not shifted left since the product bits are being continually clocked to the right. Note, however, that X is still being added to the more significant n bits of the product. In this RTL algorithm, register Y initially stores the n-bit multiplier and eventually stores the less significant half of the product. In comparison, this algorithm has no need of a shift-left enable signal.

Alternatively, the rotate operation can also be used to obtain multiplication of unsigned positive numbers. The registers X, Y, and P require n bits to store the multiplicand, multiplier, and partial product, respectively. P and Y will display the product. In addition, a single flip-flop, R, is used for temporarily storing the carry-out. This flip-flop is unnecessary if the shift/add operation is done in the same clock period. The RTL algorithm for this alternative multiplication scheme

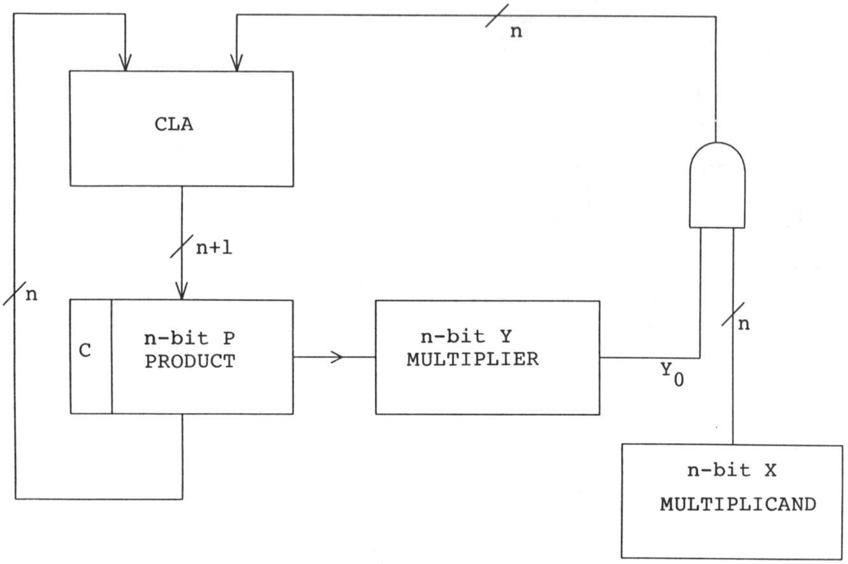

FIGURE 9.22 The logic block diagram for the improved multiplication scheme.

is obtained as follows:

T_1: P ← 0; N ← 0; X ← (multiplicand); Y ←
(multiplier);
T_2: R,P ← P plus Y_0X; R,P ← rr R,P;
T_3: R,Y ← rr R,Y ; N ← N plus 1;
IF (N < n minus 1) T_2 ← 1;
T_4: HALT;

Multiplication involving signed complement numbers is relatively complex. One reasonable scheme may consider the signs separately, take the absolute value of both the multiplicand and the multiplier, and then use a sequential multiplication technique that is valid for positive numbers. The product is finally complemented or not depending on the sign of the product. The RTL algorithm for such 2's complement sign and magnitude sequential multiplication is listed below:

T_1: P ← 0; N ← 1; X ← (multiplicand); Y ←
(multiplier); C ← 0;
T_2: IF ($\overline{X}_{n-1}$ $\overline{Y}_{n-1}$ = 1) S ← 0;
IF ($\overline{X}_{n-1}$ Y_{n-1} = 1) S ← 1, Y ← $\overline{Y}$ plus 1;
IF (X_{n-1} $\overline{Y}_{n-1}$ = 1) S ← 1, X ← $\overline{X}$ plus 1;
IF (X_{n-1} Y_{n-1} = 1) S ← 0, X ← $\overline{X}$ plus 1,
Y ← $\overline{Y}$ plus 1;
T_3: C,P ← P plus XY_0; N ← N plus 1;
T_4: C,P,Y ← sr C,P,Y ; IF (N = n plus 1) T_3 ← 1;
T_5: IF (S = 1) P,Y ← ($\overline{P,Y}$) plus 1;
T_6: HALT;

At T_2 a test is conducted for the various combinations of sign-bits. The flag S is reset if the signs are equal; otherwise it is set to a 1. At S_5, the sign of the product is eventually restored. Figure 9.23 shows the logic block diagram corresponding to the sign and magnitude multiplication algorithm. When compared to the logic circuit of Fig. 9.22, the current logic circuit needs additional modules for performing the 2's complement conversion.

In the scheme just discussed, the signs of the operands are handled separately. Furthermore, the resultant product has to be transformed back to its equivalent signed-complement form. To overcome these shortcomings, a new variation of the standard shift-and-add scheme, called Booth's algorithm, may be employed. It accommodates 2's complement numbers without any pre- or post-adjustments.

The presence of m consecutive 1's in a multiplier corresponds to m additions and m shifts. Since a string of m consecutive 1's from bit weights 2^1 to 2^k (where $k > 1$) is equivalent to $2^{k+1} - 2^1$, the corresponding operation is equivalent to a subtraction in the beginning, m subsequent shifts, and a final addition step. For example, the multiplier (0001111001110) would generally involve seven add-and-shift operations. However, if the same multiplicand is rewritten as (0010000010000 − 0000001000010), the multiplication would involve only four add-and-shift operations. In summary, strings of zeros may be ignored while each string of 1's may be replaced by a single subtraction of the multiplicand at the position of the least significant 1 and an addition at one bit to the left of the most significant 1.

Testing a string of 1's is accomplished by adding an extra bit, Y_{-1}, next to the Y_0 bit. As bits of Y are continually shifted to the right, the pair of bits $Y_0 Y_{-1}$ may be tested for identifying either the beginning or continuation or ending of a string of 1's. The multiplicand is subtracted from the partial product when the right end of the string is detected while the multiplicand is added to the partial product upon

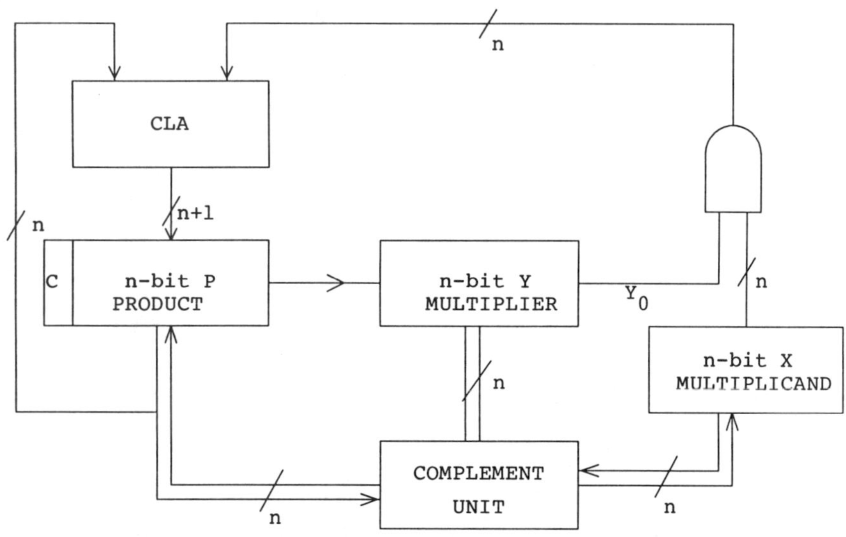

FIGURE 9.23 Logic block diagram for sequential sign and magnitude multiplication.

encountering the left end of the string. Otherwise, the partial product is only shifted. The RTL program for Booth's sequential multiplication scheme can be summarized as follows:

```
T1: X ← (multiplicand); Y ← (multiplier); P ← 0;
    N ← n; Y-1 ← 0;
T2: IF ((Y0Y-1 = 00 OR 11) AND N = 0) T5 ← 1;
    IF (Y0Y-1 = 00 OR 11) T3 ← 1;
    IF (Y0Y-1 = 01) P ← P plus X , T3 ← 1;
    IF (Y0Y-1 = 10) P ← P plus X̄ plus 1;
T3: P,Y,Y-1 ← scr P,Y,Y-1; N ← N minus 1;
T4: IF (N ≠ 0) T2 ← 1;
T5: HALT;
```

Figure 9.24 shows the basic logic block diagram for this multiplication algorithm. This circuit includes a test module and a dual adder/subtracter unit. The more significant part of the product is contained in register *P* while the less significant part of the product is contained in register *Y*.

The sequential multiplication schemes as expected are relatively slower. The major amount of multiplication time is lost by the carries while they ripple through the adders. One of the ways to overcome this problem is to use the CLA adders

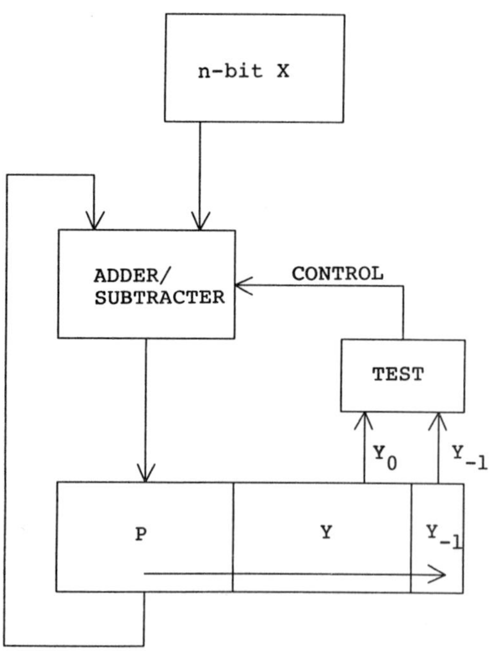

FIGURE 9.24 Logic block diagram for Booth's multiplication.

in place of the regular ripple adders. Using the carry-save multiplication, the carries from the addition of the partial products are initially stored in a set of FFs and used as one of the adder inputs during the next addition cycle. It prevents the carries from rippling through different adders.

The carry-save scheme consists of two phases: carry-save (CS) and carry-propagate (CP). The multi-step CS phase involves the standard add- and-shift operation, during which the carry is temporarily stored in a FF. The carry is not allowed to ripple from one bit position to the next. The true product however, is, obtained in the final single-step CP phase when the carries are allowed to propagate. The corresponding RTL multiplication algorithm is given as follows:

```
T1: X ← (multiplicand); Y ← (multiplier); CP ← 0;
    CS ← 1; P ← 0; C ← 0; N ← n;
T2: Pi ← Pi plus Ci plus XiY0; Ci ← (carry-out); N
    ← N minus 1;
T3: P,Y ← sr P,Y ; IF (N ≠ 0) T2 ← 1 ELSE CS ← 0,
    CP ← 1;
T4: P ← P plus C;
T5: HALT;
```

where both CS and the CP phases are mutually exclusive. At T_2, bit-by-bit addition is performed and the carry-outs are stored in register C. This CS phase is continued until add-and-shift has been executed n times. Finally, at T_4, the classical addition involving carry-propagation is executed. Figure 9.25 shows the corresponding logic circuit of a carry-save adder.

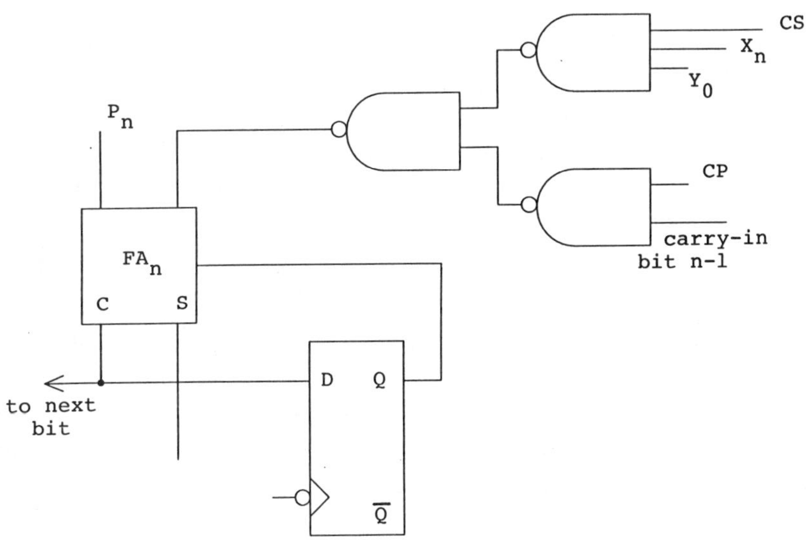

FIGURE 9.25 A carry-save adder circuit.

9.7.3 Sequential Division

In principle, division is the inverse of multiplication. The standard division process involves repeated comparison, subtraction, and shift operation. In comparison, however, in binary division, it is not necessary to determine how many times the divisor fits within the dividend. If the divisor is larger than the partial remainder, the divisor is moved to the right one bit before the next comparison is made. This shift of the divisor may also be accomplished by shifting the dividend to the left while the divisor is kept fixed.

Figure 9.26 shows the logic block diagram for division. In this division technique, also known as the comparison method, the subtraction is achieved by adding the 2's complement of the divisor to the partial remainder. Initially, the n-bit register D is loaded with the divisor and the $2n$-bit dividend is divided between the two registers X and Y. A comparison test is performed by the comparator unit to determine whether or not the subtraction should be performed. When $X \geq D$ the GO/NO-GO output is a 1, the quotient bit 1 is loaded in Y_0, and the partial remainder is shifted to the left one bit. On the other hand, if the GO/NO-GO output is a 0, the new quotient bit is set to 0 when the subtraction is not performed and the partial remainder is shifted to the left only. Finally, after n successive iterations, the remainder is obtained in X while the register Y holds the quotient.

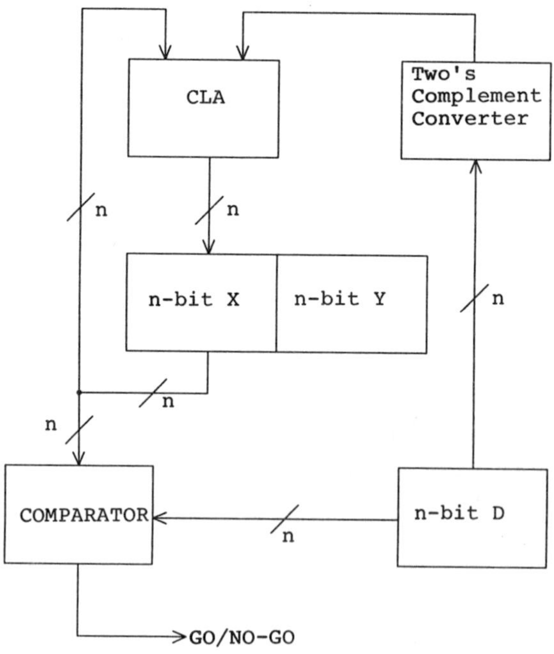

FIGURE 9.26 The block diagram of a comparison division.

The RTL algorithm corresponding to this division scheme involving only positive numbers is obtained as follows:

T_1: X,Y ← (Dividend); D ← (Divisor); N ← 0; E ← 0;
T_2: IF (D_{n-1} ≠ 1 OR D = 0) E ← 1, T_9 ← 1;
T_3: IF (X_{n-1} ≠ 1) X,Y ← sl X,Y , Y_0 ← 0, N ← N plus 1, T_3 ← 1;
T_4: IF (GO/NO-GO ≠ 1 OR N = n) T_9 ← 1;
T_5: X ← X + $\overline{D}$ + 1; Y_0 ← 1;
T_6: X,Y ← sl X,Y ; N ← N plus 1;
T_7: IF (N = n) T_9 ← 1;
T_8: IF (GO/NO-GO = 1) T_5 ← 1 ELSE Y_0 ← 0, T_6 ← 1;
T_9: HALT;

where E is an error flag that is turned on whenever it encounters a division by zero or when the final quotient requires more than n bits. The divide overflow must be avoided since this will force the quotient to be too long for transfer into other units.

An alternative division algorithm, referred to as the restoration division, can be formulated in which the divisor is always subtracted from the partial remainder. Whenever the difference turns out to be negative, the remainder is restored. The RTL algorithm for the corresponding sign and magnitude division is obtained as follows:

T_1: X,Y ← (dividend) ; N ← n minus 1;
T_2: Q_n ← X_n ⊕ D_n; E,X ← X plus $\overline{D}$ plus 1;
T_3: IF (E = 1) E,X ← X plus D, OFF_0 ← 1, T_{11} ← 1;
T_4: E,X ← X plus D, OFF_0 ← 0;
T_5: E,X,Y ← sl E,X,Y;
T_6: IF (E = 0) E,X ← X plus $\overline{D}$ plus 1 ELSE T_8 ← 1;
T_7: IF (E = 0) E,X ← X plus D, T_{10} ← 1 ELSE T_9 ← 1;
T_8: X ← X plus $\overline{D}$ plus 1;
T_9: Y_0 ← 1; N ← N minus 1;
T_{10}: IF (N ≠ 0) T_5 ← 1;
T_{11}: HALT;

where the sign-bits of both the numerator and the denominator are used to determine the sign of the quotient. The basic logic block diagram for the corresponding division scheme is shown in Fig. 9.27. The flip-flops, X_n and D_n, store the signs of the dividend and the divisor, respectively, while the flip-flop Q_n takes the sign of the quotient. The overflow flip-flop OFF_0 is turned on whenever there is a divide-overflow and flip-flop E is used to store the carry-outs.

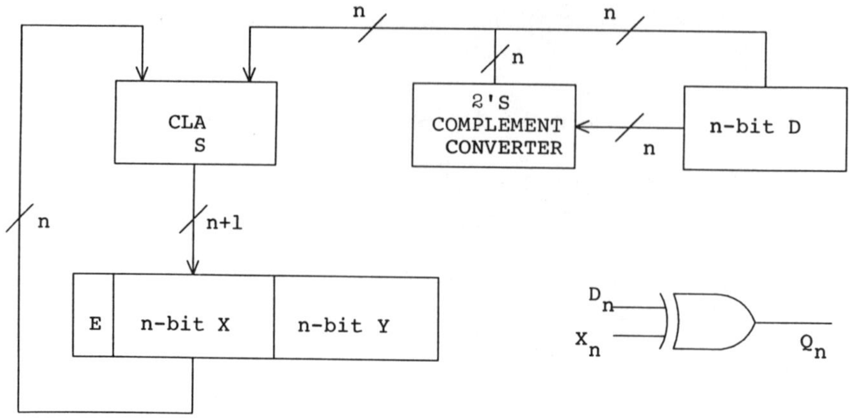

FIGURE 9.27 The logic diagram for restoration division.

9.8 PROBLEMS

1. Design a 3-bit counter that counts up when a control variable $E = 0$ and counts down when $E = 1$.
2. Design synchronous sequential circuits using J-K FFs that generate a 1 whenever the following sequence occurs: (a) 0010, (b) 1101, (c) 1010, (d) 1011 and (e) 10110.
3. Using D FFs and assorted gates, design a sequential comparator capable of determining which of the two multi-bit numbers, X and Y, of equal length is larger.
4. Design a synchronous sequence detector, using J-K FFs, which produces an output of 1 whenever any one of the three sequences 1100, 1010, and 1001 occur. The circuit resets to its initial state after a 1 has been generated.
5. Design a register capable of performing the 2's complement operation on its content.
6. Obtain a register using J-K FFs having control inputs for the increment, complement, and parallel transfer micro-operations. Show how the 2's complement can be implemented in this register.
7. Use 4-bit binary counters in parallel and assorted gates to realize a divide-by-39 counter.
8. **(a)** Design a 256 × 32 bit RAM from 256 × 8 bit RAMs.
   (b) Design a 256 × 8 bit RAM from 32 × 8 bit RAMS.
   (c) Design a 4096 × 2 bit RAM from 1K × 1 bit RAMs.
   (d) Design a 1K byte RAM from 256 × 4 bit RAMs.
9. Obtain a RTL multiplication algorithm so that 1's complement numbers can be handled.

9.9 SUGGESTED READINGS

Aagard, R. C., T. C. Lee, and D. Chen, Advanced Optical Storage Techniques for Computers, *Appl. Opt.*, **11,** 2133 (1972).

Ammon, G. J., Archival Optical Disc Data Storage, *Opt. Eng.*, **20,** 394 (1981).

Basu, D., and T. Jayashree, On a Simple Post-correction for Nonrestoring Division, *IEEE Trans. Comput.*, **C-24,** 1019 (1975).

Dadda, L., Composite Parallel Counters, *IEEE Trans. Comput.*, **C-29,** 942 (1980).

Dhurkadas, A., Faster Parallel Multiplier, *Proc. IEEE.*, **72,** 134 (1984).

Hwang, K., *Computer Arithmetic*, Wiley, New York, 1979.

Johnson, E. L., A Digital Quarter Square Multiplier, *IEEE Trans. Comput.*, **C-29,** 258 (1980).

Johnson, E. L., and M. T. Jong, An Implicit Division Technique Using a Digital Multiplier, *IEEE Circuits Syst. Magazine*, **4**(2), 16 (1982).

Johnson, E. L., and M. A. Karim, *Digital Design: A Pragmatic Approach*, PWS-Kent Publishing, Boston, 1987.

Jong, M. T., Binary Fraction Multiplication Algorithms, *IEEE Circuits Syst. Magazine*, **1**(4), 8 (1979).

Kline, R. M., *Structured Digital Design Including MSI/LSI Components and Microprocessors*, Prentice-Hall, Englewood Cliffs, 1983.

Lai, H. C., and S. Muroga, Logic Networks of Carry-Save Adders, *IEEE Trans. Comput.*, **C-31,** 870 (1982).

Oklobdzija, V. G., and M. D. Ercegovac, An On-line Square-Root Algorithm, *IEEE Trans. Comput.*, **C-31,** 70 (1982).

Strader, N. R., and V. T. Rhyne, A Canonical Bit-Sequential Multiplier, *IEEE Trans. Comput.*, **C-31,** 791 (1982).

Swartzlander, E. E., The Quasi-serial Multiplier, *IEEE Trans. Comput.* **C-22,** 317 (1973).

Swartzlander, E. E., Merged Arithmetic, *IEEE Trans. Comput.* **C-29,** 946 (1980).

Verber, C. M., Integrated Optical Architectures for Matrix Multiplication, *Opt. Eng.*, **24,** 74 (1985).

DIGITAL OPTICAL COMPUTING

CHAPTER TEN

Devices

10.1 INTRODUCTION

This chapter reviews some of the existing devices that can be used to realize digital optical computing schemes. The idea of the optical computer was first put forward with the observation of optical nonlinearity in nonlinear material. Scientists envision the development of optical nonlinear devices that eventually will replace electronic transistors of a conventional electronic computer. Nonlinear optical devices are discussed in this chapter from that perspective. It should be emphasized that nonlinear operations are deemed necessary for implementing intelligent decision-making operations. One of the major stumbling blocks in nonlinear optics of the early 1960s was the tremendous power requirement to achieve any reasonable nonlinear behavior. The higher the size of the device, the higher the power requirement. Therefore one way to overcome this problem is to go to smaller dimensions. This has contributed to miniaturized integrated optics, which make things work with less total power. Another device that will be covered is the threshold device. A correlator, for example, can make intelligent decisions as to whether or not a particular pattern or a string of numbers is present in an input. A threshold detector forms an integral component of such correlators. There are other related device issues. For example, to introduce or retrieve a real world data in either electronic or optical form into or from an optical system, an interface device is necessary. Spatial light modulators may serve as input devices that can convert input information into and out of an optical information processor. This chapter provides a description of several of these spatial light modulators.

10.2 NONLINEAR DEVICES

All the devices are characterized by their output–input relationship. Most physical phenomena are observed usually in the linear range (of the output–input relationship), although strictly speaking linearity is only an approximation of a finite range of a nonlinear behavior. Optical applications, such as imaging and information processing, involve only a limited range of linearity because they involve very low power. Nonlinear optical behavior, on the other hand, is observed when the intensity of the laser beam is extremely high. This was one of the reasons why although optical computing was prophesized in the early 1960s with the invention of optical nonlinearity it did not materialize into working devices until recently. To perform any logical operation, such as those mentioned in Chapter 7, it is necessary that the logical device be associated with nonlinearity. Again, when the nonlinearity is associated with feedback, it results in bistability. The term bistability, however, has been used rather synonymously with nonlinearity. For a combinatorial type of logic operation, however, it is not necessary to have bistability. Bistability is only required for a memory operation like that in a flip-flop.

Consider a nonlinear input–output relationship as that shown in Fig. 10.1, where the output is defined as HIGH or logical 1 when the output is in the upper branch, and as LOW or logical 0 when the transmitted power is low. Provided the presence of an input A is given by a value equal to or greater than the threshold value, the output due to A alone is sufficient to produce a HIGH output. Under such a constraint, the device functions as an OR gate. Again, by redefining the input level of a two-input device slightly above 50% of the threshold, both inputs will be required for generating a HIGH output. The resulting device thus functions as an AND gate.

10.2.1 Optical Bistable Devices

An optical system that possesses two different transmission states for the same input intensity is said to exhibit optical bistability. A flip-flop is a bistable device. Consider, for example, a J-K flip-flop, which can hold an output of either 0 or 1 when the J-K inputs are both 0. To achieve bistability, in general, some form of feedback is necessary to make the intensity transmission dependent on the output intensity. Strictly speaking, a physical system can never produce two different

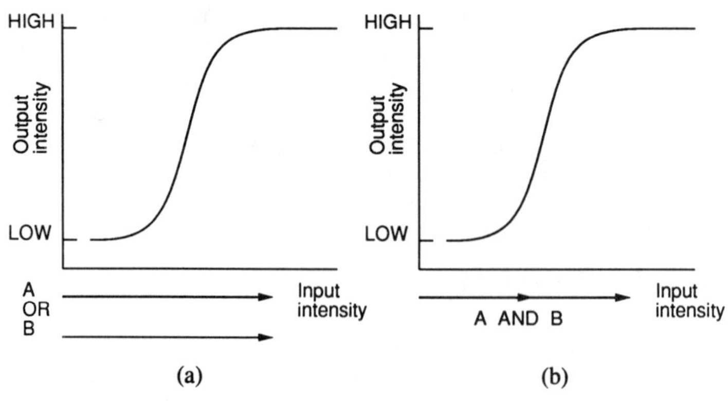

FIGURE 10.1 (a) An OR logic using nonlinearity, (b) an AND logic using nonlinearity.

outputs with the same input. However, the trick is the use of feedback, which makes the output not only a function of the input but also a function of the past output. Thus the past output acts as a hidden input to the system. If the previous output was different, then the next output can be different even with the same input. Depending on the feedback, bistable devices can be classified as electrical, optical (intrinsic), or hybrid. Bistable devices can also be classified according to the source of their nonlinearity, such as excess charge carrier generation (electronic) or lattice heating. Yet another type of classification may depend on the physical characteristics, such as polarization, absorption/gain, or refractive index, which are modified by nonlinearity.

An optical feedback may be provided in a resonator structure by means of simple mirrors. A Fabry–Perot resonator, for example, consisting of two parallel partially transmitting mirrors separated by a spacer (usually a dielectric) may be utilized for this purpose. A ray of light incident on it is partially reflected and partly transmitted repeatedly through the resonator as shown in Fig. 10.2a. As the light propagates through the resonator, it will have an amplitude change and a phase change. The amplitude change is due primarily to insufficient reflection and refraction at the mirrors. The phase change, on the other hand, is caused by the path difference of different partial reflection and refraction of waves, before they reach the output boundary. When the two phases meet in (out of) phase, constructive (destructive) interference occurs. The amount of transmittance can be estimated exactly from the reflection characteristics of the mirrors, their spacing, and the medium refractive index. As shown by the transmission characteristics of Fig. 10.2b, a Fabry–Perot resonator acts as an optical bandpass filter, which has peak transmission values at some optical frequencies. The corresponding center frequency of transmission is a function of the optical path length. The nonlinearity is introduced into the Fabry–Perot system by means of a material that has nonlinear dispersive or absorptive characteristics such that the refractive index of the material changes with the incident light intensity. Thus by changing the incident light intensity, the optical path length can be altered. Initially, the Fabry–Perot resonator

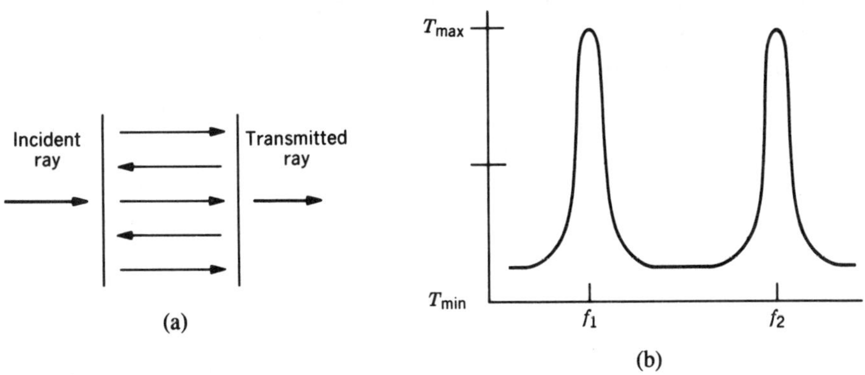

FIGURE 10.2 (a) A Fabry–Perot resonator and (b) its transmission characteristics.

is slightly detuned such that it is operated very close to the resonant frequency with less than the peak transmission. When the intensity is increased, the refractive index is changed in a way that the resonator suddenly attains its peak transmission at the operating laser frequency. Consequently, by changing the input intensity, the transmission characteristics are changed. If the input intensity is reduced, the resonator—due to internal feedback—still retains a high transmission state.

The feedback effect of the Fabry–Perot resonator may be enhanced by creating stacks of multiple filters providing multiple feedbacks. This is analogous to designing a cascaded resonator or distributed electrical transmission line or simply a high-order filter. It is a well-known fact that increasing the order of the filter causes the filter to have a sharper tuning characteristic, that is, narrower bandwidth. The multiple resonator is generally made of stacks of optical material in HLHL · · · HHH · · · LHLH form, where H stands for $\lambda/4$ thickness of high-index material and L stands for $\lambda/4$ thickness of low-index material. The high–low stacks act as high reflection mirrors around the central thicker HHH spacer, as shown in Fig. 10.3. Nonlinearity is introduced into these layers by introducing a material that possesses an intensity-dependent refractive index. The change of the refractive index could take place either electronically or thermally. The simplest intensity-dependent refractive index, however, is of thermal origin. In such a case, when the temperature of the material is changed, it changes the optical path, and hence the refractive index. This temperature change brings about a change in the tuning or resonant characteristics of the resonator containing the material. The nonlinearity in ZnSe interference filters, for example, is of thermal origin. Such filters have already been used for exhibiting bistability and logical operations.

10.2.2 Polarization Bistability

Instead of using the intensity levels as the two binary states, a proposal has been made to use two orthogonal polarizations as the two states. The major advantage of using polarization is that both 1 and 0 involve flow of energy; thus a change of state between 0 to 1 does not necessarily mean energy is being destroyed or lost. Even in electronics, in the generation and transmission of 1's and 0's a transfer of energy must take place. For example, in transistor–transistor logic, when a driver of a stage is driving the second stage to a LOW state, then energy flows in the reverse direction from the second stage to the first stage, thus sinking current in the driver. Alternately, when a device drives another device to a HIGH state, the

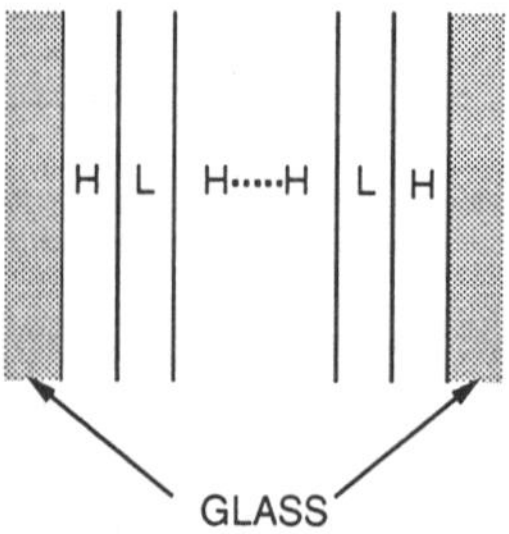

FIGURE 10.3 A multiple stack nonlinear resonator filter.

driver directs current into the second stage. Thus energy in the cascaded structure of logic elements either flows in or out of the device while, respectively, transferring a 0 or 1 logic level. Polarization logic shares a similar basis for defining binary logic. It is determined by the direction of the electric field of the electromagnetic wave. Thus logic levels involve a flow of energy, which is vital for cascadability. In an intensity coded logic, cascadability is achieved by encoding inputs such that the position of the illuminated pixel for binary 1 is reversed as in that for 0. Such schemes have been used in dual-rail coding, which is discussed in more detail in Chapter 11.

10.2.3 Self Electro-optic Effect Device

Self electro-optic effect (SEED) devices depend on changes in the optical absorption characteristics in multiple quantum well (MQW) material initiated by the application of an electric field normal to the thin semiconductor layers. It is constructed by placing a MQW structure inside the intrinsic region of a p-i-n photodiode, as shown in Fig. 10.4. A MQW is created by placing stacks of alternate low bandgap (called wells) and high bandgap (called barriers) materials, where electrons and holes are generally confined within the well layer. Due to the short thickness of the well layers, quantum size effect governs the optical and electrical characteristics of these structures, which are therefore named quantum well structures.

The optical absorption spectrum demonstrates impulse-like absorption spikes and room temperature exciton resonance. The spikes occur due to the confinement of electrons and holes in the wells, which results in quantization of the motions of holes and electrons normal to the well layers, which in turn results in discrete energy states. As a result, the optical intraband absorption no longer remains a slowly varying function of wavelength but demonstrates distinct peaks.

A room temperature excitonic effect occurs when the excitons modify their structure to fit inside the well structure. As a result, electrons and holes are forced to orbit closer, and the binding energy increases by a factor of 2 or 3. The excitonic effect also enhances the electro-absorption. With an electric field perpendicular to the well, the field pulls the electrons and holes apart. However, well barriers pre-

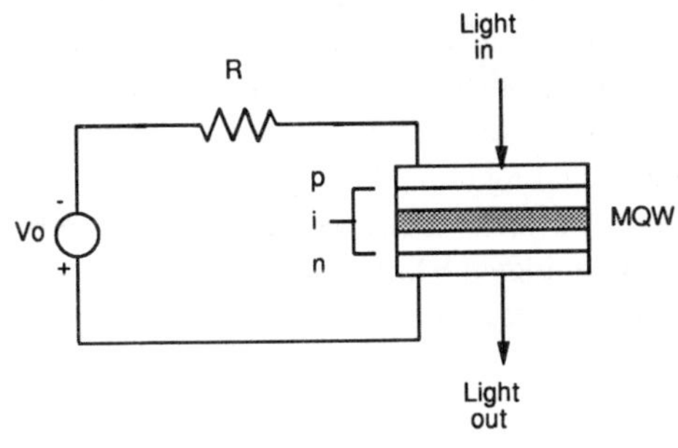

FIGURE 10.4 A multiple quantum well SEED device.

vent the separation of the electron–hole pair. Even after more than double the regular ionization potential has been applied, the well material still exhibits excitonic resonance due to the electric field. This polarizing effect, known as the quantum confined Stark effect, shifts energy to the electron–hole pair, which causes a highly absorbing peak (edge) to appear in the spectral region where the sample, is normally transparent.

In the original SEED device, the p-i-n diode was kept at a reverse bias, and the MQW acted as both a detector and electro-absorptive material. To make the device switch, the incident light wavelength is chosen to be near the exciton resonance position for zero voltage across the diode. When the incident optical power is low, all the supply voltage is dropped across the diode in the absence of any photocurrent. Due to the voltage, the MQW material shifts the exciton absorption to longer wavelengths and the optical absorption is low. When the power of the incident beam increases, so does the photocurrent (diode working as detector), consequently reducing the reverse bias voltage across the MQW material. As the excitation resonance moves back to the operating laser frequency due to reduced voltage, it results in increased absorption (electro-absorptive effect). Increased absorption increases the photocurrent, which causes a positive feedback on the absorption by reducing the MQW voltage, thus moving closer to peak absorption conditions. Suddenly, the output switches to a lower state. When the input power is subsequently lowered, the switching to a higher state is achieved at a lower power.

The function of the electrical circuit is to provide an opto-electronic feedback. Replacing the resistive load by a photodiode, a diode based SEED (D-SEED) can be obtained. This development is very much analogous to the way diode transistor logic was developed from resistor transistor logic in electronics. Symmetric SEED (S-SEED), on the other hand, consists of two quantum well p-i-n diodes electrically connected in series. It has a gain that is rather insensitive to optical power fluctuations and it provides effective input–output isolation.

10.2.4 Optical Phase Conjugate Device

Optical phase conjugation is a mechanism of creating a beam from another beam with equal frequency but opposite phase. In particular, four-wave mixing may be used to generate a phase conjugate wave. In resonant degenerate four-wave mixing, the resonant interactions of four waves with the same frequency take place. The four waves are distinguishable by the different propagation and polarization directions. A common setup for four-wave mixing is shown in Fig. 10.5. A forward and backward propagating pump beam interacts in a nonlinear medium, which has at least a cubic type of nonlinearity. When the third probe beam is introduced at an angle, a reflected conjugate beam is generated as a result. When three waves are mixed in a cubic nonlinear medium, the total field of the nonlinear medium becomes

$$E(t) \propto \tfrac{1}{2}\,[E_{\mathrm{f}} + E_{\mathrm{b}} + E_{\mathrm{p}} + \text{complex conjugate}]^3 \qquad (10.1)$$

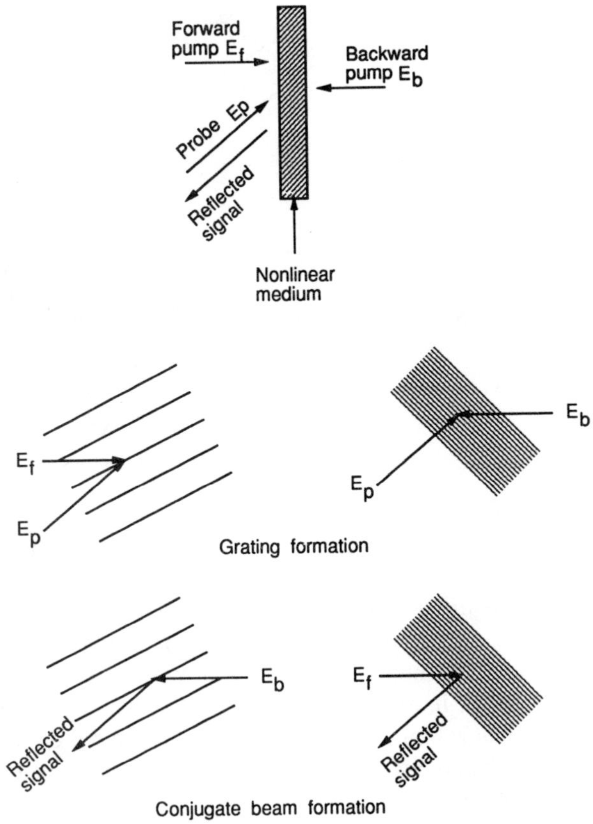

FIGURE 10.5 A setup for four-wave mixing.

Usually the pump beam has high power while the probe is generally weak. The induced cubic polarization term generates a triple product of the conjugate of the input probe beam (E_p^*) with the other two input beams (E_f and E_b). This results in the desired beam, which is nothing but the conjugate beam output. Here the energy from the pump beam is transferred to the conjugate probe beam as a multiplicative gain term.

There is another way to explain the physical process involved. When the three beams are mixed in a nonlinear medium, they interfere with each other and form gratings. These are equivalent to a thick holographic grating. The forward pump beam E_f and probe beam E_p form a lower frequency grating. This grating has the correct orientation to generate a reflected beam E_r. The input beam E_p and E_b also form another grating, which converts E_f into E_r. Usually the pump beam thus transfers its high power to the reflected beam. When the two pump beams are made collinear, the reflected beam exactly retraces the path of the probe beam. Thus a phase conjugate device is such that when light falls on it, the light retraces exactly the original path but in a reverse direction. The output is nothing but the conjugate

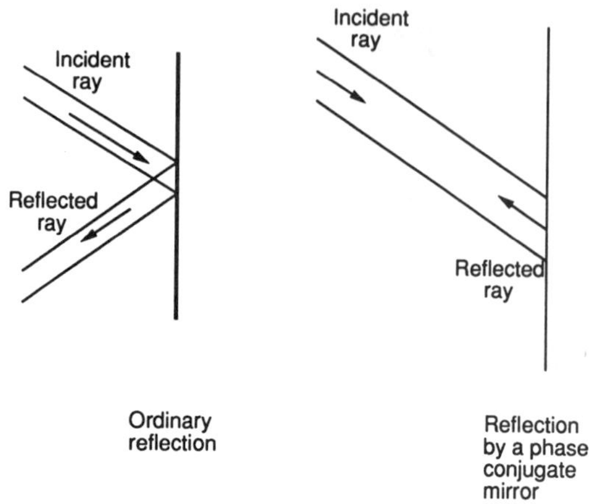

FIGURE 10.6 Comparison of a phase conjugate reflector with an ordinary mirror.

of the original input, which is equivalent to having the same amplitude wave with a reverse phase as shown in Fig. 10.6

An optical phase conjugate (OPC) system may be used in performing a three-input AND operation. This is realized in a degenerate four-wave mixing setup where the three beams are multiplied. Accordingly, one can use optical phase conjugation, for example, in an AND based optical symbolic substitution (OSS) operation. Since the realization is direct, only the noninverted or white pixels of the dual-rail encoded inputs are used. To recognize an input pattern, the input image is first divided by means of a beam splitter, and then the image is relatively shifted by mirrors and projected onto a third-order nonlinear material. A third beam encoded with the recognition mask is directed at the material. These are the two inputs of the three-input AND gate. Now the decoder mask, which is supposed to identify the recognized bright output, is directed toward the nonlinear material as the probe beam. The conjugate signal is then the product of the three signals, which counterpropagates with respect to the third probe beam. Finally, using a second beam splitter, this ANDed signal is extracted as the system output. The complete OPC based OSS system is shown in Fig. 10.7. OPC can also be used to correct phase distortion in images.

10.3 INTEGRATED OPTICS

Silicon VLSI technology is a very mature technology, and optical devices fabricated using such technology have a similar potential for large-scale production. Except for the source, all circuit components are made out of silicon. Silicon detectors are already in use in many applications. The use of GaAs material may allow light sources, modulators, detectors, and other useful components to be integrated into single monolithic structures.

In digital optical computing, many arithmetic circuits have already been realized which use large optical tables, lenses, mirrors, and free space communication

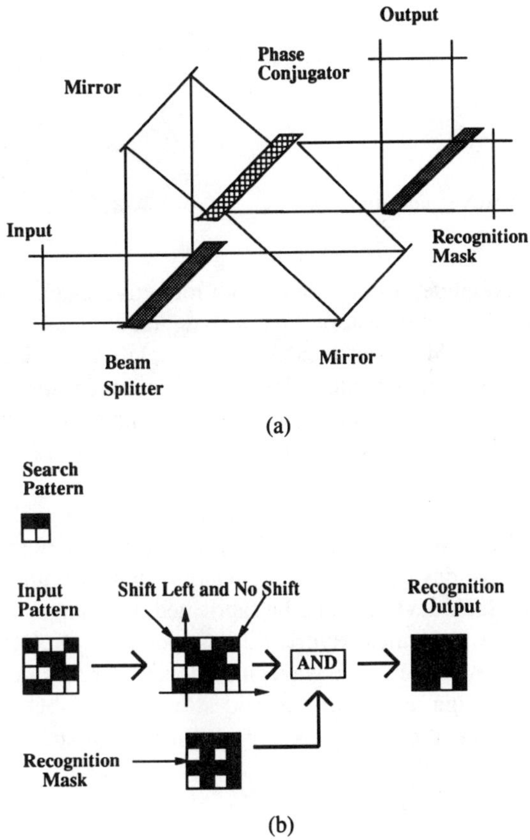

FIGURE 10.7 Recognition in an OPC based optical symbolic substitution: (a) system and (b) operations.

to perform simple logic operations. The power expended may be too high, making all these schemes impractical for any real use. By controlling the passage of light through a waveguide over a long interaction length, it may be possible to enhance the nonlinear effects with less power. Obviously, therefore, the incentive is to try to miniaturize all the components so that the power and space requirements are minimized and mechanical stability is achieved. Mechanical stability is an important advantage of integrated devices. Along with other advantages, it is expected to play an increasing role in advancing the digital optical computing.

10.3.1 Waveguides

In a miniaturized world, there is not much free space left to utilize the free space communication capability of optics. Therefore the light must be confined within a small area so that the light may be transferred easily to other appropriate areas for errorless operation. The waveguide serves as wiring for connecting two or more optical integrated circuits.

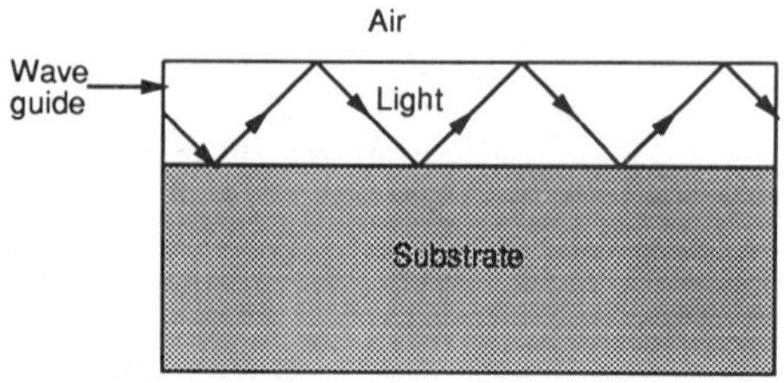

FIGURE 10.8 A waveguide structure.

In an optical waveguide, a thin film of high refractive index material is deposited on a low refractive index material in a rectangular structure, as shown in Fig. 10.8. Inside the waveguide, light incident at an angle on the side of the low refractive index material undergoes total internal reflection because of the appropriate index difference. The reflected light beam thus remains confined within the high refractive index material with a rather small loss into the surrounding low refractive index air or substrate.

Several possibilities of fabricating waveguide structure exist. In fiber optic applications, GaAs injection lasers are already in use. It may be possible to fabricate all the necessary active devices, lasers, modulators, switches, and detectors, from GaAs or its alloys. Such devices could be fabricated on a single wafer because of the similarity in lattice structures. Hybrid lithium niobate ($LiNbO_3$) is another possibility. Using $LiNbO_3$ as the active waveguide device, GaAs sources, and silicon detectors, one can design almost any hybrid structure. $LiNbO_3$ substrates have already been demonstrated to have superior optical performances.

10.3.2 Mirrors

Reflectors may be made by having a very fine grating etched into a substrate before the film is deposited. The periodic variation in refractive index causes a portion of the incident light to be reflected by each grating groove, as shown in Fig. 10.9. If the groove period is half that of the wavelength of the incident light, the reflected

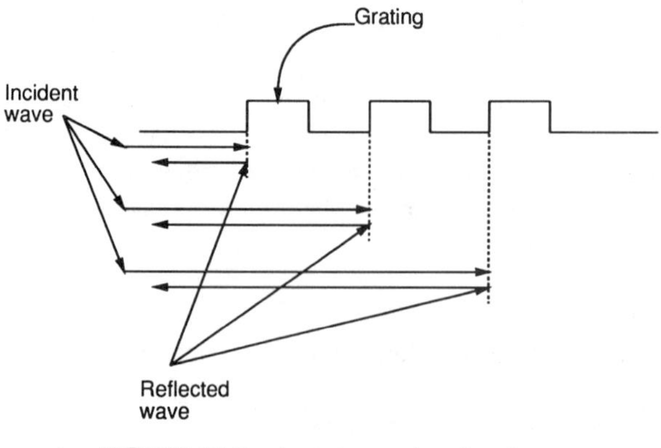

FIGURE 10.9 An integrated optic mirror.

signal adds in phase. The band of wavelength over which a strong reflective effect is observed may be narrowed by selecting a long distributed reflector structure with small reflectivity per groove. In fact, a structure of cascade filters consisting of slightly different periodicity of gratings could be used as a demultiplexer for the different frequencies of optical signals.

10.3.3 Couplers

Couplers are necessary to interface between an external source and an IC or to couple between different segments of an IC or between different ICs. Very often, one may have all the integrated optics circuit components on one single chip except for the source. Alternatively, the output of one device may have to be coupled to another circuit; then one is certainly in need of providing coupling between the source and the other components. Couplers can be of several different types.

End Coupler A very small aperture, short focal length lens or microscope objective may be used for end coupling, as shown in Fig. 10.10a. The alignment of the coupling is very critical for efficient transfer of power to the waveguide. Lateral or transverse misalignments have much more deteriorating effect than longitudinal or angular misalignments. The coupling efficiency is also a function of the nature of the waveguide: such as the step index or graded index.

Prism Coupler A prism coupling is formed by placing a high refractive index prism in close contact with a waveguide, as shown in Fig. 10.10b. In prism coupling, light enters through the surface of the waveguide. An incident beam is fed into the prism placed perpendicular to one of the surfaces. When it hits the surface

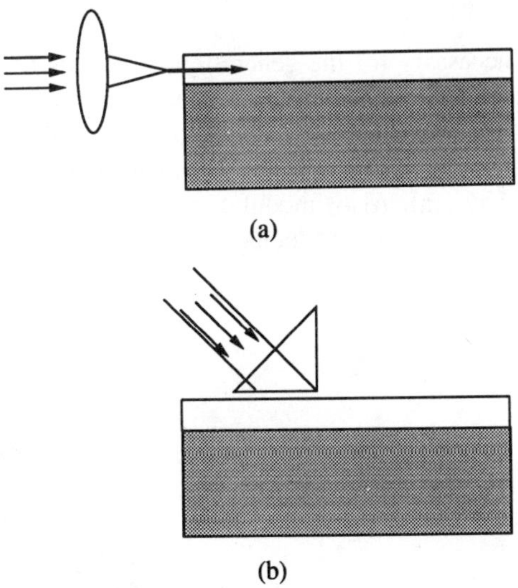

FIGURE 10.10 (a) End coupler and (b) prism coupler.

in contact with the waveguide at an angle larger than the critical angle, the light passes through the base and is tunneled to the accompanying waveguide. With some of the configurations, coupling efficiency could reach up to 100%. A prism coupler could also be used to couple light out of the waveguide. The mechanism involved in such coupling is known as frustrated total internal reflection.

Grating Coupler A grating is first deposited on the film's surface, which may cause an incident beam to suffer a periodic phase modulation. As a result, a set of diffracted patterns may be generated along different directions. As shown in Fig. 10.11, one of these diffracted beams travels in a direction parallel to the waveguide film. Although the coupling efficiency of this arrangement is low, it is rather easy to fabricate. A grating coupler could also be used to transfer the power of a forward traveling mode to that of a backward traveling mode in the same waveguide. Here, the grating etched on the waveguide structure causes a periodic phase modulation on the reflected wave, which adds in phase to create a backward propagating wave. Figure 10.9 depicts a similar situation of backward coupling.

Directional Coupler When two waveguides are laid on the same substrate side by side, as shown in Fig. 10.12, a coupling occurs between them. As a result, these two waveguides periodically exchange power between them as a function of distance. Under proper conditions, 100% coupling can take place between them if the two waveguides are identical. The degree of coupling is a function of the distance separating them and the interaction length, which in turn is a function of waveguide spacing and the velocity of light along the guides. By controlling the velocity using an electro-optic effect, one can modify the coupling efficiency between the two waveguides.

10.3.4 Active Devices

Active devices are necessary for the generation, modulation, and gain of optical signals. Although the total power may be very low, the presence of a high power concentration in a very small volume is able to cause efficient device operation. Usually, gas lasers occupy a huge volume and, depending on the power, may require a cooling subsystem. Also, modulators need extra electronics for control and interface circuitry. Therefore reducing the size of the lasers and modulators

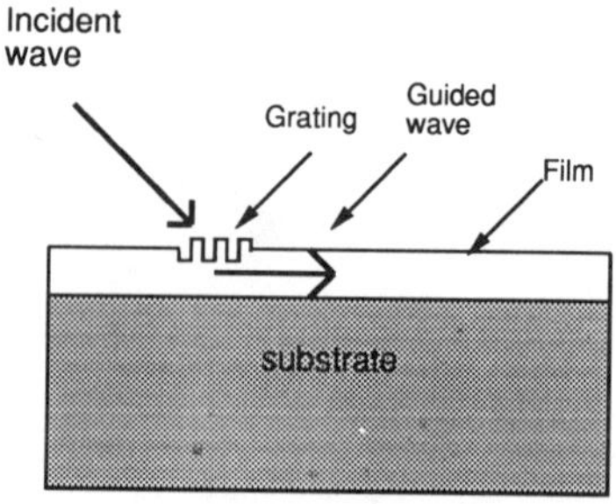

FIGURE 10.11 Grating coupler.

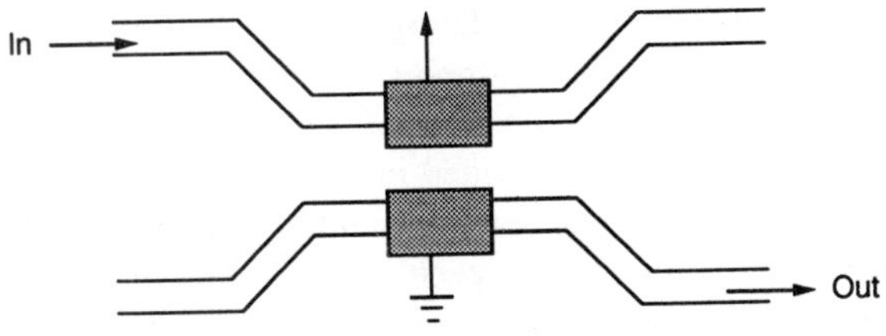

FIGURE 10.12 Directional coupler.

will have the maximum impact on the size reduction of the overall optical computing system.

Lasers Solid state laser sources may be fabricated inside an integrated device. Injection lasers, for example, may be deposited on a waveguide structure by forming a p-n junction of two dissimilar materials (Fig. 10.13). The laser action is produced by the injection of current to the forward biased junction and consequent population inversion. The wavelength of the laser radiation is a function of the bandgap of the lasing material. The necessary feedback may be provided by polishing both surfaces perpendicular to the junction or etched mirrors or distributed reflectivity of material so that the combination may act as a distributed feedback laser. The small transmission coefficient of the cavity mirrors allows the laser power to escape from the cavity to other circuit components.

Modulators For optical processing, it is necessary to transfer information from an external source to the laser frequency by means of modulation. Light may be modulated by modulating the power supplied to the source or the output of a source by means of an external modulator placed outside the source. The external modulators may utilize either electro-optic or magneto-optic effects for providing the required modulation.

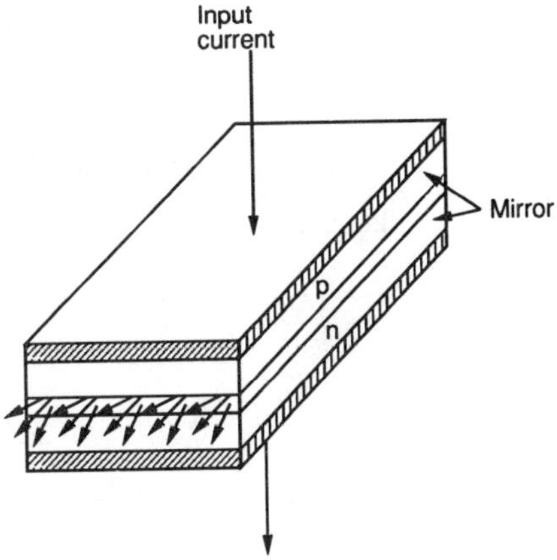

FIGURE 10.13 An integrated laser source.

There are several options for modulating light. Direct amplitude modulation by any optical effect usually requires high power. Alternatively, one can make use of phase modulation coupled with interference to realize amplitude modulation. In such a setup, a waveguide is first divided into two sections and then in one of the sections a voltage is applied to cause a predetermined amount of phase shift. When two waves shifted by 180° relative to each other are combined by recombining the sectionalized waveguides, a destructive interference may occur, resulting in no output. On the other hand, constructive interference occurs when light in neither section is modulated, so that they both may mix in phase. Such an arrangement can be suitably used for a binary type of modulation. Another approach for modulation is by controlling the leakage of light into the substrate, thereby modulating the light passing through the waveguide itself.

Surface acousto-optic waves (SAWs) can be used to create a grating on the waveguide surface. The SAW is generated by a comb-like interdigital transducer deposited on the surface of piezoelectric material by an alternating electricity. As the light travels, it is diffracted by the grating. Here the modulation is synonymous with diffraction or altering the direction of light propagation. Diffraction causes the light to be binary modulated as a portion of the diffracted light beam due to its angular orientation may or may not appear at a particular location. An acousto-optic cell is shown as a modulator in Fig. 10.14.

Switching Devices Optical bistable devices are capable of performing switching, which is necessary for logical operation. Again, to create high-density optical logic circuits, they must be fabricated in integrated form. It may be noted that to obtain a nonlinear behavior, a large amount of power is necessary. Since the nonlinear coefficient is too small, a large volume of nonlinear material will be necessary. Therefore the length of the device when compared to its width should be significantly large. As stated earlier, bistability requires a feedback, which could be optical or electronic. Also, for a logic operation, different inputs are expected to interact among each other; the interaction is later translated to a particular form of logical expression. Hybrid type of switching devices use feedback while interac-

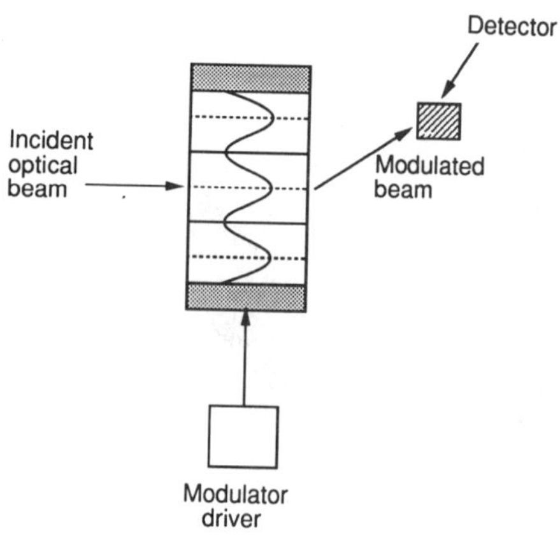

FIGURE 10.14 An acousto-optic binary modulator.

tions between different inputs are used in bidirectional couplers and interferometric switches.

Interferometric Switch The working of an interferometric switch is similar to the realization of amplitude modulation by phase modulation. By splitting a single beam into two beams using a divided waveguide structure, one may build an interferometric switch. Under normal circumstances, the two path lengths can be made equal, and hence consequent constructive interference will result in a high output. By supplying a modulation voltage to the modulator of one of the branches, a change in optical path length can be introduced such that one of the waves is delayed by a π phase resulting in a destructive interference. Consequently, using voltage controlled coupling, the device could be made to act as a logical gate.

The particular waveguide arrangement as illustrated in Fig. 10.15 can be worked as an X-OR gate. One of the branches has two control variables, while the other has only one, and each control variable is capable of producing a phase shift of π. Consider that V produces a phase shift bias of π constantly in the lower branch. Consequently, when either V_1 or V_2 is turned on, the output power is present. But if both V_1 and V_2 are turned on or off, the output is totally diminished. Thus the device behaves like an X-OR logic. When the bias voltage V is turned off, the device behaves as an X-NOR gate.

Hybrid Devices Integrated optical bistable devices may consist of a waveguide constructed of a nonlinear material exhibiting an electro-optic nonlinear effect. The detected output is fed back using an electronic detector. Then the feedback signal is applied to the nonlinear medium by means of a metallic electrode also integrated on the same substrate. A possible setup of such hybrid devices is shown in Fig. 10.16.

Directional Coupler Switch By controlling the coupling between two waveguides, switches may be fabricated (Fig. 10.17). An electrically controlled coupler can be used to control the coupling efficiency. By varying the coupling efficiency, such devices can be used to act as logic devices. Electrical feedback may also be

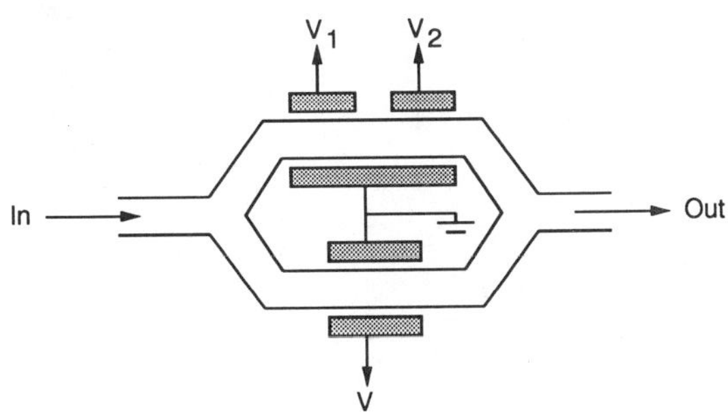

FIGURE 10.15 An interferometric X-OR gate.

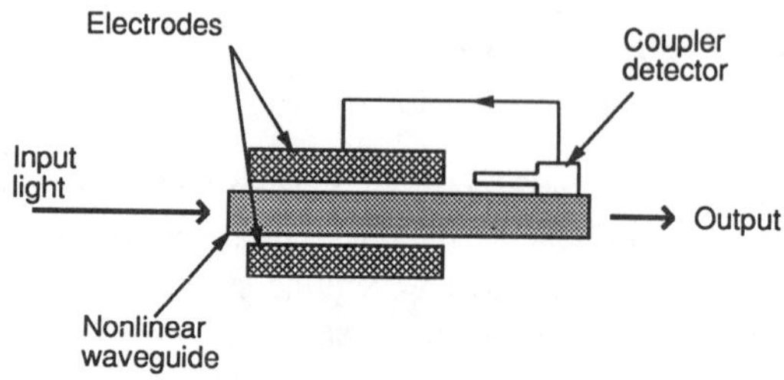

FIGURE 10.16 Hybrid bistable device.

introduced by means of a detector detecting the output light from one of the branches. Accordingly, it can exhibit bistability due to the feedback.

10.3.5 Analog Integrated Devices

Analog optical computing processes generally require a large optical setup. However, integrated optics may provide a tolerable alternative. Exploration of such possibilities may lead to very powerful portable devices, which will not require any accurate alignment and thus will eliminate the need for vibration isolation systems. Analog optical processing frequently utilizes Fourier transform lenses and modulators. Again, by suitably combining lenses and modulators, one may realize efficient spectrum analyzers.

Lenses A lens is the most common processing element needed in analog optical computing. Usually it is fabricated using a higher index material inside the lower index substrate. The effective refractive index depends on the thickness of the substrate. Note that an ordinary lens is created by a material of higher refractive index by giving a specific shape to it. Making a two-dimensional cross-section similar to a lens of another substrate material, one can fabricate an integrated optics lens. The refractive index could also be varied gradually by varying the thickness of the guiding layer to obtain an aberration-free lens. By creating indentations in

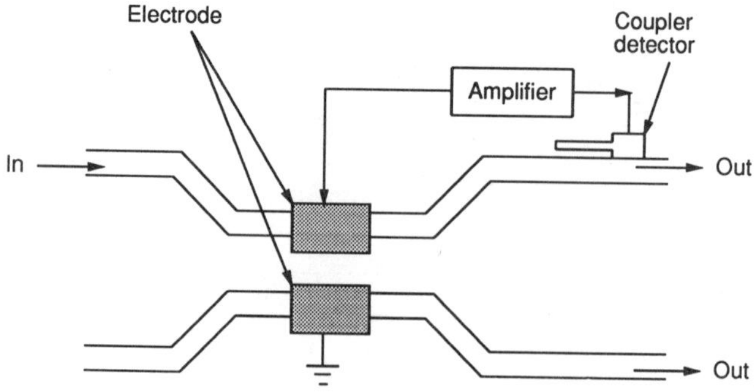

FIGURE 10.17 A bidirectional coupler switch.

the substrate under a constant thickness waveguide film, one may introduce varying path length, which in turn will result in a lens action. Gratings similar to a Fresnel lens can also be deposited or etched. Binary approximation of the one-dimensional quadratic phase distribution may be used to create binary phase pads. In a modified Fresnel lens, the phase indentations or protrusions may be tilted inward such that they act as tiny mirrors to force all the specular reflection to converge to a focus. All these lenses are shown in Fig. 10.18.

Spatial Light Modulators Spatial light modulators (SLMs) are important for conveying information to the integrated devices. The difference between a SLM and an ordinary modulator is that in a SLM two-dimensional modulation of a cross-section of the beam takes place. In a normal modulator, only the sequential type of bit strings are modulated. Thus a SLM is a rather parallel type of device. Both one-dimensional and two-dimensional integrated SLMs are now becoming increasingly popular. The acousto-optic SLMs, in particular, may be implemented using integrated optics. In a spatial light modulator, voltage is used to cause a deflection of the beam while the absence of voltage implies no deflection. The waveform is spatially modulated by the modulating voltage in the direction of deflection, as

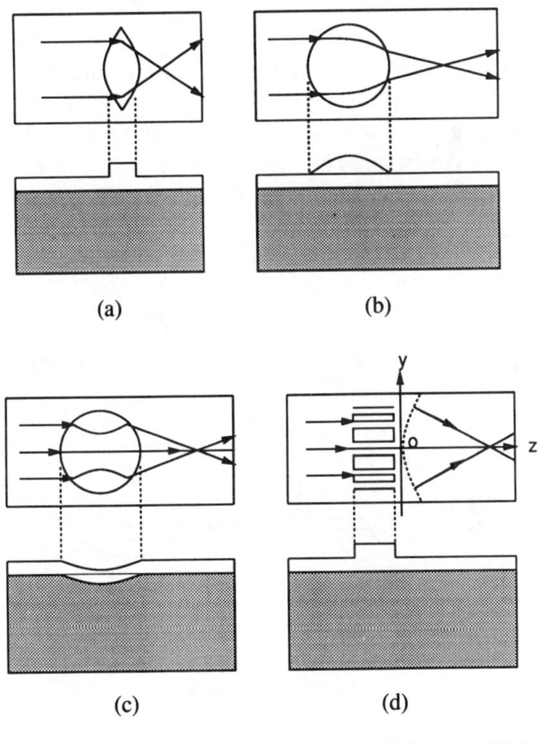

FIGURE 10.18 Integrated optics lens: (a) lens shaped dielectric, (b) Luneburg lens, (c) geodesic lens, and (d) grating lens.

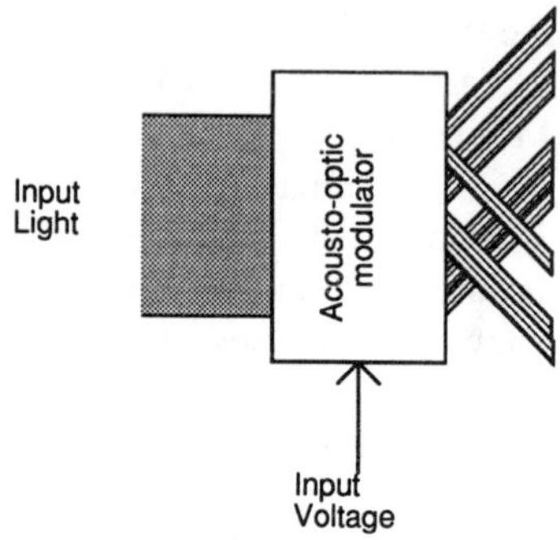

FIGURE 10.19 An acousto-optic modulator as a SLM.

shown in Fig. 10.19. An integrated circuit coupled with liquid crystals results in cost-effective SLMs. Specific types of SLMs and the principles involved in their operation are described in detail in Section 10.5.

Spectrum Analyzers Radio-frequency spectrum analyzers are a natural choice for optics since it is very easy to perform a Fourier transform of any input information by means of a simple lens. However, before applying the lens, the input must be transformed into an optical signal via modulation. As shown in Fig. 10.20, lens L_1 is used to collimate a diode laser beam. The SAW modulated by the signal voltage introduces a periodic strain and as a result modulates spatially the refractive index of the guiding layer. The modulated strain wave is then Fourier transformed by another integrated lens and then different spectral components may be gathered on arrays of detectors. The higher the frequency of the modulating voltage, the more deflection; hence the higher frequency component is recorded by the detector arrays.

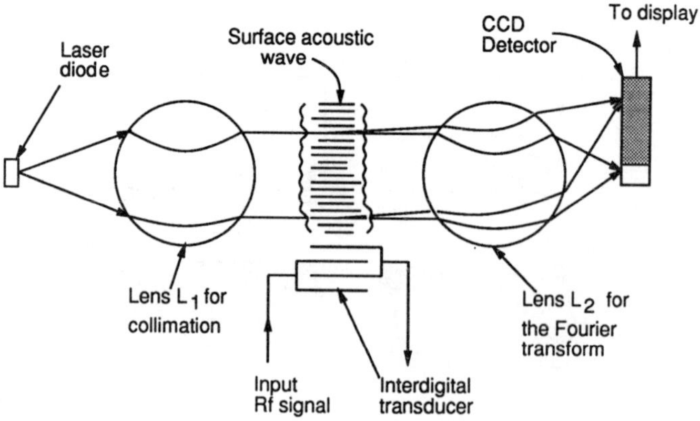

FIGURE 10.20 Integrated optic spectrum analyzers.

10.4 THRESHOLD DEVICES

All logic gates are a subset of more general threshold logic gates described in Section 7.9. In general, a threshold logic gate of n inputs has weights w_n associated with it. The weighted sum of the inputs is compared to a certain threshold. If the sum is more than the threshold, the output in HIGH, and if it is less than the threshold, the output is LOW. A threshold logic gate can be used to implement complex logic functions. A single threshold gate can also perform two-class decision-making. A bistable or multistable device can be used to perform threshold logic.

Optics has already proven to have superior analog capability; thus multiplication, free space communication, interconnection, and integration capability can be used in combination to perform weighted summation. It can be shown that a set of threshold logic operation can be decomposed as a vector matrix multiplication, and thus, a free space vector matrix multiplier can be utilized to implement a threshold logic device, as shown in Fig. 10.21. One of the major problems in the threshold logic is that the findings of proper weights to execute a logic operation may take rather a long time. However, threshold logic devices use a reduced number of gates and less interconnections than the conventional Boolean logic devices.

10.5 SPATIAL LIGHT MODULATORS

Most of the previously discussed optical computing devices are either logical or interconnection type. Optics is expected to overwhelm other computing techniques due to its inherent two-dimensional parallel data processing capability. When data are processed in a two-dimensional format, optics can outperform electronics in the battle of fast computing. The single most important optical element that would enable two-dimensional data processing is the spatial light modulator. A SLM can be used as an input device that can load serial data in a two-dimensional form to an optical system. Technically, a SLM is a device that is capable of spatially modulating a two-dimensional optical signal. As a result, the light wave itself becomes the carrier of information. The SLM can be referred to as a controlled transparency. In optical computing systems, a SLM could be used as an input

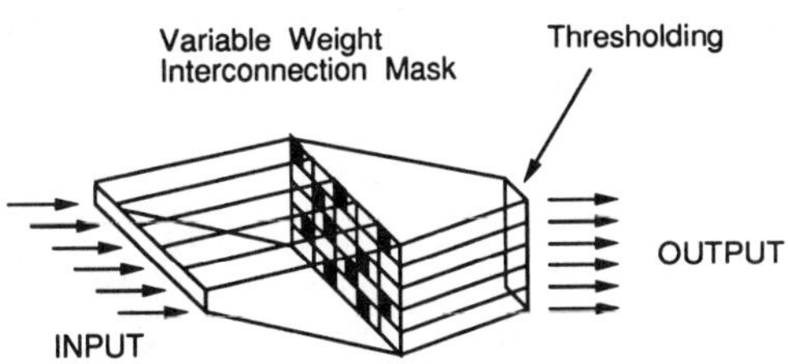

FIGURE 10.21 A possible optical threshold logic setup.

device, a holographic filter, an optical mask, a serial to parallel data converter for storage, and a detector.

As an input device, the SLM can accept two types of information: electronic and optical. Based on the source of input, the SLMs are divided into optically addressed and electronically addressed types. The optically addressed SLM can take the input either from an incoherent source or a coherent one. In the former case it is referred to as an incoherent-to-coherent converter. If the input device can be made to operate in real-time, real-time optical computing will become a reality. The most primitive form of SLM as input device, however, is a photographic film. No real-time SLMs have been capable of beating the performance of a film in terms of its spatial resolution. However, they are seriously handicapped because of their large processing time requirement. With a real-time SLM, one may expect tremendous speed enhancement.

Historically, Fourier domain filters such as those described in Chapter 3 through 6 are made of high-resolution films. With advancement of SLM technology, it is now becoming more possible to encode complex two-dimensional filters using SLMs. For example, electronically addressable magneto-optic spatial light modulators are used rather frequently to encode binary as well as trinary phase-only filters. Thus the real-time matching operation of the type described in Chapter 6 can be done using filters fabricated by SLMs.

SLMs could also be used as a spatial mask for selectively attenuating or allowing certain spatial frequencies in the Fourier plane. Such a possibility can be used in edge enhancement by deemphasizing the lower frequency component of an input image. Serial electronic data can be converted into two-dimensional data, which could then be processed or could be stored two dimensionally.

10.5.1 Electronically Addressed SLM

An electronically addressed SLM consists of a two-dimensional array of elements, which are individually addressed by an electronic interface. Two-dimensional electronic information can be imprinted on the light beam by using modulation via the opto-electronic birefringence effect. This is accomplished by using liquid crystals (LCs). LCs are anisotropic crystals unique in that they can rotate polarization direction of an incoming light. The rod-shaped LC molecules have a natural tendency to align in the same direction, as shown in Fig. 10.22. As a result, it exhibits two important effects: an electric field effect and birefringence. Due to birefringence, when light passes through a LC material, a phase shift between the two

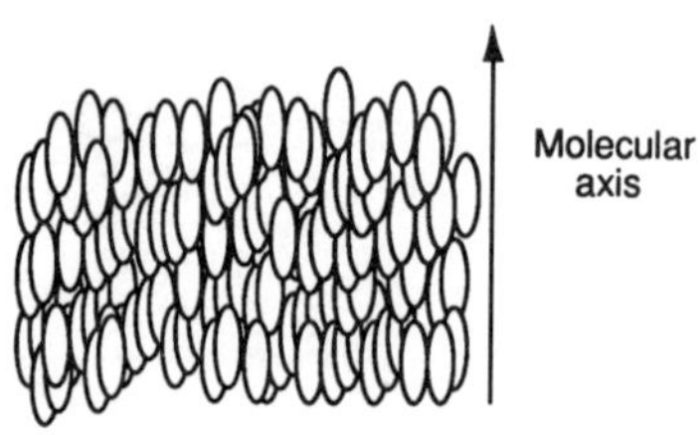

FIGURE 10.22 Liquid crystal.

beams parallel and perpendicular to the molecular axis takes place. This phase shift is a function of the thickness and the difference of refractive index of the parallel and perpendicular beams. The difference of the refractive index, however, is controllable by an applied voltage. Thus by changing the applied voltage one can change the polarization of the transmitted light and then use it along with a combination of two of the polarizers for spatial light modulation.

In a typical SLM structure, shown in Fig. 10.23, the LC cell sandwiched between two transparent electrodes is designed to act like a quarter-wave plate. Usually, the incident light falls on the input surface at an angle of 45° with the axis of the molecules. Then in the absence of any applied voltage, the output light suffers a 45° rotation of polarization. When it is reflected by a mirror and passes through the same LC plate again, it suffers further rotation, resulting in a total of 90° phase shift compared to the original beam. This reflected beam after passing through the LC a second time is blocked by the polarizer. When a voltage is applied across the LC plate, the phase shift is reduced due to reduction in birefringence. Thus the resulting light beam is no longer blocked completely. By changing voltage, the transmission characteristics of the SLM can be changed. Gray level modulation, in particular, can be realized by selecting intermediate operating voltages. In fact, LC televisions work on the same principle.

Silicon/VLSI ferroelectric liquid crystal SLMs use the silicon CMOS technology to produce low cost SLMs. The corresponding SLM consists of a silicon–

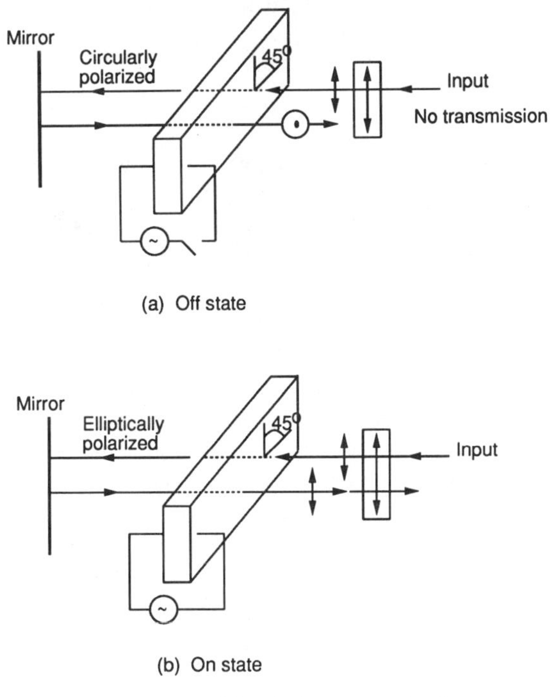

FIGURE 10.23 A liquid crystal cell.

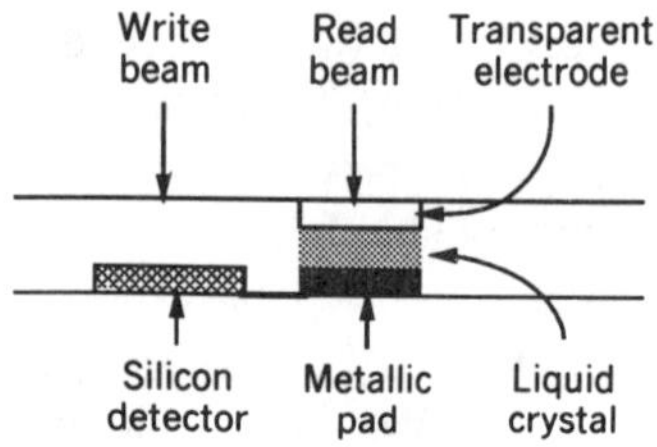

FIGURE 10.24 A typical VLSI/LC cell operation.

CMOS backplane and ferroelectric liquid crystal (FLC) in direct contact with the detector silicon chip. A FLC cell is created by a transparent conductor, which is realized by depositing indium tin oxide on the inner side of a glass plate, while the bottom of the cell consists of a metallic pad as shown in Fig. 10.24. Using one-dimensional register addressing or two-dimensional static random access memory addressing, the FLC cells can be modulated by means of metal pads and the transparent conductor. Using silicon junction phototransistors as detectors and silicon inverters, such a device can also be made to perform a two-dimensional threshold logic operation. Alternately, an optically addressed SLM consists of a CMOS backplane containing a phototransistor detector, analog/digital processors, and metal pads to modulate a liquid crystal material. The photodetector serves as the input plane while the FLC serves as the output plane. By incorporating specialized digital/analog processors on the CMOS plane, specific computations on the input light intensity can be performed on the two-dimensional optical input. Binarization or threshold logic operation or logarithmic processing can then be used on an incident image by specific design of the electronics.

10.5.2 Magneto-optic SLM

A magneto-optic SLM consists of electronically controllable two-dimensional pixel arrays. Each cell is composed of a magnetic element whose magnetization direction can be controlled by means of an electric current, as shown in Fig. 10.25. As polarized light passes through each cell, its polarization direction is changed with the magnetization orientation of the pixel. The switching of the magnetic field takes place in two steps. It is initiated by the applied current through the junction of two conductors situated near the pixel under consideration while the remaining part of the switching action takes place by pushing a domain wall that separates the two magnetization states. A binary modulation may be attained by selecting the polarization of the magnetic field such that one of the modulator cells produces a polarization rotation by 45° while the other is rotated by −45°. By placing a polarizer with an appropriate direction configuration, one of these two beams could be blocked, while the other could be allowed to pass. Consequently, the outgoing light is binary modulated.

10.5.3 Optically Addressed SLM

Liquid crystal light valves are made to act as optically addressed SLMs capable of imprinting a two-dimensional pattern on a beam of light in real-time. A LC light

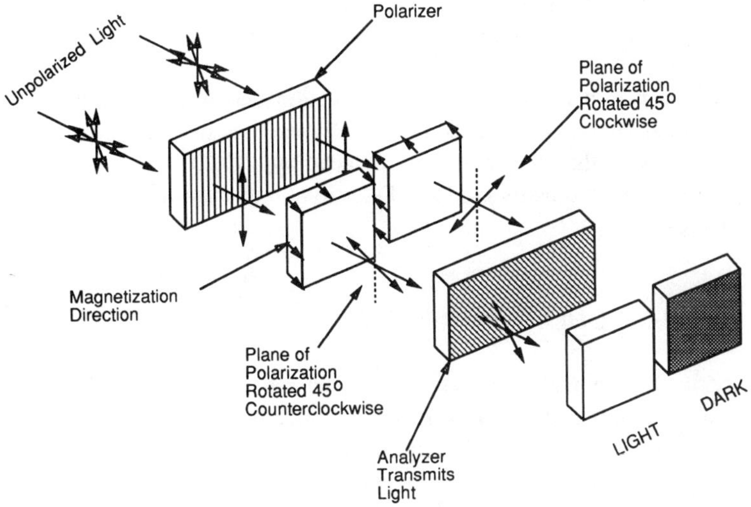

FIGURE 10.25 Operation of a magneto-optic SLM.

valve consists of a read and write side, as shown in Fig. 10.26. The principal component of such a SLM is a photoconductor composed of a liquid crystal material that is used to transfer the optical input to an equivalent voltage on the LC layer. The photoconductor has a very high resistivity, and it utilizes almost all the voltages when there is no light. A negligible amount of voltage may be applied to the LC layer when there is no light to limit the transmission of the read-out light. In the presence of incident light, the resistivity decreases; as a result, a voltage variation that directly corresponds to the input beam pattern is transferred onto the

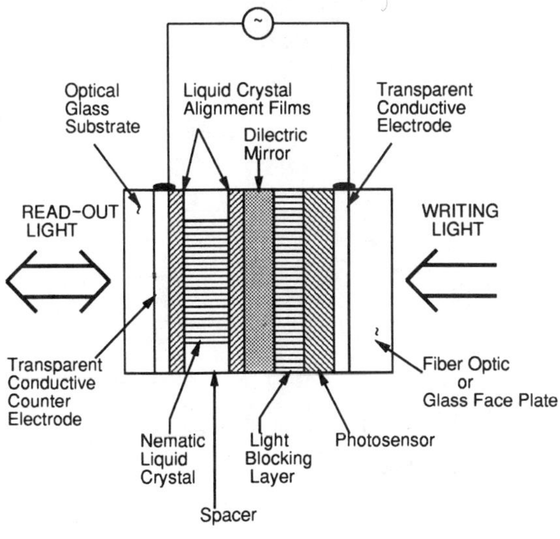

FIGURE 10.26 Cross-section of a LCLV.

LC layer. A coherent read-out beam is then reflected back but modulated by the birefringent LC layer. Consequently, the input optical image is transferred to the read-out beam.

A dielectric mirror is used to provide isolation between the input and output beam. For an ac-driven LCLV, CdS is usually used as the photoconductor. In a more recent version the input has been replaced by a CCD array. Accordingly, it has high speed, high sensitivity, less distortion, high resolution, and low noise characteristics.

10.6 THETA MODULATION DEVICES

Analog optics has demonstrated powerful transform capabilities. With this in the background, if an intensity radiation could be converted into a frequency variation, then linear Fourier optics could be exploited for nonlinear logic operations. Such a desirable conversion can be achieved by a variable grating mode (VGM) device, which can convert an intensity to a local variation of spatial frequencies over a two-dimensional plane.

The heart of a VGM device is a thin layer of liquid crystal that produces periodic intensity variation stripes when a voltage is applied to the layer. This results in a phase grating characterized by the spatial frequency, which is again a function of the applied voltage across the LC layer. A typical VGM effect is shown in Fig. 10.27a. Now by placing a photoconductor is series with the LC structure, as in Fig. 10.27b, it can work as an image transducer by creating a voltage proportional

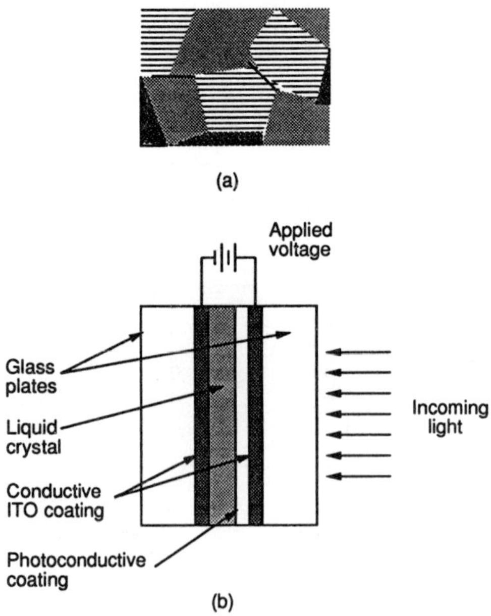

FIGURE 10.27 (a) A typical VGM effect. (b) A cross-section of a VGM device.

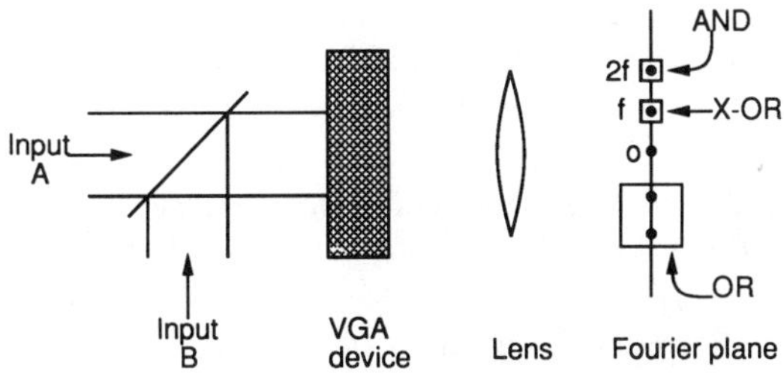

FIGURE 10.28 A typical setup for theta modulation optical computing.

to the image intensity. The photoconductor is sandwiched between two transparent electrodes. A constant dc bias voltage is applied to the electrode. When no input light is present, most of the voltage drops across the thin photoconductive layer. As a result, the voltage across the LC layer turns out to be insufficient to induce a VGM effect. As light impinges on the photoconductor, its resistance decreases; hence more voltage appears across the LC layer, which initiates the VGM effect. Thus the intensity variation is converted into a proportional local phase grating. A high-resolution image is observed due to the isolation of individual constant gray level regions by means of the high lateral impedance photoconductive layer.

Nonlinear encoding of data values by means of a grating nonlinear logic operation may be achieved using VGM devices. For example, when the intensity of one incoherent input is on the write side of the VGM, its Fourier transform f passes through a distance s_1 from the zero-order frequency. Now if two beams of equal intensity are superimposed on the input side, then the spatial frequency will become $2f$. Then by placing a spatial filter in the Fourier plane, various space variant logic processing can be performed. For example, an aperture just at a distance s_1 will produce an X-OR output. Figure 10.28 shows the setup for several space variant logic operations.

10.7 CONCLUSION

As this book is being written more devices are being invented. The unrealistic price of many such devices has retarded the growth of the parallel optical computing area. Yet most of the computing devices built so far are rather primitive in comparison to what is expected of the ultimate computer. Even so, such devices have shown remarkable performance. Most of the optical computing algorithms and architecture will ultimately have to be tested by the devices that will implement them. Legitimate breakthroughs in this area will be necessary to pave the way for the optical supercomputer.

10.8 SUGGESTED READINGS

Abraham, E., C. T., Seaton, and S. D. Smith, The Optical Computer, *Sci. Am.*, **248,** 85(1983).

Agrawal, G. P., and G. Flytzanis, Bistability and Hysteresis in Phase-conjugated Reflectivity, *IEEE J. Quant. Electron.*, **QE-17,** 374(1981).

Armitage, J. D., and A. W. Lohmann, Theta Modulation in Optics, *Appl. Opt.*, **4,** 399(1965).

Athale, R. A., and S. H. Lee, Development of an Optical Parallel Logic Device and a Half-Adder Circuit for Digital Optical Processing, *Opt. Eng.*, **18**(5), 513(1979).

Austin, J. W., and L. G. Deshazer, Optical Characteristics of a Saturable Absorber Inside a Fabry–Perot Interferometer, *J. Opt. Soc. Am.*, **61,** 650(1971).

Awwal, A. A. S., and M. A. Karim, Low Frequency Nonlinear Fabry–Perot Resonator, *Microwave Opt. Technol. Lett.*, **2**(3), 88–91(1989).

Bartelt, H. U., A. W. Lohmann, and E. E. Sicre, Optical Logical Processing in Parallel with Theta Modulation, *J. Opt. Soc. Am.*, **1,** 944(1984).

Becker, A., and W. S. C. Chang, Electrooptical Switching in Thin Film Waveguides for a Computer Communication Bus, *Appl. Opt.*, **18,** 3296(1979).

Bowden, C. M. (Ed.), *Optical Bistability 1*, Plenum Press, New York, 1981.

Bowden, C. M. (Ed.), *Optical Bistability 2*, Plenum Press, New York, 1983.

Brenner, K. H., 3D-Integration of Digital Optical Systems, *Opt. Comput. Tech. Digest Ser.*, **6,** 25–28(1991).

Casasent, D., Spatial Light Modulator, *Proc. IEEE*, **65,** 143(1977).

Cathey, W. T., and R. A. Schmidt, Low Loss Polarization-Based Optical Logic Gates, *SPIE Proc.*, **813,** 5(1987).

Chavel, P., A. A. Sawchuk, T. C. Strand, and A. R. Tanguay, Jr., Optical Logic with Variable-Grating-Mode Liquid-Crystal Devices, *Opt. Lett.*, **5,** 398–400(1980).

Chen, Y. C., and J. M. Liu, Polarization Bistability in Semiconductor Lasers, *Appl. Phys. Lett.*, **46,** 16(1985).

Choudhury, A. H., F. Wood-Jezercak, B. J. Ackerson, M. A. Karim, and A. A. S. Awwal, All Optical Logic Gates Using Crossed Laser Beams in Colloids, *Microwave Opt. Technol. Lett.*, **1**(5), 175(1988).

Clark, N. A., and S. T. Lagerwall, Physics of Ferroelectric Fluids: The Discovery of a High Speed Electro-Optic Switching Process in Liquid Crystals, *Recent Dev. Condensed Matter Phys.* **4,** 309–319(1981).

Cotter, L. K., T. J. Drabik, R. J. Dillon, and M. A. Handschy, Ferroelectric-Liquid-Crystal/Silicon-Integrated Circuit Spatial Light Modulator, *Opt. Lett.*, **15,** 291–293(1990).

Cuykendall, R., Three-Port Reversible Logic, *Appl. Opt.*, **27,** 1772(1988).

Drabik, T. J., and M. A. Handschy, Silicon VLSI Liquid Crystal Technology for Micropower Optoelectronic Computing Devices, *Appl. Opt.*, **29,** 5220–5223(1990).

Fatehi, M. T., K. C. Wasmundt, and S. A. Collins, Jr., Optical Flip-Flops and Sequential Logic Circuits Using a Liquid Crystal Light Valve, *Appl. Opt.*, **23,** 2163(1984).

Feitelson, D. G., *Optical Computing: A Survey for Computer Scientists*, MIT Press, Cambridge, MA, 1988.

Fukushima, S., and T. Kurokawa, Real-Time Image Processor with Stacked Liquid-Crystal Panels Based on Array Logic, *Opt. Lett.*, **14,** 157(1989).

Garmire, E., S. D. Allen, J. H. Marburger, and C. M. Verber, Multimode Integrated Optical Bistable Switch, *Opt. Lett.*, **3,** 691(1978).

Gibbs, H. M., S. L. McCall, and T. N. C. Venkatesan, Differential Gain and Bistability Using a Sodium Filled Fabry–Perot Interferometer, *Opt. Phys. Rev. Lett.*, 1135(1976).

Gibbs, H. M., S. L. McCall, T. N. C. Venkatesan, A. C. Gossard, A. Passner, and W. Weigmann, Optical Bistability in Semiconductors, *Appl. Phys. Lett.*, **35,** 451(1979).

Gibbs, H. M., S. L. McCall, and T. N. C. Venkatesan, Optical Bistable Devices: The Basic Component of All Optical Systems?, *Opt. Eng.*, **19,** 463(1980).

Gomes, C. M., S. Tsujikawa, H. Maeda, H. Sekine, T. Yamazaki, M. Sakamoto, F. Okumura, and S. Kobayashi, Gray-Scale Controllable Ferroelectric Liquid Crystal Spatial Light Modulator, *Opt. Comput. Tech. Digest Ser.* **6,** 72–75(1991).

Gravesteijn, D. J., H. M. Tongeren, M. Sens, T. Bertens, and C. J. van der Poel, Phase-Change Optical Data Storage in GaSb, *Appl. Opt.*, **26,** 4772(1987).

Gu, C., and P. Yeh, Optical Thresholding and Max Operation, *Opt. Comput. Tech. Digest Ser.*, **6,** 68–71(1991).

Haug, H., R. Marz, and S. Schmitt, Dielectric Function for Semiconductors with a High Exciton Concentration, *Phys. Lett.*, **77A,** 287(1980).

IEEE Journal of Quantum Electronics, Vol. **QE-21,** Special issue on Optical Bistability, 1985.

Imai, Y., and Y. Ohtsuka, Optical Multiple-Output and Multiple-Valued Logic Operation Based on Fringe Shifting Techniques Using a Spatial Light Modulator, *Appl. Opt.*, **26,** 274(1987).

Jager, D., W. Niessen, F. Forsmann, H. Thienpont, and R. V. Geen, Optical Bistability and Switching in Nonresonant GaAs:Cr Self-Electrooptic Effect Device, *Appl. Opt.*, **27,** 1769(1988).

Janossy, I., M. R. Tagizadeh, J. G. H. Mathew, and S. D. Smith, Thermally Induced Optical Bistability in Thin Film Devices, *IEEE J. Quantum Electron.*, **QE-21,** 1447–1452(1985).

Jared, D. A., R. Turner, and K. M. Johnson, Design and Fabrication of VLSI Ferroelectric Liquid Crystal Spatial Light Modulators, *Opt. Comput. Tech. Digest Ser.*, **6,** 55–58(1991).

Jeon, H., and A. A. Sawchuk, Optical Crossbar Interconnections Using Variable Grating Mode Devices, *Appl. Opt.*, **26,** 261(1987).

Johnson, K. M., M. R. Surette, and J. Shamir, Optical Interconnection Network Using Polarization-Based Ferroelectric Liquid Crystal Gates, *Appl. Opt.*, **27,** 1727(1988).

Kaplan, A. E., Hysteresis Reflection and Refraction by a Nonlinear Boundary—A New Class of Effects in Nonlinear Optics, *JETP Lett.*, **24,** 114(1976).

Karim, M. A., *Electro-Optical Devices and Systems*, PWS-Kent Publishing, Boston, 1990.

Kastasl'skii, A. A., Optical and Electrical Effects in Circuits Containing a Fabry–Perot Resonator, *Sov. Phys. Semicond.*, **7,** 653(1973).

Khan, A. H., and U. R. Nejib, Optical Logic Gates Employing Liquid Crystal Optical Switches, *Appl. Opt.*, **26,** 270(1987).

Khoo, I. C., and Y. R. Shen, Liquid Crystals: Nonlinear Optical Properties and Processes, *Opt. Eng.*, **24,** 579(1985).

Lee, D. L., *Electro-magnetic Principles of Integrated Optics*, Wiley, New York, 1986.

Lentine, A. L., G. J. Przybylek, L. E. Smith, L. A. D'Asaro, and D. A. B. Miller, Integrated Array of Self Electro-Optic Effect Device Logic Gates, *Opt. Comput. Tech. Digest Ser.*, **6,** 6–9(1991).

Li, Y., G. Eichmann, R. Dorsinville, and R. R. Alfano, Parallel Digital and Symbolic Optical Computation via Optical Phase Conjugation, *Appl. Opt.*, **27,** 2025–2032(1987).

Liu, H. K., J. A. Davis, and R. Lilly, Optical-Data-Processing Properties of a Liquid-Crystal Television Spatial Light Modulator, *Opt. Lett.*, **10,** 635(1985).

Liu, J. M., and Y. C. Shen, Optical Flip-Flop, *Electron. Lett.*, **21,** 236(1985).

Lizuka, K., *Engineering Optics*, 2nd ed., Springer-Verlag, New York, 1986.

Lohmann, A. W., and J. Weigelt, Digital Optical Adder Based on Spatial Filtering, *Appl. Opt.*, **25,** 3047(1986).

Ma, J., Y. Taketomi, Y. Fainman, J. E. Ford, and S. H. Lee, Enhanced Photorefractive Effects with a DC Field and Moving Grating in GaP at 633 nm, *Opt. Comput. Tech. Digest Ser.*, **6,** 64–67(1991).

Mathey, P., G. Pauliat, J. C. Launay, and G. Roosen, Huge Optical Amplification by Applying Pulse Electric Fields to Photorefractive Crystals, *Opt. Comput. Tech. Digest Ser.*, **6,** 60–63(1991).

McAulay, A. D., *Optical Computing Architectures*, Wiley, New York, 1991.

McCall, S. L., H. M. Gibbs, G. G. Churchill, and T. N. C. Venkatesan, Optical Transistor and Bistability, *Bull. AM. Phys. Soc.*, **20,** 636(1975).

McCall, S. L., H. M. Gibbs, and T. N. C. Venkatesan, Optical Transistor and Bistability, *J. Opt. Soc. Am.*, **65,** 1184(1975).

McCall, S. L., H. M. Gibbs, and T. N. C. Venkatesan, U. S. Patent Nos. 4,012,699 (1975) and 4,121,167(1976).

McGraw, J., D. Rogovin, and W. Ho, Four-wave Mixing in Liquid Suspensions of Microparticles, *SPIE Nonlinear Opt. Applications*, **613,** 77(1986).

McManus, J. B., and R. S. Putnam, Construction of an Optical-Carry Adder, *Appl. Opt.*, **26,** 1557(1987).

Meyer-Arendt, J. R., *Introduction to Classical and Modern Optics*, 2nd ed., Prentice-Hall, Englewood Cliffs, NJ, 1984.

Miller, D. A. B., M. H. Mozolowski, A. Miller, and S. D. Smith, Nonlinear Optical Effects in InSb with a cw CO Laser, *Opt. Commun.*, **27,** 133(1978).

Miller, D. A. B., C. T. Seaton, M. E. Prise, and S. D. Smith, Band-gap Resonant Nonlinear Refraction in III-V Semiconductor, *Phys. Rev. Lett.*, **47,** 197(1981).

Ojima, M., A. C. Pirson, Y. H. Lee, J. F. Morhange, H. M. Gibbs, N. Peyghambarian, F. Y. Juang, P. K. Bhattacharaya, and D. A. Weinberger, Optical NOR Gate Using Diode Laser Sources, *Appl. Opt.*, **25,** 2311(1986).

Okuda, M., K. Onaka, and T. Sakaii, Transient Behavior of a Saturable Optical Resonator with Distributed Bragg Reflector, *Jpn. J. Appl. Phys.*, **17,** 2133(1978).

Perez, O., and M. A. Karim, An Efficient Implementation of Joint Fourier Transform Correlation Using a Modified LCTV, *Microwave Opt. Tech. Lett.*, **2,** 193–196(1989).

Reintjes, J. F., *Nonlinear Optical Parametric Processes in Liquids and Gases*, Academic Press, Orlando, FL, 1984.

Ross, W. E., D. Psaltis, and R. H. Anderson, Two-Dimensional Magneto-optics Spatial Light Modulator for Signal Processing, *SPIE Proc. Real-Time Signal Processing V*, **341,** 191(1982).

Roux, J. D., and F. W. Leuschner, Polarization-Based Optical Computing Using Liquid Crystals, in *Technical Digest Optical Computing*, p. 276, Topical Meeting, Salt Lake City, Utah, Feb. 1989.

Seaton, C. T., S. D. Smith, F. A. P. Tooley, M. E. Prise, and M. R. Tagizadeh, Realization of an InSb Bistable Device as an Optical AND Gate and Its Use to Measure Carrier Recombination Time, *Appl. Phys. Lett.*, **42**(2), 131(1983).

Shen, Y. R., *The Principles of Nonlinear Optics*, Wiley, New York, 1984.

Smith, P. W., I. P. Kaminov, P. J. Maloney, and L. W. Stulz, Self-contained Integrated Bistable Optical Devices, *Appl. Phys. Lett.*, **34,** 62(1979).

Smith, P. W., and E. H. Turner, A Bistable Fabry–Perot Resonator, *Appl. Phys. Lett.*, **30,** 280(1977).

Smith, P. W., E. H. Turner, and B. B. Mumford, Nonlinear Electro-optic Fabry–Perot Devices Using Reflected-Light Feedback, *Opt. Lett.*, **2,** 55(1978).

Smith, S. D. (Ed.), *Optical Bistability III*, Springer-Verlag, New York, 1986.

Smith, S. D., J. G. H. Mathew, M. R. Tagizadeh, A. C. Walker, B. S. Wherrett, and A. Hendry, Room Temperature, Visible Wavelength Optical Bistability in ZnSe Filters, *Opt. Commun.*, **51,** 357–362(1984).

Smith, S. D., A. C. Walker, B. S. Wherrett, F. A. P. Tooley, J. G. H. Mathew, M. R. Tagizadeh, and I. Janossy, Cascadable Digital Optical Logic Circuit Elements in the Visible and Infrared: Demonstration of Some First All-Optical Circuits, *Appl. Opt.*, **25,** 1586–1593(1986).

Soffer, H., G. J. Dunning, Y. Owechko, and E. Marom, Associative Holographic Memory with Feedback Using Phase Conjugate Mirrors, *Opt. Lett.*, **11,** 118(1986).

Suzuki, Y., J. Shimada, and H. Yamashita, High Speed Optical–Optical Logic Gate for Optical Computer, *Electron. Lett.*, **21,** 161–162(1985).

Szöke, A., V. Daneu, J. Goldhar, and N. A. Kurnit, Bistable Optical Elements and Its Application, *Appl. Phys. Lett.*, **15,** 376(1969).

Tomlinson, W. J., J. P. Gordan, P. W. Smith, and A. E. Kaplan, A Reflection of a Gaussian Beam at a Nonlinear Interface, *Appl. Opt.*, **21,** 2041(1982).

Tooley, F. A. P., N. C. Craft, S. D. Smith, and B. S. Wherrett, Experimental Realization of an All Optical Single-Gate Full Adder Circuit, *Opt. Commun.*, **63,** 365(1987).

Venkatesan, T. N. C., and S. L. McCall, Optical Bistability and Differential Gain Between 85 and 296 K in a Fabry–Perot Containing Ruby, *Appl. Phys. Lett.*, **30,** 282(1977).

Verdeyen, J. T., *Laser Electronics*, 2nd ed., Prentice-Hall, Englewood Cliffs, NJ, 1989.

Walker, N. P., and M. A. Fiddy, Efficient Optical Bistable Device Based on Self-Focusing, *Appl. Opt.*, **28,** 201(1989).

Warde, C., H. L. Rivera, and H. K. Liu, High-Gamma Spatial Light Modulator for Nonlinear Optical Processing,, *SPIE*, **613,** 255(1986).

Warde, C., A. M. Weiss, A. D. Fisher, and J. I. Thackara, Optical Information Processing Characteristics of the Microchannel Spatial Light Modulator, *Appl. Opt.*, **20,** 2066(1981).

Watrasiewicz, B. M., Optical Digital Computers, *Opt. Laser Technol.* **7,** 213(1975).

Winful, H. G., and J. H. Marburger, Hysteresis and Optical Bistability in Degenerate Four Wave Mixing, *Appl. Phys. Lett.*, **36,** 631(1980).

Yariv, A., *Optical Electronics*, 3rd ed., Holt, Rinehart and Winston, New York, 1985.

Yu, F. T. S., S. Jutamulia, and D. A. Gregory, Real Time Liquid Crystal TV XOR and XNOR Gate Binary Image Subtraction Technique, *Appl. Opt.*, **26,** 2738(1987).

CHAPTER ELEVEN

Shadow-Casting and Symbolic Substitution

11.1 INTRODUCTION

Conventional digital computations are plagued by the bottleneck of sequential processing. This arises due to limited interconnections between memory and processing elements and among different computing modules, interactions among electrons and consequent inability to support multichannel communication through a single bus, and inability to reduce the propagation delay of digital electronic devices via miniaturization. The parallel processing capabilities of optical signals are sought to overcome these shortcomings. Optical signals propagate in parallel, cross each other without interference (whereas charged electrons interact heavily with each other), have the lowest possible propagation delay of all signals, and can provide a million channel free-space interconnections with primitive lenses (whereas an unrealistic number of connections are necessary for electrons), and thus they are excellent candidates for implementation in parallel computing machines. The communication bottleneck forces electronic computers to update the computation state space sequentially, whereas optics allows the whole state to change in parallel.

The optical shadow-casting (OSC) system is considered to have remarkable promise for optical digital computing because of the relative ease with which it can exploit the parallelism offered by photons. In the lensless optical shadow-casting system, once the inputs are encoded, the computation speed is limited only by the speed with which light travels from the input plane to the output plane. The parallelism is evident from the fact that 10^6 uniprocessors can operate in an area of 1 square inch.

The OSC system originally proposed by Tanida and Ichioka uses a two-dimensional spatially encoded pattern as data input and light emitting diodes (LEDs) as light sources. In the original shadow-casting system, the inputs were represented by vertical and horizontal stripes of opaque and transparent bars (strictly speaking, this coding is nothing but dual-rail coding, as shall be seen later). With this system, programmability is attained by changing the LED pattern and it is possible to realize 16 logical operations between two variables as well as half addition as an example of an arithmetic operation. Later on, polarized codes were introduced into the shadow-casting system, which enabled complex combinational logic units like binary and trinary full adders to be realized; this could not have been achieved using an unpolarized encoding scheme. This improved technique, called the polarization-encoded optical shadow-casting (POSC) scheme, uses horizontally polarized (H) and vertically polarized (V) codes as additional design variables for encoding the input, in addition to transparent (T) and opaque (F), for the sources and output decoding mask, thereby increasing the flexibility of designing complex arithmetic logic units. Since OSC can be considered as a subset of the POSC, from now on, in this chapter, only the more general POSC is discussed. In addition to POSC, the latter half of this chapter introduces and discusses optical symbolic substitution (SS) schemes.

Considering that the truth table is a valid way of representing a digital logic, generation of the output may be looked on as substituting a set of outputs for each set of input combinations. The very powerful optical symbolic substitution extends the same concept to the optics domain, where the truth table entries may be represented by a combination of two-dimensional pixel patterns. In symbolic substitution a combination two-dimensional pixel pattern is identified from a two-dimensional signal input and another pattern is substituted in that location. Thus symbolic substitution consists of two phases: a recognition and a corresponding substitution phase. Since the recognition and substitution phases can be performed using optics in parallel, it can be used to update in a parallel fashion the computation states of a classical finite state machine.

11.2 SHADOW-CASTING SYSTEM AND DESIGN ALGORITHM

The lensless POSC system, as shown in Fig. 11.1, consists of spatially encoded two-dimensional binary pixel patterns as the inputs. The input patterns are kept in close contact at the input plane and the resulting input overlapped pixel pattern is illuminated by a set of light emitting diodes (LEDs) from the source plane. The light originating from each of the LEDs located at the source plane produces a shadow of the input overlap pixel pattern at the output plane. The overlap of these projected shadows results in an output overlap pixel pattern at the output plane. A decoding mask placed at the output plane is then used to spatially filter and detect the logical output. To prevent crosstalk, the required spacing among the different processing elements should be as follows:

$$d = \frac{ab}{b - a} \tag{11.1}$$

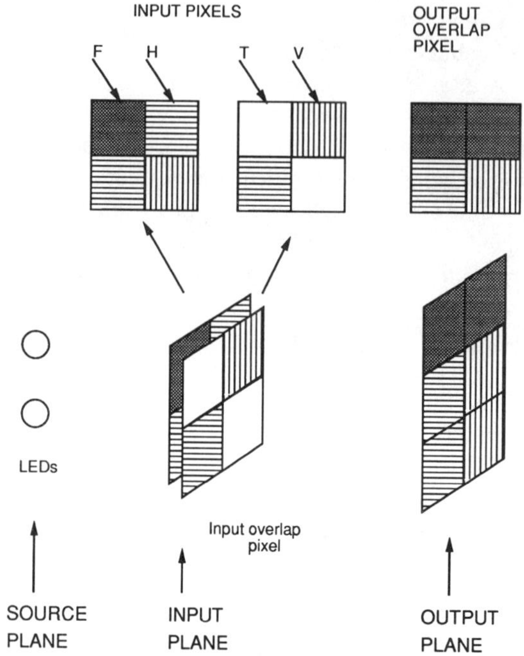

FIGURE 11.1 An optical shadow-casting setup.

such that

$$\frac{z_0}{z_1} = \frac{a}{b - a} \tag{11.2}$$

where a is the width of the pixel subcell, b is the width of the shadow of each pixel subcell, d is the LED spacing, z_0 is the distance between the source and the input overlap plane, and z_1 is the distance between input and output overlap planes. These parameters are shown in Fig. 11.2. Also, to obtain proper geometric shad-

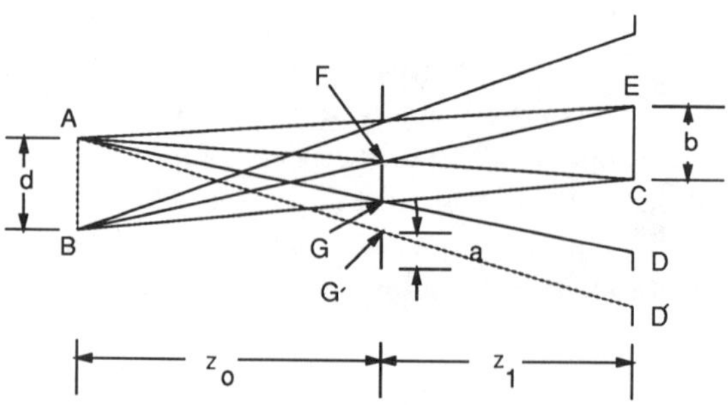

FIGURE 11.2 Distance parameters of an optical shadow-casting setup.

ows, the distance between the input and the observation plane must maintain the constraint

$$\lambda < z_1 \ll \frac{a^2}{\lambda} \tag{11.3}$$

In the OSC system, the only design variable that was considered was switching LED configuration. The idea was later extended to allow for changes in the output decoding mask and finally polarized codes for implementing a shadow-casting based full adder by changing the encoding of the LED source pattern and also the output decoding mask. Fortunately, a generalized POSC design algorithm for designing binary as well as trinary arithmetic/logic units has recently been developed. Consequently, a variety of arithmetic units, carry-free adder, programmable logic array, sequential logic element such as a J-K flip-flop, multiprocessor, and associative logic units have already been designed. The POSC design algorithm has the capability to implement several image processing operations such as median filtering and edge detection.

The POSC design algorithm enables one to determine the input encoding, source encoding, and output decoding mask. For the purpose of encoding, each input pixel is divided into a number of pixel subcells to represent a particular combination of the POSC code set {T, V, H, F} where $\overline{\mathrm{T}} = \mathrm{F}$ and $\overline{\mathrm{H}} = \mathrm{V}$. The POSC design methodology resembles a digital design technique. In a POSC system, the LEDs in the source plane project the minterms to produce a desired output and the overlap of the encoded input patterns is responsible for generating the required minterm. The POSC overlap equation represents the relationship of the overlapped inputs (minterm) placed in a pixel subcell with the corresponding code of the expected output bit.

Unlike the more elementary OSC, a POSC system is capable of being used for designing arithmetic/logic units with two or more outputs. These outputs may appear at one particular location or at several spatially different locations. To differentiate between two outputs appearing at the same location, however, one may choose orthogonal codes for the outputs. Since the number of orthogonal codes is limited to only two, multiple (more than two) outputs may be generated by directing the outputs to spatially different locations. The use of more than one orthogonal code for the same output enables one to increase the memory efficiency of a POSC system by storing multiple minterms in the same input pixel location. Moreover, noting that two different outputs may have one or more minterms common among themselves, more reduction in the memory size (i.e., input pixel size) can be attained by carefully placing the minterms in the memory map. This can be accomplished by having minterms shared between two or more outputs.

The following POSC algorithm is general enough to design any arbitrary sequential or combinatorial logic circuit.

Step 1 The truth is characterized. Multiple outputs resulting from a single processor or different uniprocessor outputs are dictated by its different output columns.

Step 2 The groups of minterms corresponding to a common output are first listed together, and then classified as sharing (minterms common to more than one output) and nonsharing types.

Step 3 The minterms are grouped in pairs so that the total number of minterm pairs is a minimum, starting with the sharing ones first. This can be accomplished in two ways. First, one can exhaustively search all possible pairing and then choose the minimum pair condition. A tree-structured search may systematically yield the desired pairing of minterms. The minimum number of pairs is obtained only when there is a maximum number of common pairs among different outputs. In the second method, the set of the input conditions corresponding to each output bit is divided into two subsets: (1) nonsharing, that is, input conditions that generate only that output bit, and (2) sharing, that is, input conditions that generate more than one output bit. Input conditions are paired first among the first kind and then among the second kind. The total number of pixel subcells required to encode the inputs is equal to the sum of the number of pairs and the number of singles.

Step 4 The output bits having the maximum number of input conditions are mapped first by placing the corresponding input conditions among the pixel subcells of the input overlap pixel in close proximity. It is desirable to place the nonsharing ones in the peripheral of the input overlap pixel and the sharing ones toward the interior so that they can be shared with those output bits that share the same input condition(s). The output bit having the next highest number of independent input conditions (i.e., of those that have not yet been considered) is then mapped in the input overlap pixel next to the already placed dependent (sharing) input conditions so that the resulting perimeter of the mapping block (a geometric enclosure occupied by the set of all input conditions for a particular output bit) is a minimum. In placing the minterms, the locations may be iteratively changed to find the best solution whereby the mapping block can have a minimum perimeter and a minimum number of sides. All input conditions are mapped likewise in the input overlap pixel. The placement of a particular group is also influenced by the fact that the OR result of a mapping block must have a unique location in the output overlap plane.

Step 5 The input coding pattern is obtained from the address map described in Step 4 by solving the pair of POSC overlap equations (corresponding to each minterm pair) when the uncomplemented literals are assigned T code and the complemented literals are assigned the code of the output.

Step 6 The source LED pattern for the output is determined next. The size of the LED pattern is equal to the size of the largest mapping block. If the outputs are generated sequentially, then the number of LEDs is usually equal to the number of the pixel subcells in the pixel. In that case, depending on the location of the minterm, LEDs are turned ON, and depending on the code of the output, they are either unpolarized or polarized.

Step 7 The space-variant output decoding mask is designed next so that all outputs may be decoded in parallel. Each of the mapping blocks represents an OR operation that is performed between its pixel subcells. The location of the output can be found by noting that the function of each LED is to produce a relatively shifted version of the input overlap pixel at the output plane. Consequently, the output at each pixel subcell in the output overlap pixel can be related to the input overlap pixel and the LED. By explicitly writing these relationships, one can obtain the location of the ORed output of different segments of the input overlap pixel. The pixel subcell at the output overlap pattern (for each mapping block) where the ORed output will appear is made transparent. Each of the transparent openings of the decoding mask therefore yields a particular output. If only one output is needed at a time, then the central pixel is chosen for the output; in this case, the outputs are chosen solely by switching the source LED pattern.

11.3 POSC LOGIC OPERATIONS

Based on the design algorithm summarized in Section 11.2, it is possible to design a POSC system capable of generating logical outputs. When two variables are overlapped, the corresponding truth table consists of only four input combination entries. The four minterms are given by 00, 01, 10, and 11. These four minterms $\overline{A}\overline{B}$, $\overline{A}B$, $A\overline{B}$, and AB are then placed in a 2×2 pixel structure at locations $\{11\}$, $\{12\}$, $\{21\}$, and $\{22\}$. Assigning an output of T for each minterm, one may write the POSC overlap equation as

$$\overline{a_{11}} \wedge \overline{b_{11}} = \mathrm{T} \tag{11.4}$$

where $\wedge$ indicates an overlap operation. Solution of Eq. (11.4) yields $a_{11} = F$ and $b_{11} = F$. By solving POSC equations for the other minterms, the POSC codes for the remaining pixel subcells may be obtained. The resulting codes for A and B thus become

$$A = \begin{bmatrix} \mathrm{F} & \mathrm{F} \\ \mathrm{T} & \mathrm{T} \end{bmatrix} \quad \text{and} \quad B = \begin{bmatrix} \mathrm{F} & \mathrm{T} \\ \mathrm{F} & \mathrm{T} \end{bmatrix}$$

For the decoding mask the central pixel subcell in the output plane is usually made transparent. The equation relating the central pixel subcell output to the input overlap data and LEDs (organized in a 2×2 format) is given by

$$L_{11}I_{11} + L_{12}I_{12} + L_{21}I_{21} + L_{22}I_{22} = O_{22} \tag{11.5}$$

where I_{ij} (O_{ij}) represents the minterm stored at the $\{i, j\}$ input (output) pixel subcell location. Note that the logical sum of all four minterms appears at the center of each 3×3 output pixel, which is the reason for choosing the central pixel subcell for output. By selecting 16 different combination of L_{ij}, one can perform up to 16 logical operations between two binary input variables.

11.4 POSC MULTIPROCESSOR

The POSC design algorithm can be used to design an arithmetic unit that is capable of simultaneously producing, for example, both full addition and full subtraction of binary numbers. Such a device may be referred to as a multiprocessor. The number of encoded inputs that are to be introduced in the corresponding POSC system input plane is three. The truth table for the full adder and subtracter is shown in Table 11.1, where A, B, and C are the three input bits, addend (minuend), augend (subtrahend), and carry-in (borrow-in) of the adder/subtracter unit; S and C', respectively, represent the sum and carry-out, while D and B represent, respectively, the difference and borrow-out.

In this design, all output bits can be generated simultaneously in parallel. First, all the output bits are listed along with the corresponding input conditions:

$$S\text{:}\quad \{2, 3, 5, 8\}$$

$$C'\text{:}\quad \{4, 6, 7, 8\}$$

$$D\text{:}\quad \{2, 3, 5, 8\}$$

$$B\text{:}\quad \{2, 3, 4, 8\}$$

It is obvious that the two outputs S and D are identical. Therefore the generation of only three output bits will suffice for this design. To find the optimum pairing for involving only a minimum memory, all possible pairs are formed from four input conditions taking two at a time. This provides a total of ${}^4C_2$ possible pairs. Thus the three bits $S(=D)$, C', and B with their pairs of input conditions can be listed as follows:

$$S\text{:}\quad \{(2, 3), (5, 8)\}, \{(2, 5), (3, 8)\}, \{(2, 8), (3, 5)\}$$

$$C'\text{:}\quad \{(4, 6), (7, 8)\}, \{(4, 7), (6, 8)\}, \{(4, 8), (6, 7)\}$$

$$B\text{:}\quad \{(2, 3), (4, 8)\}, \{(2, 4), (3, 8)\}, \{(2, 8), (3, 4)\}$$

Note that each of the output bits can be generated in three different ways. For generating all three outputs, one may combine with each group of S, one of three

TABLE 11.1 Truth Table for Adder/Subtracter

Input Conditions	Input			Output			
	A	B	C	S	D	C'	B
1	0	0	0	0	0	0	0
2	0	0	1	1	1	0	1
3	0	1	0	1	1	0	1
4	0	1	1	0	0	1	1
5	1	0	0	1	1	0	0
6	1	0	1	0	0	1	0
7	1	1	0	0	0	1	0
8	1	1	1	1	1	1	1

different groups of C' and one of three different groups of B. Thus the total number of such groupings that can be considered is 27. All these groupings are listed in Table 11.2. From the table it is obvious that only grouping number 7 involves four pairs of nonrepeating minterms, while the others are not that optimum. Thus four pairs of input combinations (2, 3), (5, 8), (4, 8), and (6, 7) may now be used for obtaining the address maps.

Alternatively, to find the proper minterm groupings, the input conditions of Table 11.1 can be divided into two groups: sharing (s) and nonsharing (ns). Accordingly, the outputs may be listed as follows:

S: s (2, 3, 8); ns (5)

C': s (4, 8); ns (6, 7)

B: s (2, 3, 4, 8);

TABLE 11.2 Possible Groupings for Adder/Subtracter Unit for Minimum Memory

Sets of Input Conditions	Common Pairs	Number of Nonrepeating Pairs
1. (2, 3) (5, 8) (4, 6) (7, 8) (2, 3) (4, 8)	(2, 3)	5
2. (2, 3) (5, 8) (4, 6) (7, 8) (2, 4) (3, 8)		6
3. (2, 3) (5, 8) (4, 6) (7, 8) (2, 8) (3, 4)		6
4. (2, 3) (5, 8) (4, 7) (6, 8) (2, 3) (4, 8)	(2, 3)	5
5. (2, 3) (5, 8) (4, 7) (6, 8) (2, 4) (3, 8)		6
6. (2, 3) (5, 8) (4, 7) (6, 8) (2, 8) (3, 4)		6
7. (2, 3) (5, 8) (4, 8) (6, 7) (2, 3) (4, 8)	(2, 3) (4, 8)	4
8. (2, 3) (5, 8) (4, 8) (6, 7) (2, 4) (3, 8)		6
9. (2, 3) (5, 8) (4, 8) (6, 7) (2, 8) (3, 4)		6
10. (2, 5) (3, 8) (4, 6) (7, 8) (2, 3) (4, 8)		6
11. (2, 5) (3, 8) (4, 6) (7, 8) (2, 4) (3, 8)	(3, 8)	5
12. (2, 5) (3, 8) (4, 6) (7, 8) (2, 8) (3, 4)		5
13. (2, 5) (3, 8) (4, 7) (6, 8) (2, 3) (4, 8)		6
14. (2, 5) (3, 8) (4, 7) (6, 8) (2, 4) (3, 8)	(3, 8)	5
15. (2, 5) (3, 8) (4, 7) (6, 8) (2, 8) (3, 4)		6
16. (2, 5) (3, 8) (4, 8) (6, 7) (2, 3) (4, 8)	(4, 8)	5
17. (2, 5) (3, 8) (4, 8) (6, 7) (2, 4) (3, 8)	(3, 8)	5
18. (2, 5) (3, 8) (4, 8) (6, 7) (2, 8) (3, 4)		6
19. (2, 8) (3, 5) (4, 6) (7, 8) (2, 3) (4, 8)		6
20. (2, 8) (3, 5) (4, 6) (7, 8) (2, 4) (3, 8)		6
21. (2, 8) (3, 5) (4, 6) (7, 8) (2, 8) (3, 4)	(2, 8)	5
22. (2, 8) (3, 5) (4, 7) (6, 8) (2, 3) (4, 8)		6
23. (2, 8) (3, 5) (4, 7) (6, 8) (2, 4) (3, 8)		6
24. (2, 8) (3, 5) (4, 7) (6, 8) (2, 8) (3, 4)	(2, 8)	5
25. (2, 8) (3, 5) (4, 8) (6, 7) (2, 3) (4, 8)	(4, 8)	5
26. (2, 8) (3, 5) (4, 8) (6, 7) (2, 4) (3, 8)		6
27. (2, 8) (3, 5) (4, 8) (6, 7) (2, 8) (3, 4)	(2, 8)	5

The input conditions 2 and 3 are present in both S and B while input condition 4 is present in both C' and B and input condition 8 is present in S, C', and B. Grouping is done by taking two input conditions at a time from the nonsharing ones. The only nonsharing pair that can be formed is (6, 7), belonging to the C' output. This leaves the pair (4, 8) to be considered for realizing C'. But since the pair (4, 8) is also common to B, we then need to consider the pair (2, 3) only to generate B. Again, since the pair (2, 3) is common to both S and B, the only pair left out to generate S is (5, 8). The pairs for S are thus (2, 3) and (5, 8), while that for C' are (4, 8) and (6, 7) and that for B are (2, 3) and (4, 8). The total number of groups is therefore four: (2, 3), (5, 8), (4, 8), and (6, 7). Accordingly, one could choose a 2×2 matrix for the input overlap pixel. The four pairs of input conditions are configured in the matrix such that the shape of the mapping block is regular (square or rectangular) and its perimeter is a minimum.

We first place (2, 3) and (5, 8) at the locations $\{1, 1\}$ and $\{1, 2\}$ of the input matrix, such that

$$S \text{ address map } = \begin{bmatrix} 2, 3 & 5, 8 \\ - & - \end{bmatrix}$$

Since the output B shares minterm (2, 3) with output S, the pair (4, 8) required to generate B is placed next to (2, 3) so that

$$B \text{ address map } = \begin{bmatrix} 2, 3 & - \\ 4, 8 & - \end{bmatrix}$$

Now for output C', there is only one pixel subcell left next to (4, 8), and placing (6, 7) in that location makes it easy to generate C' from the OR operation of the neighboring pixel subcells. The overall mapping block is thus given by

$$\text{overall mapping block } = \begin{bmatrix} 2, 3 & 5, 8 \\ 4, 8 & 6, 7 \end{bmatrix}$$

The maximum size of each mapping block is 2 in either the horizontal or vertical direction. Consequently, the output overlap pixel pattern can be generated by shifting the input overlap pattern once to the right and once downward.

Each output bit is obtained as a result of OR logic performed among the corresponding minterms (within the mapping block) shown in Fig. 11.3a by means of source LEDs. Consequently, the output bits $S(D)$, C', and B are obtained, respectively, at $\{1, 2\}$, $\{2, 1\}$, and $\{3, 2\}$ locations of the output overlap pixel. These particular locations can be determined from the projection equations of the output pixel subcells in terms of LED source, L, and input overlap pixel, I. One

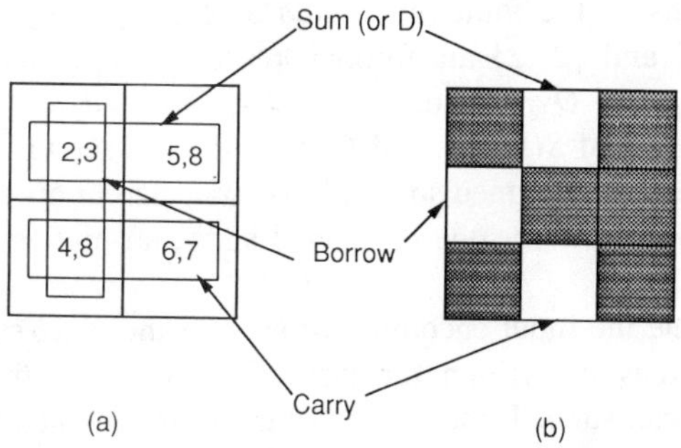

FIGURE 11.3 POSC multiprocessor: (a) input pixel minterm assignment and (b) output decoding mask.

can write the projection equations to relate input and output as

$$O_{11} = L_{23}I_{11}$$
$$O_{12} = L_{22}I_{11} + L_{23}I_{12}$$
$$O_{13} = L_{21}I_{12} + L_{22}I_{12} + L_{23}I_{13}$$
$$\vdots$$
$$O_{21} = L_{23}I_{21} + L_{13}I_{11}$$
$$\vdots$$
$$O_{25} = L_{21}I_{23} + L_{11}I_{13}$$
$$O_{31} = L_{13}I_{21}$$
$$\vdots$$

and so on where O_{ij} is the ijth pixel subcell of the output pixel. When all LEDs are ON, the above equations reduce to

$$\vdots$$
$$O_{13} = I_{11} + I_{12} + I_{13}$$
$$O_{21} = I_{11} + I_{21}$$
$$O_{25} = I_{23} + I_{13}$$
$$O_{31} = I_{21}$$
$$\vdots$$

Thus OR operations of the minterms stored at {1, 2} and {1, 2}, {1, 1} and {2, 1}, and {1, 2} and {2, 2} are formed, respectively, at the following output locations: O_{12}, O_{21}, and O_{32}. Accordingly, O_{12}, O_{21}, and O_{32}, respectively, correspond to the location of $S(D)$, B, and C output bits as shown in Fig. 11.3b. The output decoding mask is designed so that it has transparent openings at only those output locations. To allow for one horizontal shift and one vertical shift all four LEDs must be ON.

Now to determine the input encoding for each of the pixel subcells, the POSC overlap equations may be written for each of the minterm pairs. To store two minterms at each pixel subcell, the nonzero outputs are denoted by two orthogonal polarization directions. Accordingly, the POSC overlap equations for minterm pair (2, 3), for example, can be written

$$\overline{a_{11}} \wedge \overline{b_{11}} \wedge c_{11} = \mathrm{V} \tag{11.6}$$

$$\overline{a_{11}} \wedge b_{11} \wedge \overline{c_{11}} = \mathrm{H} \tag{11.7}$$

where x_{ij} represents the code of the ijth pixel subcell of input X. Solving Eqs. (11.6) and (11.7), we obtain $a_{11} = \mathrm{F}$, $b_{11} = \mathrm{H}$, and $c_{11} = \mathrm{V}$. To determine the input encoding for the other pixel subcells, the POSC overlap equations are written for each of the other minterm pairs with outputs denoted by two orthogonal polarization directions. After solving all the pairs of POSC overlap equations for the remaining locations, the pixel encoding for all the inputs are obtained, as shown in Table 11.3.

Now a comparison of this adder/subtracter with its electronic counterpart is in order. Unlike the all-electronic digital adder/subtracters, this multiprocessor provides the outputs of both addition and subtraction simultaneously. Figure 11.4a shows an implementation of an electronic adder/subtracter. The control input is A/S; when A/S = 1, the input performs a subtraction, and when A/S = 0, the unit performs an addition. Figure 11.4b shows a POSC equivalent circuit that is capable of generating all the outputs in parallel. In the electronic circuit only two outputs are generated at a time, which amounts to utilizing 6 gates/output bit, while in the later case only 2 gates/output bit are involved. The OR gates may not be counted in POSC since OR operations are done merely by overlapping the

TABLE 11.3 Input Encodings for Adder/Subtracter

Input	1	0
A	$\begin{bmatrix} \mathrm{F} & \mathrm{T} \\ \mathrm{H} & \mathrm{T} \end{bmatrix}$	$\begin{bmatrix} \mathrm{T} & \mathrm{F} \\ \mathrm{V} & \mathrm{F} \end{bmatrix}$
B	$\begin{bmatrix} \mathrm{H} & \mathrm{H} \\ \mathrm{T} & \mathrm{H} \end{bmatrix}$	$\begin{bmatrix} \mathrm{V} & \mathrm{V} \\ \mathrm{F} & \mathrm{V} \end{bmatrix}$
C	$\begin{bmatrix} \mathrm{V} & \mathrm{H} \\ \mathrm{T} & \mathrm{V} \end{bmatrix}$	$\begin{bmatrix} \mathrm{H} & \mathrm{V} \\ \mathrm{F} & \mathrm{H} \end{bmatrix}$

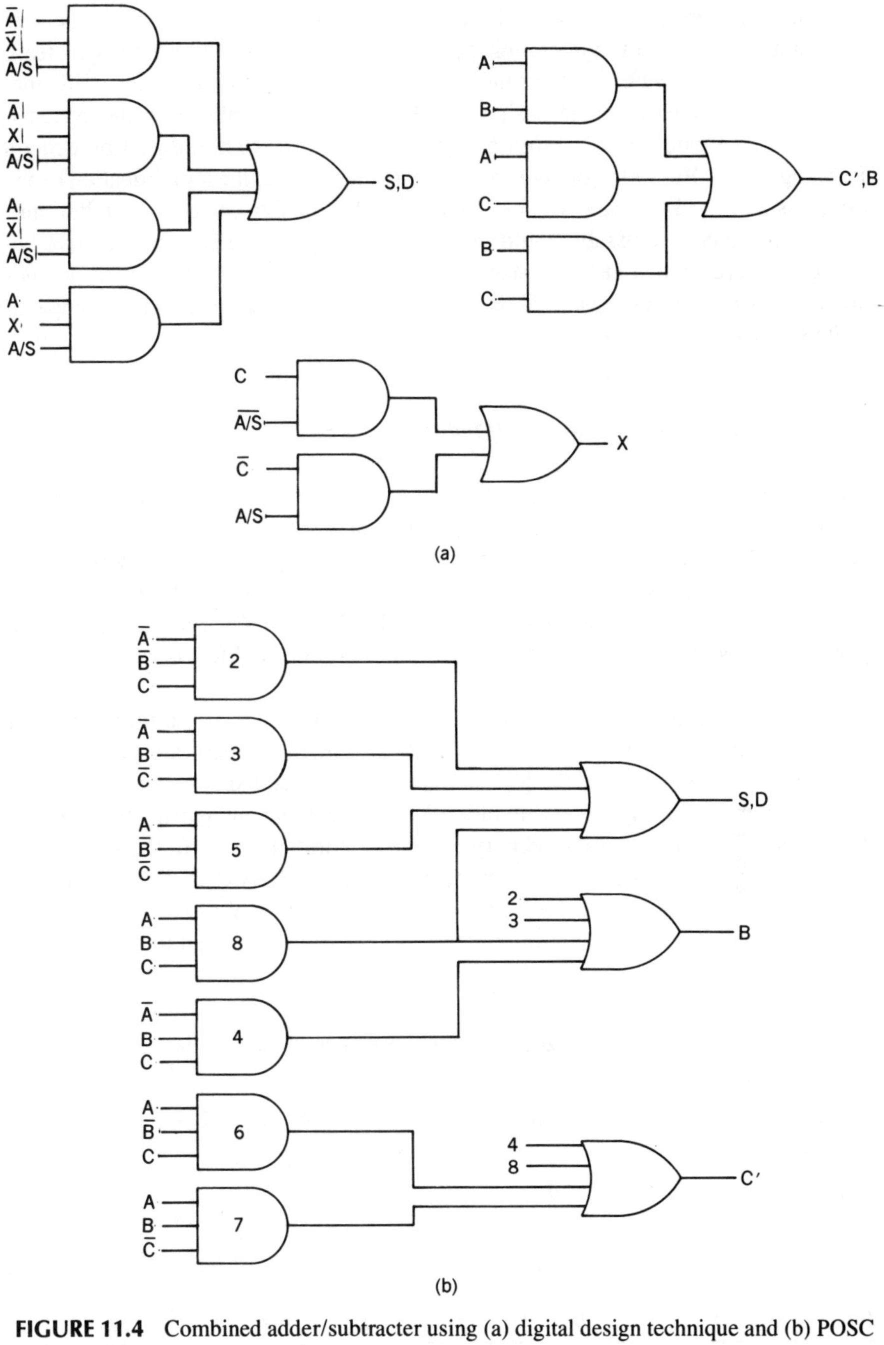

FIGURE 11.4 Combined adder/subtracter using (a) digital design technique and (b) POSC design technique.

shadows at a common pixel subcell; no actual hardware is involved. Note also that this simultaneous generation of multiple functions is achieved without any increase in the system hardware. Furthermore, this design points out an important distinction of optical digital system design as follows. In digital all-electronic systems, the number of gate inputs as well as the number of gates are reduced by using a Karnaugh map. But in the optical shadow-casting scheme, the simplification of the circuit is achieved by the sharing of minterms. The design of such a parallel multiple-output dual-addressable multiprocessor memory device is very attractive because it would reduce memory size by 75% or more. The POSC device just designed not only gives out parallel output bits but is highly efficient when compared to its electronic counterpart.

11.5 PARALLEL ALU USING POSC

The multiprocessor design algorithm of the last section may be extended to enable the design of an arithmetic logic unit having parallel output. A novel design of an optical ALU is discussed here which is capable of performing both arithmetic as well as logical operations. The adder/subtracter truth table shown in Table 11.1 represents the arithmetic part of this ALU, while Table 11.4a represents its logical part. It may be noted that the functional truth table of the ALU is rather massive in size.

Two separate approaches to finding an optimized design of the multi-function ALU are examined. In the first design, all arithmetic outputs are generated in parallel, and at least half of the logical outputs are generated simultaneously. It is shown that more intelligent consideration in the placement and pairing of minterms could enable one to design a memory-optimized parallel ALU. In the second design, to be covered in the next section, a conventional ALU architecture is followed where the outputs of the ALU are generated separately. The resulting size of the encoded input pixels are minimized by a technique referred to as truth table

TABLE 11.4a Truth Table for 16 Logical Operations of Two Binary Variables

Input Conditions	Input A	B	Output AND	NAND	OR	NOR	XOR	$\overline{\text{XOR}}$
1	0	0	0	1	0	1	0	1
2	1	1	1	0	1	0	0	1
3	0	1	0	1	1	0	1	0
4	1	0	0	1	1	0	1	0

Input Conditions	Input A	B	Output A	$\overline{A}$	B	$\overline{B}$	$A\overline{B}$	$\overline{A}+B$	$A+\overline{B}$	$\overline{A}B$	TRUE	FALSE
1	0	0	0	1	0	1	0	1	1	0	1	0
2	1	1	1	0	1	0	0	1	1	0	1	0
3	0	1	0	1	1	0	0	1	0	1	1	0
4	1	0	1	0	0	1	1	0	1	0	1	0

partitioning. To facilitate the ALU design, the arithmetic part and logical part are treated separately.

The truth table of the arithmetic unit is identical to the multiprocessor example designed in the last section. Thus the pairing of the minterms and the placement of the minterms in the pixel map can also be done identically. Basically, the arithmetic unit of the ALU is designed the same way as the multiprocessor unit. The pixel map of the arithmetic unit of the ALU is repeated and shown in Fig. 11.5a, along with the corresponding output map of Fig. 11.5b. Next, an attempt is made to use the already placed minterms to generate as much of the logical unit as possible. More pixel subcells are added to the existing pixel structure, to enable the generation of the remaining logical part of the truth table. However, it is important to note that the coding and decoder mask may have to be changed to accommodate these alterations in the pixel structure. Therefore the coding and output decoder design is postponed until the pixel map for the logical section is obtained.

Unlike the arithmetic section, the logical unit should also produce all the outputs in parallel. To reduce the size of the input pixels, as before, the outputs of

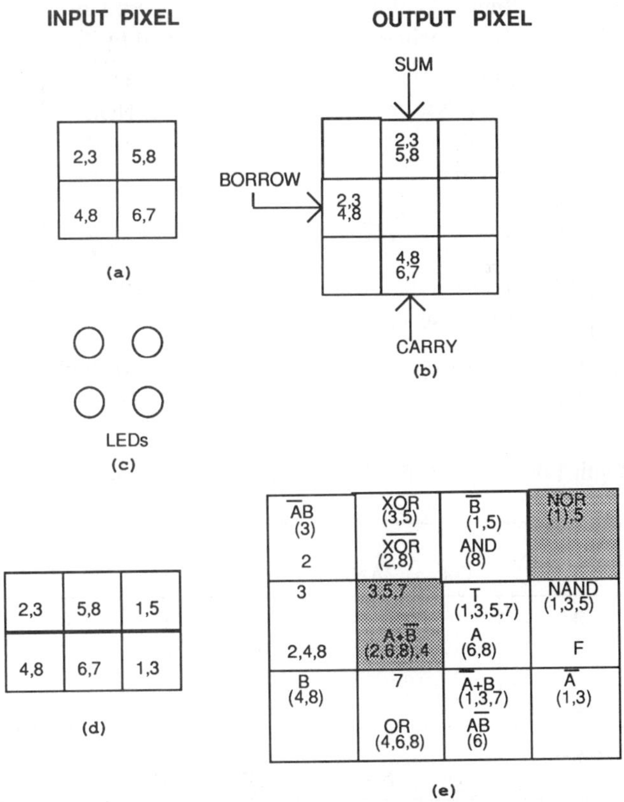

FIGURE 11.5 Parallel generation of eight logical outputs: (a) input overlap address mapping for arithmetic part; (b) output overlap pixel for arithmetic part; (c) the LED pattern; (d) overlap address mapping (for both arithmetic and logical part); and (e) the output decoding mask.

those logical operations that are complements of one another are paired together. Accordingly, one output of each of these pairs can be controlled by a control input C while the others are generated by the inverted C (i.e., $\overline{C}$). It is then possible to perform eight of the 16 logical operations simultaneously by selecting a code for either C of $\overline{C}$. The operations obtained by C and $\overline{C}$ are as follows:

Control Input	Logical Operations							
C	OR	AND	$A+\overline{B}$	$A\overline{B}$	$\overline{\text{XOR}}$	A	B	FALSE
$\overline{C}$	NOR	NAND	$\overline{A}B$	$\overline{A}+B$	XOR	$\overline{A}$	$\overline{B}$	TRUE

Using the above assignment, the modified truth table, which includes the control input C, is obtained as shown in Table 11.4b. The input condition numbers of Tables 11.1 and 11.4b are kept the same, for simplicity. Table 11.4 groups the eight logical outputs together, such that the logical value of the control input C is the same for those eight outputs. Next, all the outputs are listed along with the corresponding input conditions.

Logical Operation Obtained by C	Input Conditions (Even Minterms)	Logical Operation Obtained by $\overline{C}$	Input Conditions (Odd Minterms)
AND:	{8}	NAND:	{1, 3, 5}
OR:	{4, 6, 8}	NOR:	{1}
$\overline{\text{XOR}}$:	{2, 8}	XOR:	{3, 5}
A:	{6, 8}	$\overline{A}$:	{1, 3}
B:	{4, 8}	$\overline{B}$:	{1, 5}
$A\overline{B}$:	{6}	$\overline{A}+B$:	{1, 3, 7}
$A+\overline{B}$:	{2, 6, 8}	$\overline{A}B$:	{3}
FALSE:	{none of the even minterms}	TRUE:	{1, 3, 5, 7}

The first column of logical operations listed above involves only even input conditions. Note that all the even input conditions (2, 4, 6, 8) have already been used

TABLE 11.4b Truth Table for 16 Logical Operations of Two Binary Variables Including the Control Input

Input Conditions	Input A	B	C	Output OR	AND	$\overline{\text{XOR}}$	$A+\overline{B}$	$A\overline{B}$	A	B	FALSE
2	0	0	1	0	0	1	1	0	0	0	0
4	0	1	1	1	0	0	0	0	0	1	0
6	1	0	1	1	0	0	1	1	1	0	0
8	1	1	1	1	1	1	1	0	1	1	0
Input Conditions	**Input A**	**B**	**C**	**Output NOR**	**NAND**	**XOR**	**$\overline{A}B$**	**$\overline{A}+B$**	**$\overline{A}$**	**$\overline{B}$**	**TRUE**
1	0	0	0	1	1	0	0	1	1	1	1
3	0	1	0	0	1	1	1	1	1	0	1
5	1	0	0	0	1	1	0	0	0	1	1
7	1	1	0	0	0	0	0	1	0	0	1

to generate the sum, carry-out, and borrow-out corresponding to the ALU arithmetic section. With a carefully designed decoder, it may be possible to extract the logical outputs of the first column from the minterms already stored for the arithmetic operations. But the odd input condition 1 listed in the second column is not used by the arithmetic part of the ALU. Interestingly, this particular input condition alone is responsible for generating nonzero output for six of the logical operations, NAND, NOR, $\overline{A}$, $\overline{B}$, $\overline{A}B$, and TRUE. Consequently, it is essential to place the minterm 1 in the input overlap pixel to cover those six logical operations. But all the locations of the 2 × 2 input pixel have already been filled by other minterms when solving for the arithmetic operations. The only way to incorporate the minterm 1 is thus to increase the pixel size. At least two pixel subcells are necessary to obtain a regular shaped (rectangular) input pixel.

The possible pairs of the odd minterms are (1, 3), (1, 5), (3, 5), (5, 7), (1), (3), and (7). Keeping the pixel assignment of the arithmetic part shown in Fig. 11.5a unchanged, one notes that the pairs (3, 5) and (5, 7) can be formed by the OR operation between the two to-be-added neighboring pixel subcells $\{I_{11}, I_{12}\}$ and $\{I_{12}, I_{22}\}$. However, logical operations, such as $\overline{A}$ and $\overline{B}$ cannot be generated with the already assigned coding since the input condition 1 can no longer be placed as a neighbor to both 3 and 5. Accordingly, the pairs (1, 3) and (1, 5) may be placed along with the previously stored minterms for generating all the remaining logical operations. For example, (1, 3) and (1, 5) with a proper decoder can generate NAND, NOR, $\overline{A}$, $\overline{B}$, and $\overline{A}B$. Furthermore, the placement of the (1, 3) pair in the neighborhood of (6, 7) facilitates the generation of the TRUE logic output. The minterm pairs are now capable of performing all 16 logical as well as arithmetic operations. The overall mapping block, which consists of three horizontal pixel subcells and two vertical ones is shown in Fig. 11.5d. The maximum size of the OR mapping block is only 2 in the horizontal or vertical direction; consequently, the output overlap pixel pattern can be generated by shifting the input overlap pattern only once to the right and once downward. To allow for one horizontal and one vertical shift, all four LEDs at the source plane must be on as shown in Fig. 11.5c.

Using the reduced projection equations mentioned in Section 11.4, one can determine the output locations as follows:

Input Pixel Subcell	Output Pixel Subcell	Output
{1, 1}	{1, 1}	$\overline{A}B$
{1, 1}, {1, 2}	{1, 2}	XOR, $\overline{\text{XOR}}$
{1, 2}, {1, 3}	{1, 3}	$\overline{B}$, AND
{1, 3}	{1, 4}	NOR
{1, 1}, {1, 2}, {2, 1}, {2, 2}	{2, 2}	$A + \overline{B}$
{1, 2}, {1, 3}, {2, 2}, {2, 3}	{2, 3}	TRUE, A
{1, 3}, {2, 3}	{2, 4}	NAND, FALSE
{2, 1}	{3, 1}	B
{2, 1}, {2, 2}	{3, 2}	OR
{2, 2}, {2, 3}	{3, 3}	$\overline{A} + B$, $A\overline{B}$
{2, 3}	{3, 4}	$\overline{A}$

Now in order to design the intelligent decoder mask, we note from the above list that at the location $\{1, 4\}$ of the output pixel, minterm pair (1, 5) is projected. But at the same location the output of the NOR operation is generated by the minterm 1 alone. The decoding mask is thus made polarized so that it allows the light generated by minterm 1 alone to appear at the output. For a similar reason, in generating $A + \overline{B}$ at the output location $\{2, 2\}$ a horizontally polarized decoding mask is chosen to prohibit the appearance of vertically polarized light generated by minterm 4. Note here that we have already assumed a V code for the nonzero outputs corresponding to minterm 4. As a result of this choice, minterms 2, 6, and 8, which generate $A + \overline{B}$, at location $\{2, 2\}$ must be assigned an orthogonal H code. Consequently, the minterms 3, 5, and 7, which are paired with 2, 6, and 8, respectively, must be denoted by vertically polarized code V. Similarly, for the pair of input conditions (1, 5), mentioned earlier for NOR logic, minterm 1 is assigned an H code, so that the output of the NOR operation, generated at location $\{1, 4\}$, may be decoded using a horizontally polarized decoding mask. The overlap relations for the minterm pair (1, 5) are therefore

$$\bar{a}_{13} \wedge \bar{b}_{13} \wedge \bar{c}_{13} = \mathrm{H} \tag{11.8}$$

$$a_{13} \wedge \bar{b}_{13} \wedge \bar{c}_{13} = \mathrm{V} \tag{11.9}$$

where X_{ij} represents the code of the ijth pixel subcell of the input. Solving Eqs. (11.8) and (11.9), one obtains $a_{13} = \mathrm{V}$ and $b_{13} = c_{13} = \mathrm{F}$. Similarly, from the pairs of POSC overlap equations for the remaining locations, the pixels encoding for all the inputs are obtained, as shown in Table 11.5.

11.6 SEQUENTIAL ALU USING POSC

In the previous section, we had designed a parallel ALU. Here we design a conventional ALU in which all the outputs are generated separately. Note that the size of the corresponding ALU truth table is massive since it involves 20 outputs to be

TABLE 11.5 Input Pixel Encodings for Parallel ALU

Input	Code of the Pixel: 1	Code of the Pixel: 0
A	$\begin{bmatrix} \mathrm{F} & \mathrm{T} & \mathrm{V} \\ \mathrm{H} & \mathrm{T} & \mathrm{F} \end{bmatrix}$	$\begin{bmatrix} \mathrm{T} & \mathrm{F} & \mathrm{H} \\ \mathrm{V} & \mathrm{F} & \mathrm{T} \end{bmatrix}$
B	$\begin{bmatrix} \mathrm{V} & \mathrm{H} & \mathrm{F} \\ \mathrm{T} & \mathrm{V} & \mathrm{V} \end{bmatrix}$	$\begin{bmatrix} \mathrm{H} & \mathrm{V} & \mathrm{T} \\ \mathrm{F} & \mathrm{H} & \mathrm{H} \end{bmatrix}$
C	$\begin{bmatrix} \mathrm{H} & \mathrm{H} & \mathrm{F} \\ \mathrm{T} & \mathrm{H} & \mathrm{F} \end{bmatrix}$	$\begin{bmatrix} \mathrm{V} & \mathrm{V} & \mathrm{T} \\ \mathrm{F} & \mathrm{V} & \mathrm{T} \end{bmatrix}$

generated separately. This corresponds to 4 × 16 combinations of input–output relationships for the logical operations and 8 × 4 combinations of input–output relationships for the additions/subtractions. The resulting truth table will thus have 64 + 32 = 96 separate entries. In digital design, when the functional truth table is that massive in size, it is often partitioned into smaller sections so one can realize smaller circuit components for each of these sections. These components are then cascaded by exploiting some form of logical dependence so that the whole may function as a complete device. This is often a preferred method for reducing the memory size of the read-only-memory or programmable logic array.

The process of code determination of the POSC inputs resembles the programming of a programmable read-only-memory. One may thus expect savings in POSC memory requirements (i.e., smaller pixel size for the encoded inputs) when the truth table of the to-be-realized function is partitioned into primary and secondary segments. In this section, such truth-table partitioning techniques are exploited for obtaining a memory-efficient sequential ALU. The algorithm for the design of such an ALU utilizing truth-table partitioning is described as follows:

Step 1 The truth table of the arithmetic module is determined.

Step 2 The truth table is then partitioned into two sections: the primary section taking at least two input columns and the remaining inputs forming the secondary section. For multiple-output functions, each output is treated separately.

Step 3 The outputs are assigned the following polarization codes: either T, V, or H for binary 1 and F for binary 0. However, denoting a binary 1 by code T allows only one minterm to be stored at each location of the pixel subcells. Denoting a binary 1 by either V or H doubles the memory efficiency. The nonzero outputs are thus assigned either a V or H code. The output 0's are not usually considered in the treatment, because they would always occur by default whenever a 1 is not generated.

Step 4 The overlap of the primary inputs in each row of the truth table is assigned the same code as that of the corresponding output. For the purpose of identification, each of the outputs is characterized by either hold, toggle, set, or reset modes in terms of the corresponding secondary inputs.

Step 5 The overlap of the secondary inputs is assigned a POSC code so that its overlap with the code of the overlap of the primary inputs produces the code for the desired output. The overlap of secondary inputs is thus assigned either a T, F, or "don't care" code, respectively, when the output is in the hold, toggle, or set mode. A don't care means that the variable could have any one of the codes T, V, H, or F for both of its binary values.

Step 6 The input coding patterns are determined from the solution of the corresponding POSC overlap equations.

Step 7 If more than one output exists, the next output is generated from the existing codes of the overlap of the primary inputs. If this is not possible, Steps 4, 5, and 6 are repeated to obtain new assignments for the overlap of either primary and/or secondary inputs.

Step 8 The POSC overlap equations involving (i) the primary input variables and the code of their overlap and (ii) the secondary input variables and the code of their overlap are solved to provide the codes for all the inputs.

Step 9 The source LED patterns for each output are determined.

Step 10 The output decoding mask is identified.

The above steps may be sequentially followed to design an arithmetic logic unit.

To design an ALU, one is confronted with a massive truth table with 96 entries! To simplify the corresponding design process, the arithmetic part and the logical part may be treated separately. Consider, first, the arithmetic part. The truth table for the full adder and full subtracter is shown in Table 11.1, where A, B, and C are the three input bits: addend (minuend), augend (subtrahend), and carry-in (borrow-in) of the adder (subtracter) unit. S and C', respectively, represent the sum and carry-out, while D and B, respectively, represent the difference and borrow-out.

Arbitrarily, two input columns (say, addend A and augend B) may be partitioned to form the primary section. The segmented truth table in terms of the secondary input, carry-in, is obtained as in Table 11.6. for the four outputs: sum, difference, carry-out, and borrow-out. There are four distinct cases (00, 01, 10, and 11) of the primary inputs, A and B. The output type is classed in either of the four modes:

TABLE 11.6 Truth Table Partitioning for Arithmetic Part of ALU

Output	Input Conditions	Primary Input A	Primary Input B	Primary Code	Secondary Input C	Output Value	Mode
S and D	1	0	0	V	1	1	Hold
	2	1	1	H	1	1	Hold
	3	0	1	V	0	1	Toggle
	4	1	0	H	0	1	Toggle
C	1	0	0	V	X	0	Reset
	2	1	1	H	X	1	Set
	3	0	1	V	1	1	Hold
	4	1	0	H	1	1	Hold
B	1	0	0	V	1	1	Hold
	2	1	1	H	1	1	Hold
	3	0	1	V	X	1	Set
	4	1	0	H	X	0	Reset

hold, set, reset, and toggle. For the primary input conditions 00 and 11, the output sum is either 0 or 1 when the secondary input condition C is 0 or 1, respectively. So the sum holds the secondary input conditions in these two cases. But when the primary input conditions are 01 or 10, the output sum is either 0 or 1 corresponding to the secondary input condition (C) of 1 or 0, respectively; that is, the sum toggles with C. Similarly, it can be shown for the carry-out that it either holds, resets, or sets the secondary input condition C. In other words, the outputs are derived only from C, which acts as the control variable, while A and B simply supply a code for the output.

First, the coding for sum (difference) is obtained. The nonzero outputs are assigned V and H codes. Each group of primary input conditions in the partitioned truth table is identified by V and H similar to the output bit shown in Table 11.6. Since each segment of Table 11.6 has four input conditions and since two minterms can be stored in a single pixel subcell, only two pixel subcells are required for storing these two pairs of minterms. We arbitrarily start with a 2 × 1 pixel structure for A, B, and C. The pair of input conditions (1, 2) and (3, 4) are assigned pixel subcell locations {1, 1} and {2, 1}, respectively. The POSC overlap equations corresponding to input conditions 1 and 2, for example, for generating sum (difference) are given by

$$\overline{a_{11}} \wedge \overline{b_{11}} = \mathrm{V} \quad \text{(primary code)} \tag{11.10a}$$

$$\mathrm{V} \wedge c_{11} = \mathrm{V} \quad \text{(sum)} \tag{11.10b}$$

$$a_{11} \wedge b_{11} = \mathrm{H} \quad \text{(primary code)} \tag{11.11a}$$

$$\mathrm{H} \wedge c_{11} = \mathrm{H} \quad \text{(sum)} \tag{11.11b}$$

Solving Eqs. (11.10a) and (11.11a) simultaneously yields $a_{11} = b_{11} = \mathrm{H}$, and from Eqs. (11.10b) and (11.11b), one obtains $c_{11} = \mathrm{T}$. Note that the above equation pairs indicate the sum is in the hold mode. Note that C is encoded with T in the hold mode. The logical code assigned to c_{11} does not necessarily show up as the output; it only allows or blocks the primary code to appear in the output. For the remaining half segment of the truth table, one notices that V and H are assigned for two different primary input conditions. In terms of the POSC overlap equation corresponding to input conditions (3, 4) of the sum output,

$$\bar{a}_{21} \wedge b_{21} = \mathrm{V} \quad \text{(primary code)} \tag{11.12a}$$

$$\mathrm{V} \wedge \overline{c_{21}} = \mathrm{V} \quad \text{(sum)} \tag{11.12b}$$

$$a_{21} \wedge \bar{b}_{21} = \mathrm{H} \quad \text{(primary code)} \tag{11.13a}$$

$$\mathrm{H} \wedge \overline{c_{21}} = \mathrm{H} \quad \text{(sum)} \tag{11.13b}$$

Solving these POSC equations yields $a_{21} = \mathrm{H}$, $b_{21} = \mathrm{V}$, and $c_{21} = \mathrm{F}$. Note that the code for C is F as expected in a toggle mode. This completes the solution of all input conditions for the sum and difference. The input codings for generating the sum and difference are thus given by

$$A = \begin{bmatrix} \text{H} \\ \text{H} \end{bmatrix}, \quad B = \begin{bmatrix} \text{H} \\ \text{V} \end{bmatrix}, \quad C = \begin{bmatrix} \text{T} \\ \text{F} \end{bmatrix}$$

For the carry-out output, using the previous assignments of the primary section, we see that the primary inputs 00 and 01 (input conditions 1 and 3) generate a V while 11 and 10 (input conditions 2 and 4) generate an H. The code for the secondary input c_{11} is found from the pair of POSC overlap equations corresponding to input conditions 1 and 2 involving the primary code and the secondary input condition given by

$$\text{V} \wedge X = \text{F} \tag{11.14}$$

$$\text{H} \wedge X = \text{H} \tag{11.15}$$

where X stands for don't-care input C, which can be either 0 or 1. Solving Eqs. (11.14) and (11.15) simultaneously yields $X = \text{H}$ for both C and $\overline{C}$. Here the code of C is H, which acts as "don't care." Again from input conditions 3 and 4 of the carry generation, the POSC overlap equations are

$$\text{V} \wedge c_{21} = \text{V} \tag{11.16}$$

$$\text{H} \wedge c_{21} = \text{H} \tag{11.17}$$

The solution of the above pair of equations gives $c_{21} = \text{T}$. The input codings for generating the carry output are thus

$$A = \begin{bmatrix} \text{H} \\ \text{H} \end{bmatrix}, \quad B = \begin{bmatrix} \text{H} \\ \text{V} \end{bmatrix}, \quad C = \begin{bmatrix} \text{H} \\ \text{T} \end{bmatrix}, \quad \overline{C} = \begin{bmatrix} \text{H} \\ \text{F} \end{bmatrix}$$

Note that the pixel subcell c_{11} acts as a don't-care input; it is permanently tied to an H for both C and $\overline{C}$.

Similarly, for the borrow-out output, the primary input conditions 00, 11, 01, and 10 generate V, H, V, and H, respectively. Retaining the same primary code assignments, the code for C is obviously T for the pixel subcell c_{11} and V for pixel subcell c_{21}. V here represents a "don't-care" pixel subcell.

The input codings for the borrow-out output are thus

$$A = \begin{bmatrix} \text{H} \\ \text{H} \end{bmatrix}, \quad B = \begin{bmatrix} \text{H} \\ \text{V} \end{bmatrix}, \quad C = \begin{bmatrix} \text{T} \\ \text{V} \end{bmatrix}, \quad \overline{C} = \begin{bmatrix} \text{F} \\ \text{V} \end{bmatrix}$$

For the generation of the sum, difference, carry-out, and borrow-out, the code for A and B remains the same, while the code for C and $\overline{C}$ is switched. Therefore the POSC full adder/subtracter consists of three inputs, each coded with pixels of size 2×1, and only two LEDs. The outputs can be generated either sequentially (by switching the codes for C and $\overline{C}$) or in parallel by directing overlap output of primary inputs to different secondary inputs simultaneously.

Now we may concentrate on designing the logical section of the ALU. It is well known that the 16 possible functions of two binary variables serve as the basis of the arithmetic operations. These functions are listed in Table 11.4a. The input entries represent the four combinations of values for the two binary variables A and B. These two are taken as primary inputs while the other input (which controls the different operations) is taken as the secondary input.

In Table 11.4 one notices that there are some totally complementary logical operation pairs. For example, consider the output of A and $\overline{A}$ or XOR and $\overline{\text{XOR}}$, which are complementary to each other. These inverted operations may be paired together to facilitate reduction of the number of pixel subcells. Moreover, controlling one output generation by C and the other by inverted C (i.e., $\overline{C}$), one can further simplify the design process of the logical unit. As an example, consider the truth table for the NAND and AND operations. With control input C, the output can be expressed as follows:

Logic	Input Conditions	Primary Inputs A	B	Primary Code	Control Input C	Output
AND	1	0	0	V	1	0
	2	1	1	H	1	1
	3	0	1	V	1	0
	4	1	0	H	1	0
NAND	5	0	0	V	0	1
	6	1	1	H	0	0
	7	0	1	V	0	1
	8	1	0	H	0	1

Considering only the nonzero outputs of the AND and NAND operations, one can reduce the above table into the following form:

Input Conditions	Primary Inputs A	B	Primary Code	Control Input C	Output
5	0	0	V	0	1
2	1	1	H	1	1
7	0	1	V	0	1
8	1	0	H	0	1

For the first pair of input conditions (5, 2), the POSC overlap equations can be written

$$\bar{a}_{11} \wedge \bar{b}_{11} = \mathrm{V} \quad \text{(primary code)} \tag{11.18a}$$

$$\mathrm{V} \wedge \bar{c}_{11} = \mathrm{V} \quad \text{(NAND)} \tag{11.18b}$$

$$a_{11} \wedge b_{11} = \mathrm{H} \quad \text{(primary code)} \tag{11.19a}$$

$$\mathrm{H} \wedge c_{11} = \mathrm{H} \quad \text{(AND)} \tag{11.19b}$$

Solving Eqs. (11.18a) and (11.19a) simultaneously yields $a_{11} = b_{11} = \mathrm{H}$, while solving Eqs. (11.18b) and (11.19b) yields $c_{11} = \mathrm{H}$. For the other pair of input conditions (7 and 8), V and H are assigned to the respective primary input conditions. In terms of POSC overlap equations,

$$\bar{a}_{21} \wedge b_{21} = \mathrm{V} \quad \text{(primary code)} \tag{11.20a}$$

$$\mathrm{V} \wedge \bar{c}_{21} = \mathrm{V} \quad \text{(NAND)} \tag{11.20b}$$

$$a_{21} \wedge \bar{b}_{21} = \mathrm{H} \quad \text{(primary code)} \tag{11.21a}$$

$$\mathrm{H} \wedge \bar{c}_{21} = \mathrm{H} \quad \text{(NAND)} \tag{11.21b}$$

As before, solving these POSC equations yields $a_{21} = \mathrm{H}$, $b_{21} = \mathrm{V}$, and $c_{21} = \mathrm{F}$. This completes the solution of all input conditions. The input codings for the AND and NAND output are thus given by

$$A = \begin{bmatrix} \mathrm{H} \\ \mathrm{H} \end{bmatrix}, \quad B = \begin{bmatrix} \mathrm{H} \\ \mathrm{V} \end{bmatrix}, \quad C = \begin{bmatrix} \mathrm{H} \\ \mathrm{F} \end{bmatrix}, \quad \bar{C} = \begin{bmatrix} \mathrm{V} \\ \mathrm{T} \end{bmatrix}$$

Note that the C code generates AND logic, while $\bar{C}$ generates NAND logic.

Let use consider another pair of logical operations, such as A and $\bar{A}$. As before, let the output A be generated by the control input C and that of $\bar{A}$ be generated by the control input $\bar{C}$. The compact truth table for output A and $\bar{A}$ generating nonzero outputs are thus obtained as follows:

Input Conditions	Primary Inputs A	B	Primary Code	Control Input C	Output
5	0	0	V	0	1
2	1	1	H	1	1
7	0	1	V	0	1
4	1	0	H	1	1

Note that since the outputs (A and $\bar{A}$) are completely complementary, their logical sum $A + \bar{A}$ always generates exactly four nonzero outputs. To determine the encoding for each of the pixels of control input C, the POSC overlap equations are developed for each of the minterm pairs (5, 2) and (7, 4). Accordingly, the POSC overlap equations for minterm pair (5, 2), for example, are

$$\bar{a}_{11} \wedge \bar{b}_{11} = \text{V} \quad \text{(primary code)} \tag{11.22a}$$

$$\text{V} \wedge \bar{c}_{11} = \text{V} \quad (\bar{A}) \tag{11.22b}$$

$$a_{11} \wedge b_{11} = \text{H} \quad \text{(primary code)} \tag{11.23a}$$

$$\text{H} \wedge c_{11} = \text{H} \quad (A) \tag{11.23b}$$

Solving Eqs. (11.22a) and (11.23a) simultaneously, we obtain $a_{11} = b_{11} = \text{H}$, and from Eqs. (11.22b) and (11.23b), one obtains $c_{11} = \text{H}$. For the other pair of input conditions (7, 4) of the truth table, one can write the following POSC overlap equations:

$$\bar{a}_{21} \wedge b_{21} = \text{V} \quad \text{(primary code)} \tag{11.24a}$$

$$\text{V} \wedge \bar{c}_{21} = \text{V} \quad (\bar{A}) \tag{11.24b}$$

$$a_{21} \wedge \bar{b}_{21} = \text{H} \quad \text{(primary code)} \tag{11.25a}$$

$$\text{H} \wedge c_{21} = \text{H} \quad (A) \tag{11.25b}$$

The code for the pixel subcell $\{2, 1\}$ is thus obtained as $a_{21} = \text{H}$, $b_{21} = \text{V}$, and $c_{21} = \text{H}$. Therefore the code of control input C which generates A is

$$C = \begin{bmatrix} \text{H} \\ \text{H} \end{bmatrix}$$

and that of $\bar{C}$ which generates $\bar{A}$ is

$$\bar{C} = \begin{bmatrix} \text{V} \\ \text{V} \end{bmatrix}$$

It is interesting to note that the input codings for inputs A and B in the logical section are exactly the same as those in the sequential arithmetic unit.

To complete the design of control input C, it is required to combine all the pairs of the logical operations that are totally complementary to each other. In each pair, one of the logical outputs is generated by C code while the other is selected by its inverse $\bar{C}$. Accordingly, the following pair of operations may be controlled by C and $\bar{C}$.

Control Input	Logical Operations							
C	OR	AND	$A + \bar{B}$	$A\bar{B}$	$\overline{\text{XOR}}$	A	B	FALSE
$\bar{C}$	NOR	NAND	$\bar{A}B$	$\bar{A} + B$	XOR	$\bar{A}$	$\bar{B}$	TRUE

TABLE 11.7 Input Encoding of Control-Input for Different Logical Operations

Logical Operations		Code of C	Code of $\overline{C}$
OR	NOR	$\begin{bmatrix} H \\ T \end{bmatrix}$	$\begin{bmatrix} V \\ F \end{bmatrix}$
AND	NAND	$\begin{bmatrix} H \\ F \end{bmatrix}$	$\begin{bmatrix} V \\ T \end{bmatrix}$
$\overline{\text{XOR}}$	XOR	$\begin{bmatrix} T \\ F \end{bmatrix}$	$\begin{bmatrix} F \\ T \end{bmatrix}$
A	$\overline{A}$	$\begin{bmatrix} H \\ H \end{bmatrix}$	$\begin{bmatrix} V \\ V \end{bmatrix}$
B	$\overline{B}$	$\begin{bmatrix} H \\ V \end{bmatrix}$	$\begin{bmatrix} V \\ H \end{bmatrix}$
$A + \overline{B}$	$\overline{A}B$	$\begin{bmatrix} T \\ H \end{bmatrix}$	$\begin{bmatrix} F \\ V \end{bmatrix}$
$A\overline{B}$	$\overline{A} + B$	$\begin{bmatrix} F \\ H \end{bmatrix}$	$\begin{bmatrix} T \\ V \end{bmatrix}$
False	True	$\begin{bmatrix} F \\ F \end{bmatrix}$	$\begin{bmatrix} T \\ T \end{bmatrix}$

Using the POSC overlap equations for each pair of operations, one can solve the codes for the control inputs C and $\overline{C}$, as shown in Table 11.7. The POSC system for the sequential output generation is illustrated in Fig. 11.6.

Two different designs of memory-efficient POSC ALU using truth-table partitioning and minterm sharing have been illustrated in these two sections. The first design illustrates a parallel ALU, which allows one to perform up to eight logical operations in parallel. By increasing two pixel subcells to arithmetic operations, 16 logical outputs were generated. Note that there is no electronic analog of this POSC ALU architecture. In multiprocessor design we saw that a binary adder/subtracter having three inputs and four outputs requires at least four pixel subcells for encoding its inputs and four LEDs. Here in the second design, in comparison, we find that the size can, in fact, be reduced even more to only two pixel subcells, thus doubling the input resolution of the system while reducing the number of LEDs by one-half. In addition to that, the device is capable of generating 16 logical outputs of two binary inputs.

11.7 POSC IMAGE PROCESSING

A digital computer becomes easily overburdened by the vast amount of data that it has to process in today's image processing tasks. Due to the limited number of

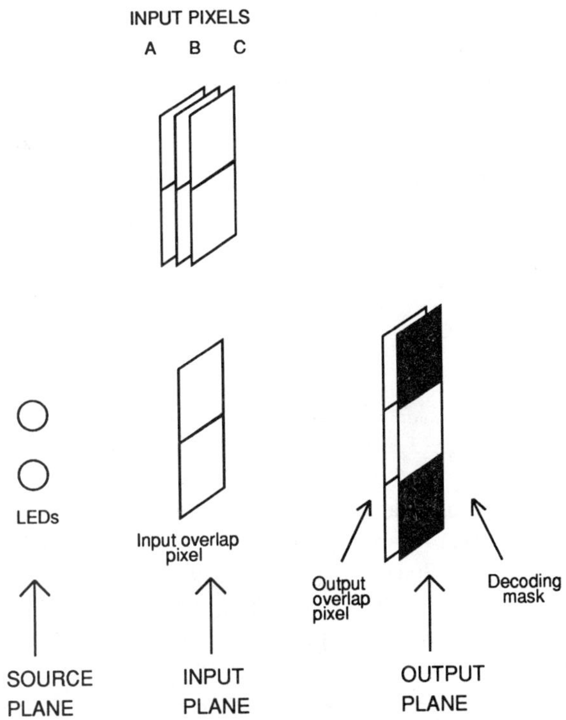

FIGURE 11.6 Implementation of sequential output generation.

communication channels, when the size of the image is very large, each pixel of information is sequentially fetched from the memory and then acted upon, and after processing the result is stored back in the memory. In the POSC optical setup, in comparison, the POSC processing and communication between data memory and the output plane are all done in parallel. Image processing tasks, especially spatial domain image processing, are therefore a natural choice for POSC. Pattern recognition operations require extraction of important features such as edges of an image. In this section, for example, we explore the use of POSC for edge detection.

For edge detection, we may consider an L-shaped window of three binary pixels. An edge is detected whenever a 0 to 1 or a 1 to 0 transition is encountered anywhere in the input image. Let a, b, and c denote the three members of the L-shaped window, as shown in Fig. 11.7a. Of the eight possible combinations of three members of the window, only two input conditions, namely, 000 and 111,

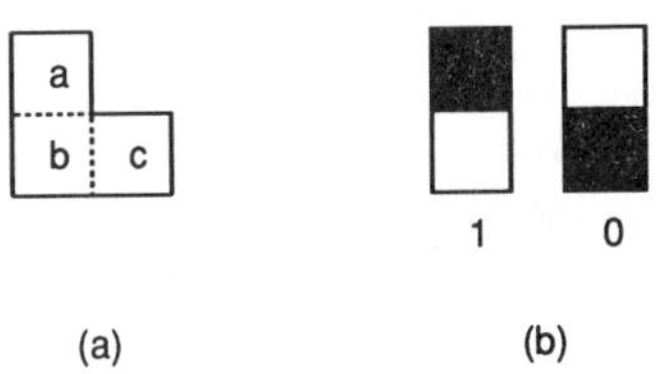

FIGURE 11.7 (a) An L-shaped window and (b) coding for 1 and 0.

do not correspond to any edge. The outputs corresponding to these input conditions may be represented as T and the rest are set to F. In other words, edges are denoted by an F output and no edge by T. Thus the constant gray level of the input image will appear as bright and the edges will appear as dark.

Assuming that each input bit is represented by two pixel subcells denoted by the subscripts 11 and 12, one can write the POSC overlap equations involving a, b, and c as follows:

$$\bar{a}_{11} \wedge \bar{b}_{11} \wedge \bar{c}_{11} = \mathrm{T} \tag{11.26}$$

$$a_{12} \wedge b_{12} \wedge c_{12} = \mathrm{T} \tag{11.27}$$

which yields $a_{11} = b_{11} = c_{11} = \mathrm{F}$ and $a_{21} = b_{21} = c_{21} = \mathrm{T}$. The corresponding solutions are shown in Fig. 11.7b. From Fig. 11.7a, it may be noted that the three members of the L-shaped window a, b, and c must interact with one another by means of overlap. Therefore one coded image a must be shifted downward by one pixel, another copy c must be shifted left by one pixel and then overlapped with the third unshifted copy b. Since the codes for a, b, and c are the same, the three shifted versions of the image will have identical characteristics. Thus three copies of one coded image will suffice for shifting and overlapping. This shifting and overlapping operation ensures that members of all L-shaped windows throughout the image will overlap with one another. Only two LEDs are needed at the source plane to perform the OR operation between two pixel subcells.

POSC schemes are implementable using many optical devices. Multi-wavelength exposure of a spatial light rebroadcaster (SLR), for example, has already been used to successfully realize AND operations, associative memory application, and dynamic matched filtering. It is therefore possible, in principle, to use a SLR device to perform the overlap operation and then generate the input overlap pattern. The SLR material when exposed to visible or blue light excites electrons, which are then trapped in an excited state indefinitely. When in this condition, the material is reexposed by infrared (IR) light; it releases those trapped electrons and an orange light is subsequently emitted. Locations of the material, which were exposed with blue light, will emit orange light, but locations that remained unexposed initially will allow the IR to pass through the material. Thus the sequential illumination by the two colored lights results in an AND operation between the two inputs. The absence of one or the other input will result in no visible light.

The optical setup is shown in Fig. 11.8 in which an argon laser is used for writing the coded image a onto the material directly. The shifted image b is projected onto the material using a beam splitter and an IR laser source. Since the IR image is invisible, the shifted image may have to be aligned using a He-Ne laser, for example. For the POSC edge detection system, an image of a ship, as shown in Fig. 11.9, is considered. Using the codes derived earlier, the image is encoded. Now the encoded image is placed in position a as shown in Fig. 11.8 and is illuminated by a collimated source and projected onto the material as shown in Fig. 11.10.

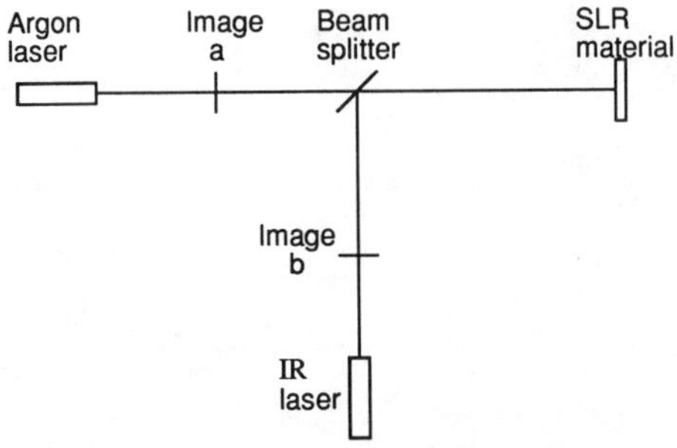

FIGURE 11.8 The optical setup for edge detection.

First, the horizontal edge may be obtained. A copy of the encoded image is inserted in position *b* and then shifted vertically, such that it is displaced relative to image *a* by two pixel subcells. The image is then illuminated by an IR source. This performs an overlap operation between images *a* and *b*. The overlapped image showing the horizontal edges is shown in Fig. 11.11. Note that the edges show up as slightly dark pixels. Note that three gray levels are produced as a result of the overlap of images *a* and *b*. The brightest gray level occurs when a region formerly illuminated by blue light is again exposed by an IR source. The darkest regions are those that were not illuminated by blue light or IR. This occurs when the pixels in *a* and *b* are exactly same, that is 11 or 00. The intermediate gray level, on the other hand, appears at the regions that were illuminated by blue light but were not exposed to IR. This stray light is the result of luminescence produced by the blue light. For extracting the vertical edges, the material is again exposed by image *a*. Next, image *b* may be shifted in the horizontal direction so that it is displaced by one pixel subcell relative to *a*. Figure 11.12 shows the corresponding vertical edges of the image. Since the SLR material acts as a two-input AND gate either only horizontal or vertical edges can be obtained at one time. When the image from the material is shifted upward and overlapped with itself (in POSC this is usually done by LEDs), resulting in an OR operation between the two pixel subcells, then the no-edges appear totally white and the edges appear totally black.

One of the major problems of the SLR material, however, is the loss of power of the image at the output. This limitation prohibits manipulation of the output image. However, by using an image intensifier and liquid crystal light valve, it is possible to provide gain to the output image, so that the output of the SLR device

FIGURE 11.9 The original image.

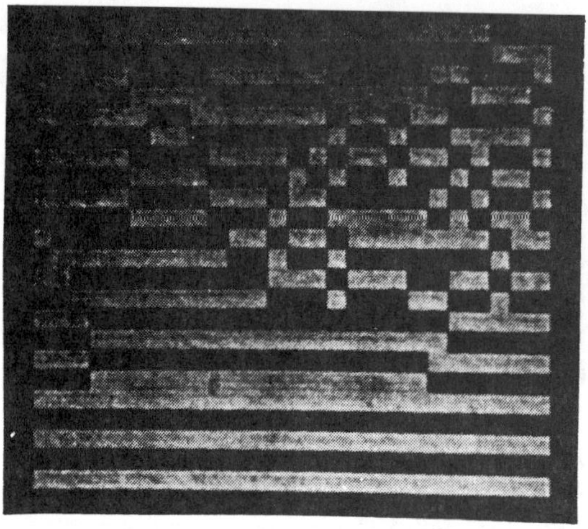

FIGURE 11.10 The coded image written on SLR material.

may be reused for further processing or cascading. Then one will be able to optically shift the image by means of mirrors and beam splitters and overlap them so that the edges can be totally extracted from the nonedge information.

11.8 SYMBOLIC SUBSTITUTIONS

To facilitate the recognition phase of symbolic substitutions (SS), the inputs need to be encoded. Two possible codings of the inputs exist: intensity and polarization. In intensity coding, a dual-rail coding is adopted where both 0 and 1 of each input have a black and a white pixel in them, and the relative positions of the black and

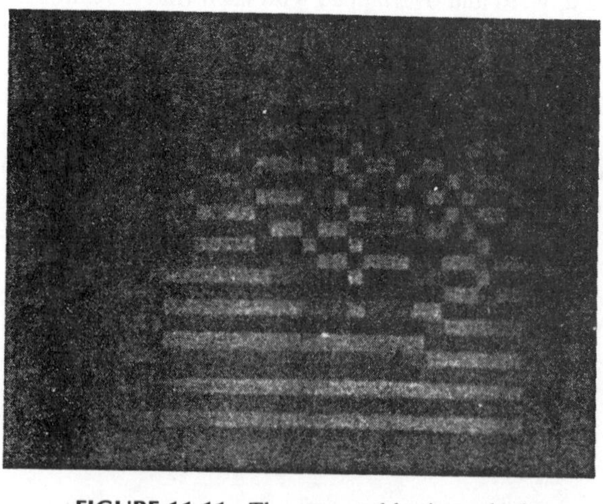

FIGURE 11.11 The extracted horizontal edge.

FIGURE 11.12 The extracted vertical edges.

white pixels indicate whether it is 1 or 0. A dual-rail coding is shown in Fig. 11.13. The shift and overlap steps of SS rely on identifying only the black pixel. Another incentive for this type of coding is the need to propagate energy when the input is either 0 or 1. The propagation of a signal represented by a dark pixel is usually associated with significant deterioration of the signal-to-noise ratio. In electronics, the logical variables propagate while energy flows either in or out of the next stage logic device. For example, a 0 is propagated when energy is being sunk in the driver gate, whereas a 1 is propagated when energy is supplied as current. In polarization coding, however, horizontal and vertical polarizations provide two distinct modes, which can be distinguished easily as 1 and 0. It also provides cascadability since both 1 and 0 can propagate as light signals. Using a polarization scheme, only one pixel subcell is required for encoding both 1 and 0 as in the POSC system. Polarization codes, when compared to the intensity codes, generally reduce the size of the inputs by half.

A logical operation in SS may be realized simply by defining a set of rules for substituting the outputs as dictated by the to-be-implemented logical function. For example, an AND operation may be implemented by substituting a 1 at each of the locations where a pattern of (11) has been recognized and substituting 0 for all other patterns. Since the code for both 1 and 0 has a bright subpixel, 0 cannot be generated automatically. Consequently, both 1 and 0 need to be reencoded, if such outputs are to be reused. Therefore it is necessary to execute the rules that generate 0, in addition to those that generate 1. Thus in dual-rail coded SS, one needs to

FIGURE 11.13 A dual-rail coding.

implement rules for all the entries of a truth table. However, if the process requires only one step of SS, then the final output need not be dual-rail coded. In that case, a bright pixel may be used to encode an output of a 1 and dark pixel may be furnished for a 0. In such situations, the number of rules are greatly reduced, since only rules for generating 1's need to be implemented. Note that only recognition suffices; no substitution is necessary.

Symbolic substitution involves implementing each input of a truth table by means of a rule. In this respect, SS is similar to the traditional digital design technique, the only difference being that the inputs are represented differently and as a consequence are recognized differently. In a digital electronic system, either 1's or 0's need to be recognized but not both. In both cases, however, the redundancy in similar bit strings are eliminated by the K-map or Quine–McClusky tabular reduction technique. For example, we are to generate an output of 1 whenever the input ABC is either 111 or 110. These input conditions could be combined together to only identify 11 for AB and we "don't care" whether C is 0 or 1. Therefore the corresponding reduced minterm is AB. Thus when implementing an arithmetic unit, for example, by means of SS, we can apply the minterm reduction technique to reduce the number of rules to be implemented.

Arithmetic operations like addition/subtraction can be carried out by defining two bit substitution rules. The first proposed SS addition was an iterative n-step process for adding two n-bit numbers. Since the outputs must be reprocessed again, it is necessary to reencode zeros for the next stage recognition purpose. Therefore both 0 and 1 need to be generated. As a result, the minimization of only generating 1's is not achieved. To derive the rules for addition, one notes that when two binary numbers are added, it results in a sum bit and a carry bit. Each of the entries of the truth table of a half adder thus becomes a rule to-be-implemented in SS. Now since the carry has more weight than the sum, it is shifted one bit to the left. Accordingly the corresponding rules are

$$\begin{array}{ccc} 0 & \rightarrow & 0 \\ 0 & & \quad 0 \end{array} \qquad \begin{array}{ccc} 0 & \rightarrow & 0 \\ 1 & & \quad 1 \end{array} \qquad \begin{array}{ccc} 1 & \rightarrow & 0 \\ 0 & & \quad 1 \end{array} \qquad \begin{array}{ccc} 1 & \rightarrow & 1 \\ 1 & & \quad 0 \end{array}$$

Note that the rules simply replace the to-be-added bits by corresponding carry and sum bits. In three of the four cases, the carry bit is zero. Accordingly, repeated applications of these rules will finally propagate the carry to the final bit position and make all the carries equal to zero, with the final sum appearing in the bottom row.

As an example, consider the example of adding 1110 + 0100, which can be carried out in four steps:

$$\begin{array}{c} 1110 \\ 0100 \end{array} \Rightarrow \begin{array}{c} 0100\emptyset \\ \emptyset 1010 \end{array} \Rightarrow \begin{array}{c} 01000\emptyset \\ \emptyset 00010 \end{array} \Rightarrow \begin{array}{c} 00000\emptyset \\ \emptyset 10010 \end{array}$$

where $\emptyset$ represents padded zero. In optics, repetition of operations is not a problem in terms of delays; however, reencoding of the output is a bottleneck in itself.

Therefore processes that require only a fixed number of steps are sought. Representing numbers in a redundant number system, however, may reduce the delay required for elementary arithmetic operations.

A modified signed digit (MSD) binary number, for example, may be represented by

$$X = \sum_{i=0}^{n} x_i 2^i \tag{11.28}$$

where the variable x_i, can take any value from the set $\{\bar{1}, 0, 1\}$. The MSD number just described is a redundant type since more than one representation is possible for numbers in this system. In the symbolic substitution-based MSD addition scheme, the numbers are first represented in MSD representation and then the to-be-added pair of numbers are mapped into another intermediate pair of numbers in a way that the latter numbers, when added, do not result in any carry.

When two single-digit binary MSD numbers are added, a carry is generated whenever the to-be-added numbers are either (1, 1) or $(\bar{1}, \bar{1})$. The carry is locally absorbed by allowing one bit propagation to the left. The generation of such a carry may be avoided by mapping the two digits in question into an intermediate sum and an intermediate carry (also known as weight and transfer bits) such that the nth intermediate sum and the $(n - 1)$th intermediate carry never form any one of the carry-generating combinations. The following steps may be used in deriving the required rule truth table for realizing carry-free addition.

1. The sets of MSD addend and augend digits are grouped into multiple classes in accordance with the value of their sums.
2. For each of the above classes, all possible two-digit MSD representations of the corresponding sum are identified.
3. Of the MSD representations derived in the previous step, the less significant digit is taken to represent the intermediate sum while the other digit is taken to represent the intermediate carry. The next step identifies the form of intermediate sum and carry.
4. A substitution rule for the addend and augend is selected from the different representations derived in Step 2 above. The substitution rule for a group of paired numbers that has the same sum is derived by examining the $(n - 1)$th pair of digits; so that the nth intermediate sum when added to the $(n - 1)$th intermediate carry does not generate a carry. For each substitution rule selected for a particular group of inputs, the allowable set of the $(n - 1)$th pair of digits must be specified so that the intermediate sum S_n and intermediate carry C_{n-1} generated from the group satisfy the equation

$$|S_n + C_{n-1}| < R \tag{11.29}$$

where R is the radix of the number system under consideration.

TABLE 11.8 Truth Table for Binary MSD Addition

Type	Addend A_n	Augend B_n	Digits at the Next Lower Order Position A_{n-1} B_{n-1}	Intermediate Carry	Intermediate Sum
1	0	0	Don't care		
	1	$\bar{1}$			
	$\bar{1}$	1		0	0
2	0	1	Both positive	1	$\bar{1}$
	1	0	Otherwise	0	1
3	0	$\bar{1}$	Both negative	$\bar{1}$	1
	$\bar{1}$	0	Otherwise	0	$\bar{1}$
4	$\bar{1}$	$\bar{1}$	Don't care	$\bar{1}$	0
5	1	1	Don't care	1	0

The pairs of the to-be-added MSD numbers may be divided into five groups in accordance with their sums as shown in Table 11.8. The sum ranges from -2 to $+2$. The two-digit MSD representations of -2 through $+2$ are as follows:

(a) $-2 = (\bar{1} + \bar{1}) = \bar{1}0$
(b) $-1 = (\bar{1} + 0, 0 + \bar{1}) = 0\bar{1}, \bar{1}1$
(c) $0 = (\bar{1} + 1, 1 + \bar{1}, 0 + 0) = 00$
(d) $1 = (0 + 1, 1 + 0) = 01, 1\bar{1}$
(e) $2 = (1 + 1) = 10$

Note that the substitution rules for (a), (c), and (e) are fixed since there is only one way to represent the numbers in those cases. Next, in order to find the substitution rules for (b) and (d), one notes that they have a possible intermediate sum digit of either 1 or $\bar{1}$. In order to find the substitution rules (i.e., for determining the allowable set of $(n - 1)$th digit pairs) for (b) and (d) we need to consider that there can be up to three cases of the intermediate sum digit at the nth digit position.

Sum 0 When the intermediate sum is a 0, the addition of no MSD trinary numbers may generate a carry. Thus at the $(n - 1)$th position, any combination of digits is allowable. This covers three of the five cases, (a), (c), and (e). Also note that if the intermediate sum at the nth position were not zero, then our carry-free addition scheme would have failed, since there is no alternate representation of these three cases.

Intermediate Sum 1 When the intermediate sum is a 1, a carry will be generated only if the intermediate carry at the $(n - 1)$th position is a 1. This implies that the carry from the previous digit position must not be a 1. Accordingly, in this case, the previous digit pairs can be anything except the pair (1, 1) or (1, 0) or (0, 1) which may generate a carry of 1. The allowed pairs of numbers are termed "all negative."

Negative Intermediate Sum $\bar{1}$ The negative intermediate sum $\bar{1}$ has the same requirement as it positive counterpart. The carries are avoided by not allowing $(\bar{1}, \bar{1})$, $(\bar{1}, 0)$, or $(0, \bar{1})$ to appear, which may generate the carry of $\bar{1}$. The allowed pairs of numbers are referred to as "all positive,"

Accordingly, the entries of Table 11.8 also include a set of A_{n-1} and B_{n-1} pairs of digits for each of the substitution rules.

There are often more than one way to satisfy the $(n - 1)$th pair of digits for a particular input condition. For example, the pairs for which $A_{n-1} + B_{n-1}$ is zero may be grouped as either "all positive" or "all negative" pairs. If such pairs are permanently grouped with "all positive," then the "all negative" term becomes anonymous with "otherwise." However, one particular grouping may lead to more simplification in minterm minimization than the other. In the following, we use Table 11.8 and a slight variation of it to demonstrate carry-free addition.

$$
\begin{array}{rl}
A = & 11010\bar{1}0 = (102)_{10} \\
B = & \bar{1}11\bar{1}\bar{1}\bar{1}1 = (-47)_{10} \\
\hline
\text{Intermediate sum} = & 00\bar{1}0101 \\
\text{Intermediate carry} = & 0110\bar{1}\bar{1}0 \\
\hline
& 011\bar{1}\bar{1}001 = (55)_{10}
\end{array}
$$

Using an alternate substitution scheme (if $A_{n-1} + B_{n-1}$ is 0), we find that any combination of intermediate sum and carry digit can be substituted at the nth bit position. Accordingly,

$$
\begin{array}{rl}
A = & 11010\bar{1}0 = (102)_{10} \\
B = & \bar{1}11\bar{1}\bar{1}\bar{1}1 = (-47)_{10} \\
\hline
\text{Intermediate sum} = & 0010101 \\
\text{Intermediate carry} = & 0100\bar{1}\bar{1}0 \\
\hline
& 011\bar{1}\bar{1}001 = (55)_{10}
\end{array}
$$

Note in the above sum that the only difference is in the substitution of the (1, 0) pair instead of a $(\bar{1}, 1)$ pair.

Adopting these rules, one can perform carry-free addition using a symbolic substitution scheme. The truth-table reduction technique, however, may be used to reduce the number of to-be-implemented rules. Such schemes can be extended to design almost any arbitrary arithmetic or logic unit, for example, fast multipliers. Let A, B, C, and D be the four inputs of such a multiplication unit. The 2-bit number AB is multiplied with another 2-bit number CD, which results in a 4-bit number $P_3P_2P_1P_0$. The truth table of the multiplier is shown in Table 11.9.

After reducing the minterms of Table 11.9, one obtains

$$P_3 = ABCD \tag{11.30}$$

$$P_2 = AXC\overline{D} + A\overline{B}CX \tag{11.31}$$

$$P_1 = XBC\overline{D} + \overline{A}BCX + AX\overline{C}D + A\overline{B}XD \tag{11.32}$$

$$P_0 = XBXD \tag{11.33}$$

where X denotes a "don't care" literal. In order to generate the outputs of the multiplier, one needs to implement these eight minterms as eight rules. Note that the requirement for the output generation is that a 1 is indicated by the presence of light, whereas a 0 is generated by default by the absence of light. By storing the minterms in a holographic memory, for example, one may implement the multiplier.

Image processing applications are also good candidates for symbolic substitution because images are naturally in the ready-to-use optical format for such optical processing. However, to implement image processing using symbolic substitution, the processing requirements must be first reduced to a truth-table form. Consider, for example, that we intend to remove "salt and pepper" noise from a given image using median filtering. Consider further that we have a binary image and suppose we want to apply a median filter on sets of three pixels, whose members are A, B, and C. Then by definition, a window of three pixels is scanned across the image and the center of the window is replaced by the median of the values of the three

TABLE 11.9 Truth Table for a 2 × 2 Binary Multiplication Unit

	Inputs				Outputs				
					Binary				
Input Conditions	A	B	C	D	P_3	P_2	P_1	P_0	Decimal
1	0	1	0	1	0	0	0	1	1
2	0	1	1	0	0	0	1	0	2
3	1	0	0	1	0	0	1	0	2
4	0	1	1	1	0	0	1	1	3
5	1	1	0	1	0	0	1	1	3
6	1	0	1	0	0	1	0	0	4
7	1	1	1	0	0	1	1	0	6
8	1	0	1	1	0	1	1	0	6
9	1	1	1	1	1	0	0	1	9

TABLE 11.10 Median Truth Table for Three Binary Numbers

Input Conditions	A	B	C	Median
1	0	0	0	0
2	0	0	1	0
3	0	1	0	0
4	0	1	1	1
5	1	0	0	0
6	1	0	1	1
7	1	1	0	1
8	1	1	1	1

pixels. The truth table of the median filter just described is given by Table 11.10. The truth table lists all the rules.

Using a K-map, the median filter truth table may be minimized to reduce the number of rules. The reduced median function is given by

$$M = ABX + XBC + AXC \tag{11.34}$$

Since M can be generated from three minterms, only the corresponding three rules need to be implemented for realizing the median filter. The same schemes can also be used for applying horizontal windows, or using a cross-shaped operator of nine pixels one can actually perform a two-dimensional median operation.

11.9 OPTICAL IMPLEMENTATION OF SYMBOLIC SUBSTITUTION

In a symbolic substitution scheme implementing binary logic, inputs are dual rail or polarization coded. A spatial domain technique could then rely on identifying either the transparent or opaque pixels. Since both 0 and 1 have both transparent and opaque pixel subcells, they may be identified by recognizing either one of them. In order to recognize a combination of input patterns, the following steps could be carried out:

1. The origin of the search pattern is selected.
2. The input data plane is copied n times, where n is the number of opaque cells present in the search pattern.
3. The copies are shifted (up, down, left, or right or a combination thereof) such that all the opaque cells present in the search pattern are superimposed on the reference cell. As a result, if and only if all the pixels superimposed on the reference cell are opaque will the result of the superposition be opaque; the presence of a single bright cell would result in a bright output. This condition could be determined uniquely by means of a NOR gate, whose output is bright if and only if all inputs are dark.

4. A decoding mask that has an opening at the location of the reference cells is used thereafter to decode the NOR gate bright output. A bright output will indicate the occurrence of the search pattern in the input. The bright output has another useful purpose. It is used subsequently to create the scribe pattern.
5. Next, in order to reconstruct the substitution pattern, the recognition pixels are shifted as many times as there are bright pixels in the scribe pattern, in all directions, such that they may translate from the origin to the corresponding position of the scribe pattern. This results in the scribe pattern being written in place of the substitution pattern.

If the above steps are analyzed, there are some basic steps that are necessary to implement SS. These operations are creating multiple copies from a single copy, shifting and overlapping, the NOR operation, and masking. Variant schemes may be implemented where each process may be executed differently. For example, one may employ beam splitter and tilted mirrors to relatively shift copies of the input pattern. In that case, an interferometric setup is employed. A phase-only hologram could also be used to generate multiple copies, and then the relatively shifted versions could be generated by wedge prisms. An example of the shift and overlap technique involved in recognition and substitution is shown in Fig. 11.14.

In SS, to search for dark pixel locations (in dual-rail encoded inputs), an OR or a collinear superposition of shifted inputs may be employed. This is followed by a threshold NOT operation and a masking operation. Thus the complete process is a NOR operation. From DeMorgan's law, however, NOR logic can be accom-

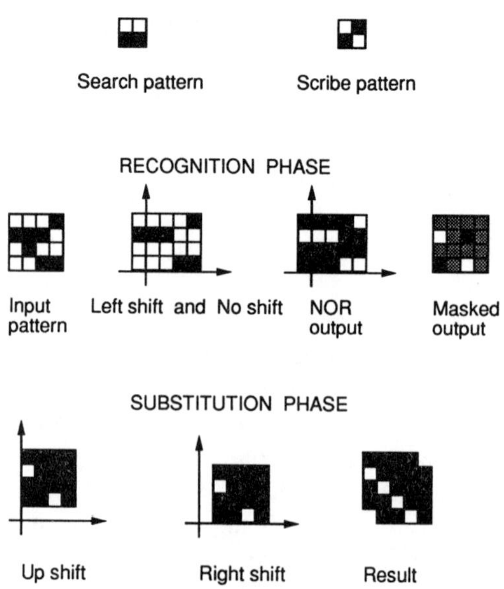

FIGURE 11.14 Symbolic substitution steps.

plished by ANDing the inverted inputs. Since in dual-rail encoding both the input and its complement are present, instead of using dark pixels, its inverted form or the white pixels could also be used. Accordingly, one can use an AND-based optical SS technique directly. The AND operation could be realized directly with the help of optical phase conjugation introduced in Chapter 10.

The above implementation techniques are based on spatial domain approaches. However, using the Fourier transform capabilities of optics, a correlator-based approach may also be used. Here, correlation may be used for both matching with a single pattern or group of patterns and also for substitution. The recognition phase of symbolic substitution is rather similar to classification of patterns into multiple classes. All the input combinations that generate the output of 1 may be considered as belonging to one class while the rest of the input combinations form the other class. Optimization techniques may be used on multiple inputs to design a matched filter so that it reduces the cross-correlation of out-of-class inputs and at the same time increases the autocorrelation of the same class of inputs. This optimization step is analogous to minterm reduction schemes in digital design. The whole process of recognizing and substituting is exactly analogous to a hetero-associative content addressable memory. NAND-based and NOR-based holographic recording schemes, which are also used to record holograms for recognition of a particular symbol, may also be employed. Before recording the hologram, a digital minimization process is first applied to reduce the number of patterns that have to be recorded. One hologram is required for recognizing one minimized rule.

As an example of minterm minimization, note the rule derived for MSD addition/subtraction in Table 11.8. Let, d_{ab} denote a don't care input, where the input may take a value of either a or b while d denotes an unconditional don't care, where the inputs may take any value possible ($\overline{1}$, 0, and 1). An observation of A_{n-1} and B_{n-1} entries of Table 11.8 reveals that they can be classified as belonging to four distinct groups. The "all-positive" ("otherwise") groups can be identified as $d_{0,1} d_{0,1} (\overline{d_{0,1} d_{0,1}})$.

11.10 LIMITATIONS AND CHALLENGES

Both POSC and SS offer very powerful optical computing schemes. Encoding, however, is one of the major bottlenecks for both of these and for almost any optical processing. POSC codes could be encoded using ferroelectric liquid crystal SLM devices. In a POSC scheme, in particular, a serious problem is the spreading of the output pixel due to the use of divergent point sources. By adopting a coherent scheme that uses holograms for image shifting and overlapping, however, this problem may be resolved. In symbolic substitution, on the other hand, as the number of to-be-implemented rules increases, the complexity of the corresponding hologram increases. Also, it may be noted that cascading is a problem in both cases. Using optical gain devices, such as the liquid crystal light valve, this problem of cascading may be alleviated. When integrated optical devices reach the maturity that the integrated electronics industry enjoys right now, many of the digital optical computing algorithms are expected to be realized.

11.11 PROBLEMS

1. Consider an optical gray adder using a POSC system that adds a binary number *AB* with another number *CD*, resulting in a sum of $S_2 S_1 S_0$. (a) Obtain the truth table showing inputs *A*, *B*, *C*, and *D*, and the corresponding output 3-bit sum. (b) Assume a 4 × 2 pixel structure for the inputs, and assuming sequential output generation show the encoding for the inputs and also the output decoder mask. (c) Next, assume a 3 × 3 pixel structure for the inputs and then show the design of a POSC system to generate all the output bits in parallel. (*Hint*: Follow the multiprocessor design)
2. Design a quaternary multiplier using a POSC scheme that multiplies a single-bit quaternary number by another single-bit quaternary number. Provide the input encoding and output decoder mask for parallel output generation.
3. Using the minterm sharing as illustrated in the multiprocessor design example, design a 2 × 2 bit multiplier that multiplies two 2-bit binary numbers *AB* and *CD* and results in a 4-bit output $P_3 P_2 P_1 P_0$. Show that as a result of applying minterm sharing it is possible to design the multiplier with only six pixel sub-cells. Each output bit is expected to be generated in parallel. Starting with the truth table, show the minterm grouping, pixel map, input encoding, and the output decoder mask.
4. Using input encoding of the multiprocessor example, show the design of an output decoder mask that will generate some of the logical outputs of an ALU in parallel. Show the effect of switching LEDs on generating different logical output.
5. Extend the two-step carry-free addition scheme as derived in Section 11.8 to the MSD trinary number system. This should result in a truth table similar to Table 11.8 but now with MSD trinary entries.
6. From the paper by Eichmann et al. (1990), derive the rules for higher order symbolic substitution for addition/subtraction using two 4-bit binary numbers at a time.
7. Incorporate the MSD number system in higher order SS (see Eichmann et al., 1990), deriving the necessary rules for carry-free addition of a 2-bit MSD number with another 2-bit MSD number.

11.12 SUGGESTED READINGS

Ahmed, J. U., and A. A. S. Awwal, Polarization-Encoded Optical Shadow Casting: An Efficient Multiplier Design, *Microwave and Opt. Technol. LETT.*, **4,** 328–331 (1991).

Awwal, A. A. S., and M. A. Karim, Edge Detection Using Polarization-Encoded Optical Shadow-Casting, *Microwave Opt. Technol. Lett.*, **1,** 377–380 (1988).

Awwal, A. A. S., and M. A. Karim, Polarization Encoded Optical Shadow-Casting Programmable Logic Array: Simultaneous Generation of Multiple Outputs, *Appl. Opt.*, **27,** 932–936 (1988).

Awwal, A. A. S., and M. A. Karim, Polarization Encoded Optical Shadow-Casting: Design of a J-K Flip-Flop, *Appl. Opt.*, **27,** 3719–3722 (1988).

Awwal, A. A. S., and M. A. Karim, Associative Polarization-Encoded Optical Shadow Casting: Gray-Level Image Encoding for Serial and Parallel Operations, *Appl. Opt.*, **28,** 284–290 (1989).

Awwal, A. A. S., and M. A. Karim, Median Filtering Using Polarization-Encoded Optical Shadow Casting, *Appl. Opt.*, **28,** 1436–1440 (1989).

Awwal, A. A. S., and M. A. Karim, Polarization Encoded Optical Shadow-Casting: Design of a Carry-Free Adder, *Appl. Opt.*, **28,** 785–790 (1989).

Awwal, A. A. S., and M. A. Karim, Multiprocessor Design Using Polarization Encoded Optical Shadow-Casting, *Appl. Opt.*, **29,** 2107–2112 (1990).

Awwal, A. A. S., and M. A. Karim, and A. K. Cherri, Polarization Encoded Optical Shadow-Casting Scheme: Design of Multioutput Trinary Combinational Logic Units, *Appl. Opt.*, **26,** 4812–4818 (1987).

Barua, S., Single-Stage Optical Adder/Subtracter, *Opt. Eng.*, **30,** 265–270 (1991).

Bocker, R. P., B. L. Drake, M. E. Lasher, and T. B. Henderson, Modified Signed Digit Addition and Subtraction Using Optical Symbolic Substitution, *Appl. Opt.*, **25,** 2456 (1986).

Bocker, R. P., B. L. Drake, M. E. Lasher, and T. B. Henderson, Photonic Computing Using Modified Signed Digit Number Representation, *Opt. Eng.*, **25,** 38 (1986).

Botha, E., D. Casasent, and E. Barnhard, Optical Symbolic Substitution Using Multichannel Correlators, *Appl. Opt.*, **27,** 817 (1987).

Brenner, K. H., New Implementation of Symbolic Substitution Logic, *Appl. Opt.*, **25,** 3061 (1986).

Brenner, K. H., A. Huang, and N. Streibel, Digital Optical Computing with Symbolic Substitution, *Appl. Opt.*, **25,** 3054–3060 (1986).

Cherri, A. K., and M. A. Karim, Symbolic Substitution Based Flagged Arithmetic Using an Efficient Symbolic Substitution, *Appl. Opt.*, **27,** 3824–3827 (1988).

Cherri, A. K., and M. A. Karim, Symbolic Substitution Based Flagged Arithmetic Using Polarization-Encoded Optical Shadow-Casting, *Opt. Commun.*, **70,** 455–461 (April 1989).

Cherri, A. K., and M. A. Karim, Symbolic Substitution Based Operations Using Holograms: Multiplication and Histogram Equalization, *Opt. Eng.*, **28,** 638–642 (June 1989).

Cherri, A. K., and M. A. Karim, Image Enhancement Using Optical Symbolic Substitution, *Opt. Eng.*, **30,** 259–264 (1991).

Eichmann, G., A. Kostrzewski, D. H. Kim, and Y. Li, Optical Higher Order Symbolic Recognition, *Appl. Opt.*, **29,** 2135–2147 (1990).

Eichmann, G., J. Zhu, and Y. Li, Optical Parallel Image Skeletonization Using Content Addressable Memory-Based Symbolic Substitution, *Appl. Opt.*, **27,** 2905 (1988).

Gaylord, T. K., and M. M. Mirsalehi, Truth-Table Lookup Processing: Number Representation, Multilevel Coding, and Logical Minimization, *Opt. Eng.*, **25,** 22 (1986).

Gaylord, T. K., M. M. Mirsalehi, and C. C. Guest, Optical Digital Truth-Table Lookup Processing, *Opt. Eng.*, **24,** 48 (1985).

Goodman, S. D., and W. T. Rhodes, Symbolic Substitution Applications to Image Processing, *Appl. Opt.*, **27,** 1708 (1988).

Guest, C. C., and T. K. Gaylord, Truth-Table Look-up Optical Processing Utilizing Binary and Residue Arithmetic, *Appl. Opt.*, **19,** 1201 (1980).

Huang, A., Parallel Algorithm for Optical Digital Computer, *Technical Digest of IEEE 10th International Optical Conference*, p. 13, 1983.

Johnson, E. L., and M. A. Karim, *Digital Design: A Pragmatic Approach*, PWS–KENT Publishing, Boston, 1987.

Karim, M. A., A. A. S. Awwal, and A. K. Cherri, Polarization Encoded Optical Shadow-Casting Logic Units: Design, *Appl. Opt.*, **28,** 2720–2725 (1987).

Kozaitis, S. P., Higher-Ordered Rules for Symbolic Substitution, *Opt. Commun.*, **65,** 339–342 (1988).

Lang, Y., D. Zhao, and H. K. Liu, Multifocus Dichromated Gelatin Hololens, *Appl. Opt.*, **22,** 3451–3456 (1983).

Li, Y., and G. Eichmann, Conditional Symbolic Modified-Signed Digit Arithmetic Using Optical Content-Addressable Memory Logic Elements, *Appl. Opt.*, **26,** 2328 (1987).

Li, Y., G. Eichmann, and R. R. Alfano, Optical Computing Using Hybrid Encoded Shadow Casting, *Appl. Opt.*, **25,** 2636–2638 (1986).

Li, Y., G. Eichmann, R. Dorisinville, and R. R. Alfano, Parallel Digital and Symbolic Optical Computation via Optical Phase Conjugation, *Appl. Opt.*, **27,** 2025 (1987).

Li, Y., G. Eichmann, R. Dorisinville, and R. R. Alfano, An AND Operation-Based Optical Symbolic Substitution Pattern Recognizer, *Opt. Commun.*, **63,** 375 (1987).

Li Y., J. Zhu, and G. Eichmann, Optical on-the-Fly Conversion of a Modified Signed Digit into Two's Complement Binary Number Representation, *Opt. Lett.*, **13,** 294 (1988).

Louri, A. Efficient Optical Implementation Method for Symbolic Substitution Based on Shadow-Casting, *Appl. Opt.*, **28,** 3264–3267 (1989).

Louri, A. Parallel Implementation of Optical Symbolic Substitution Logic Using Shadow-Casting and Polarization, *Appl. Opt.*, **30,** 540–548 (1991).

Lu, D. U., R. M. Kent, J. M. Terhove, M. A. Karim, and A. A. S. Awwal, Polarization Encoded Optical Shadow-Casting: Design of Trinary Multipliers, *Appl. Opt.*, **29,** 5242–5252 (1990).

Mait, J. N., and K. H. Brenner, Optical Symbolic Substitution: System Design Using Phase Only Holograms, *Appl. Opt.*, **27,** 1692 (1988).

McAulay, A. D., Logic and Arithmetic with Luminescent rebroadcasting Devices, *SPIE Proc.*, **936,** 321–326 (1988).

McAulay, A. D., Optical Orthogonal Neural Network Associative Memory with Luminescent Rebroadcasting Devices, *International Joint Conference on Neural Networks*, Vol. 2, pp. 483–486, Washington DC, June 1989.

McAulay, A. D., Optical Heteroassociative Memory Using Spatial Light Rebroadcasters, *Appl. Opt.*, **29,** 2067–2073 (1990).

Mirsalehi, M. M., and T. K. Gaylord, Logical Minimization of Multilevel Coded Functions, *Appl. Opt.*, **25,** 3078–3088 (1986).

Mirsalehi, M. M., and T. K. Gaylord, Truth-Table Lookup Optical Parallel Data Processing Using a Content Addressable Memory, *Appl. Opt.*, **25,** 2277 (1986).

Mirsalehi, M. M., and T. K. Gaylord, Analytical Expressions for the Sizes of Logically Minimized Truth Table for Binary Addition and Subtraction, *Appl. Opt.*, **29,** 3339–3344 (1990).

Mirsalehi, M. M., C. C. Guest, and T. K. Gaylord, Residue Number System Holographic Truth-Table Look-up Processing: Detector Threshold Setting and Probability of Error Due to Amplitude and Phase Variations, *Appl. Opt.*, **22,** 3583 (1983).

Parhami, B., Carry-Free Addition of Recoded Binary Signed-Digit Numbers, *IEEE Trans. Comput.*, **37,** 1470 (1988).

Ramamoorthy, P. A., and S. Antony, Optical Modified-Signed Digit Adder Using Polarization Coded Symbolic Substitution, *Opt. Eng.*, **26,** 821 (1987).

Rizvi, R. A., K. Zaheer, and M. S. Zubairy, Separate and Simultaneous Generation of Multioutputs in a Polarization-Encoded Optical Shadow-Casting Scheme; Design of Half- and Full-Adders and Subtractors, *Appl. Opt.*, **27,** 5176–5180 (1988).

Takagi, N., H. Yasuura, and S. Yajima, High Speed VLSI Multiplication Algorithm with a Redundant Binary Addition Tree, *IEEE Trans. Comput.*, **C-34,** 789 (1985).

Tanida J., and Y. Ichioka, Optical Logic Array Processor Using Shadowgrams, *J. Opt. Soc. Am.*, **73,** 800–809 (June 1983).

Thalman, R., G. Pedrini, B. Acklin, and R. Dandliker, Optical Symbolic Substitution Using Diffraction Grating, *Appl. Opt.*, **29,** 2126–2134 (1990).

Yatagai, T., Optical Space-Variant Logic-Gate Array Based on Spatial Encoding Technique, *Opt. Lett.*, **11,** 260–262 (1986).

Yu, F. T. S., and S. Jutamulia, Implementation of Symbolic Substitution Logic Using Optical Associative Memories, *Appl. Opt.*, **26,** 2293 (1987).

CHAPTER TWELVE

Optical Matrix Processing

12.1 INTRODUCTION

In this chapter, several discrete optical processing schemes are discussed. Most of these architectures revolve around optical matrix vector multiplication units. Many numerically useful linear algebraic and signal processing algorithms can be suitably expressed in terms of linear matrix operations, and therefore optical matrix processors can be used for such operations. Even digital Boolean logic operations can be expressed as nonlinear algebraic sums of product expression, which can easily be organized for higher efficiency and parallel evaluation in a matrix–vector format. Since digital logic involves inherent built-in self-restoring logic, these calculations can be executed with high accuracy. The bridge between the purely analog matrix computations and digital logic evaluation, however, is digital multiplication using analog convolution. In this scheme, the convolution-related property of classical optical elements is used for obtaining the product of two suitably introduced digital inputs.

12.2 MULTIPLICATION

Optically it is very simple to perform multiplication. When light passes through a modulator, it performs an analog multiplication of the input optical signal with the modulator. If light passes through a stack of multiple modulators, it performs a multiplication with all the modulators. In both cases, the to-be-multiplied numbers are represented by the transmittance of the modulator while one of the to-be-multiplied numbers could be represented by the light intensity. The multiplication result is finally obtained as the detector output. The basic equation describing the multiplication operation is the expression for intensity transmitted through successive media of transmittances $T_1, T_2, \ldots, T_n$, respectively, as shown in Fig. 12.1. The output intensity is given by

$$I_{\text{out}} = T_1 T_2 \cdots T_n I_{\text{in}} \tag{12.1}$$

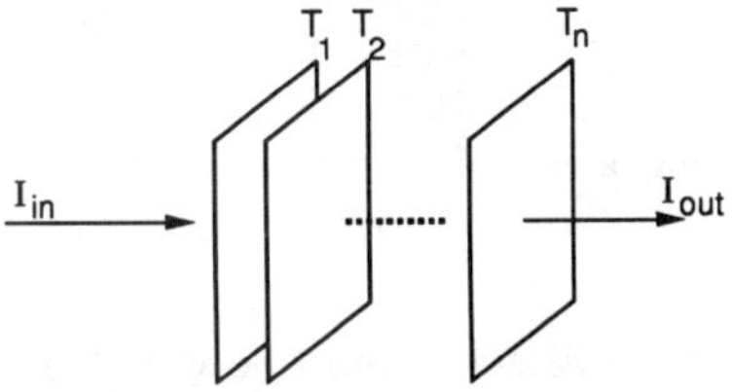

FIGURE 12.1 Multiplication via spatial modulation.

where I_{in} is the input light intensity. One of the limitations of this simple multiplication scheme is that one must be able to accurately control the transmittance of the modulators. If a photographic film is used, for example, setting the transmittance of the whole film has to be done by first characterizing the film and then developing the film with appropriate density. However, for representing a two-dimensional multiplicand, such an approach may not be very easy to perform manually, unless a computer-controlled variable aperture, capable of accurately moving in two dimensions, is available. One way this problem could be alleviated is by using a laser printer to print the modulated gray levels, and then using a camera to take its picture on a glass plate. The disadvantage, however, is that since laser printers usually utilize tiny dots of varying densities to generate gray levels, when photographed directly, it causes light to be diffracted due to the high-frequency content of gray level images. This diffraction, however, could be reduced by performing a low pass filtering on a Fourier optic setup and then imaging onto a photographic plate. The other way to modulate it is by modulating a transparent pixel by forming an opaque region inside it. Then the transmitted light intensity becomes

$$I_{out} = \int_{-a}^{a}\int_{-b}^{b} f(x, y) I_{in} \, dx \, dy \tag{12.2}$$

Now if I_{in} is a constant, it could be taken out of the integral, and thus a multiplication is performed with the total intensity and the area of the aperture.

An alternate way to represent the two to-be-multiplied numbers is by decomposing $f(x, y)$ into two orthogonal directions. The integrals could then be separated into two independent integrals such that each of the corresponding areas can be used to represent one number. Mathematically,

$$\begin{aligned} I_{out} &= \int_{-a}^{a}\int_{-b}^{b} f(x) f(y) I_{in} \, dx \, dy \\ &= I_{in} \int_{-a}^{a} f(x) dx \int_{-b}^{b} f(y) dy \end{aligned} \tag{12.3}$$

Note that the multiplier is encoded by the area of $f(x)$ and the multiplicand by that of $f(y)$. This scheme is shown in Fig. 12.2.

Note that this type of scheme can easily perform normalized floating point multiplication provided the total input light intensity represents a 1. To make this type

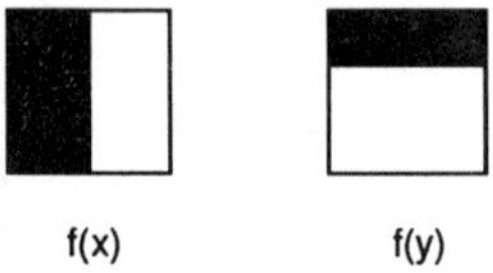

FIGURE 12.2 Multiplication of two numbers using area modulation.

of analog multiplication useful, the accuracy of the detectors must be very high. Otherwise, to achieve greater accuracy, one may prefer to use digital multiplication over the analog one. One particular technique in optics is to perform digital multiplication by analog convolution (DMAC).

12.3 MULTIPLICATION USING CONVOLUTION

Digital multiplication by analog convolution (DMAC) schemes are based on the fact that multiplying two numbers is equivalent to convolving their base-R representation so that the individual digits may have values greater than $R - 1$. Since the product is not obtained in the original base-R representation, it needs to be recorded by a subsequent conversion processor. The resulting DMAC process is thus hybrid since the associated system consists of both analog optical systems for convolution and an electronic post processor for conversion of the product into a form that is understandable.

The following theory shows how convolution may be used to find multiplication. For simplicity, the multiplier a and multiplicand b can be taken as two n-bit binary numbers:

$$a = \sum_{i=0}^{n-1} a_i 2^i \tag{12.4}$$

$$b = \sum_{i=1}^{n-1} b_i 2^i \tag{12.5}$$

where a_i and b_i may have values of either 0 or 1. The product, p, of the two binary numbers is thus given by

$$p = \sum_{k=0}^{2n-1} p_k 2^k = \sum_{i=0}^{n-1} \sum_{j=0}^{n-1} a_i b_j 2^{i+j} \tag{12.6}$$

where p_k is also binary. The product can also be expressed as

$$p = \sum_{k=0}^{2n-2} \left(\sum_{i=0}^{k} a_i b_{k-i} \right) 2^k = \sum_{k=0}^{2n-2} c_k 2^k, \tag{12.7}$$

where c_k is allowed to be mixed binary with a value in the range of 0 to n and it is understood that $a_m = b_m = 0$ when $m > n - 1$. Note that c_k is nothing but the convolution of a_k and b_k. However, the convolution of a and b in the form of analog outputs from the detector array results in a mixed binary product. Figure 12.3 shows a typical optical convolution scheme for obtaining multiplication along with a numerical example demonstrating the usefulness of Eq. (12.7). In one of the space integrating architectures, the input is applied in parallel, while a second input is channeled through acousto-optic modulators which are time multiplexed. Then the product is spatially integrated by means of optical elements. However, different convolution results are obtained sequentially at different times. Now the time and space coordinates may be switched and then the same output could be generated simply by time integration. In the example of Fig. 12.3, multiplication of the two base-10 numbers is shown to-be-obtained by convolution, multiplication with carries, and multiplication without generating any carry.

However, to have the final product recorded in binary form, the analog output c_k of the systolic convolver must be processed by an all-electronic postprocessor. The postprocessor function is therefore described by

$$\sum_{k=0}^{2n-2} c_k 2^k = \sum_{k=0}^{2n-1} p_k 2^k \tag{12.8}$$

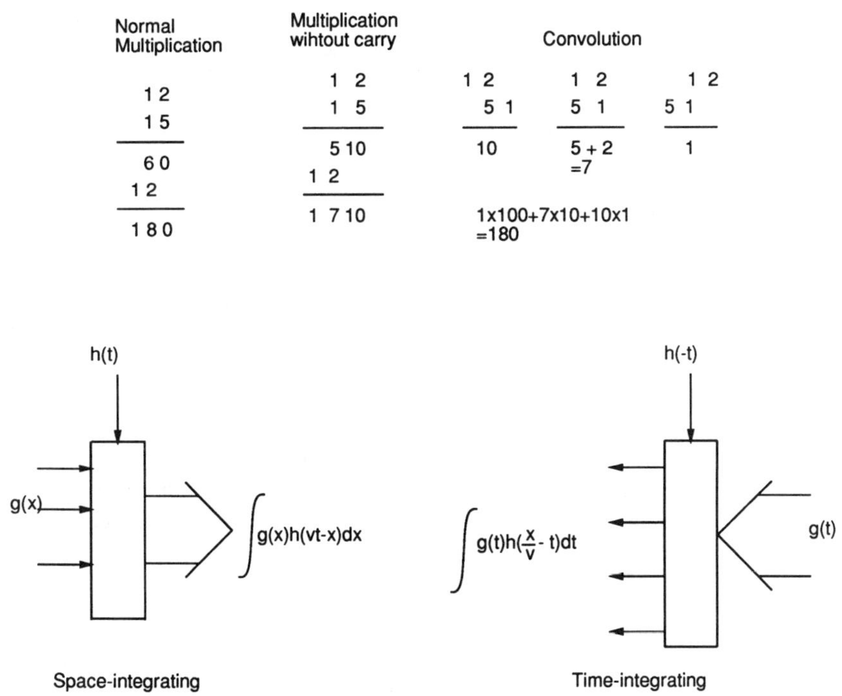

FIGURE 12.3 A typical multiplication via convolution: (a) example and (b) optical systems.

The input of the postprocessor is mixed binary while the corresponding output digits are expected to be in pure binary. In the normal postprocessing scheme, as shown in Fig. 12.4, c_k digits are sequentially passed through an analog-to-digital (A/D) converter. Note that the detector doesn't have to be a time-integrating type, in which case c_k digits are not accumulated. The A/D converter output is fed into a carry look-ahead (CLA) adder as its augend input. The addend input, on the other hand, is fed with the content of the parallel-out shift-right accumulation register, ACR. At the end of each add cycle, the least significant bit of ACR is shifted out by means of the shift-right enable. SRE. This process is continued for a total of $2n - 1$ add cycles. The postprocessing speed can be increased, however, by making use of a time-integrating detector. In that case, a q-bit CLA adder and a total of $(2n - 1)/m$ add cycles will be necessary, where m represents the number of analog columns accumulated by the time-integrating detector and

$$q = \log_2 \left[\left(\sum_{i=1}^{m} 2^{(i-1)} \right) n \right] \tag{12.9}$$

For all practical purposes, m can be either 2 or 3. With $m = 1$, this scheme reduces to the classical quasi-serial multiplication process. In any event, this postprocessor

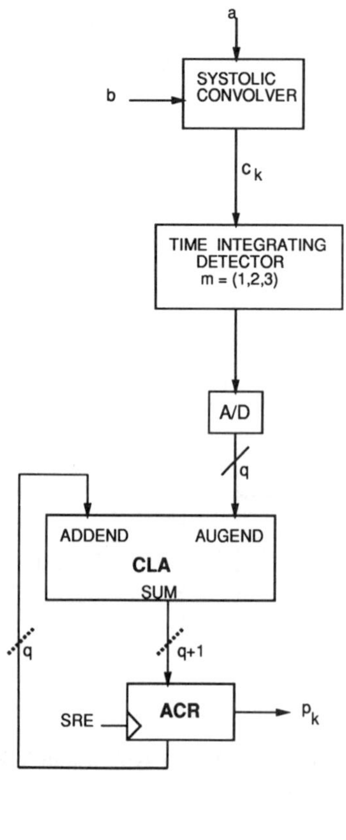

FIGURE 12.4 The quasi-serial multiplication postprocessor.

involves a large number of add cycles even for a small n. Even with a time-integrating detector, this quasi-serial multiplication scheme makes DMAC just about comparable to the all-electronic add-shift multiplication. However, it is far slower than the all-electronic speed-up schemes of which Booth's add-shift multiplication, discussed in Chapter 9, is an important example.

Alternatively, the binary equivalent of analog quantity c_k may be given by

$$c_k = \sum_{s=0}^{h-1} c_{k,s} 2^s \tag{12.10}$$

such that $h = I[(\log_2 n) + 1]$, where $I[x]$ is the integer value of x and $c_{k,s}$ is either 0 or 1. Note that when $m = 1$, $q = h$. Equation (12.10) shows the possibility of a smart postprocessor whose output can be suitably expressed as

$$\sum_{k=0}^{2n-2} c_k 2^k = \sum_{k=0}^{2n-2} \left(\sum_{s=0}^{h-1} c_{k,s} 2^s \right) 2^k$$

$$= \sum_{s=0}^{h-1} \left(\sum_{s=0}^{2n-2} c_{k,s} 2^k \right) 2^s \tag{12.11}$$

Accordingly, the number of add cycles in the postprocessor can be drastically reduced by treating the A/D converter output bit by bit and using $2n - 1$ A/D converters and $2n - 1$ parallel-in shift-right registers. Equation (12.11) makes it obvious that while the quasi-serial postprocessors (shown in Fig. 12.4) involve $2n - 1$ add cycles and an h-bit CLA adder, a smart quasi-serial (SQS) postprocessor can be designed with a $(2n - 1)$-bit CLA adder but only h add cycles. The circuit for this SQS postprocessor is shown in Fig. 12.5. This SQS speed-up postprocessor makes the complete DMAC system much more attractive from an engineering standpoint.

For comparison the all-electronic add-shift multiplication involves no mixed-binary digits and therefore, no A/D converters. The corresponding delay for an n-bit by n-bit multiplier is ndL, where d represents the propagation delay of a single NAND gate and L represents the number of NAND levels in the CLA. In comparison, Booth's all-electronic scheme saves about half of the add-shift steps and consequently the delay is about $ndL/2$. In the hybrid DMAC schemes, on the other hand, the delay must account for both the CLA adder and A/D converter. Under the best of circumstances, the A/D converter propagation delay is independent of n. Its response is proportional to the sum of the delays of its subunits, that is, the differential comparator and the priority encoder. The standard quasi-serial postprocessor delay amounts to a total delay of $[(2n - 1)L/m + 24]d$ units. In comparison, the delay for the SQS postprocessor is only $(hL + 24)d$.

Briefly stated, a CLA is composed of three successive units: propagate-generate, carry, and sum. There are three variations of the CLA adder: standard CLA, ripple CLA, and first-order CLA. They differ in their internal designs. When the internal units of an n-bit CLA adder are designed using only NAND gates, L turns out to be 8, $6 + 2z$, and 12, respectively, for these three variations, where $z = I[(n + 7)/8]$. Accordingly, standard, ripple, and first-order CLAs are used, respectively, when $n \leq 8$, $8 < n \leq 24$, and $n > 24$. The total number of NAND gates used in these CLA adders are, respectively, $(n^2 + 23n)/2$, $27n/2$, and $(n^2 + 184n)/2$.

Table 12.1 lists the relative worst-case propagation delays of the DMAC postprocessors along with those of the two all-electronic multiplication schemes. It can be inferred that while DMAC using a standard time-integrating detector (when $m = 3$) is superior to add-shift multiplication, none of the cases perform as well as Booth's method. For all practical purposes, however, parameters to Booth's scheme can be disregarded when n is small. On the other hand, DMAC based on a SQS postprocessor is exceedingly superior to all other schemes, both all-electronic and other hybrid DMACs. This logarithmic speed advantage is realized at the expense of CLA adder size, which turns out to be twice as large as that in Booth's scheme. DMAC based on the standard postprocessor involves smaller CLA adders, but this only makes them slower. However, it may be seen that while the standard postprocessor requires increasingly larger A/D converters and complicated time-integrating electronics, the SQS postprocessor is restricted to only h-bit size A/D converters and shift registers and involves no complicated time-integrating electronics.

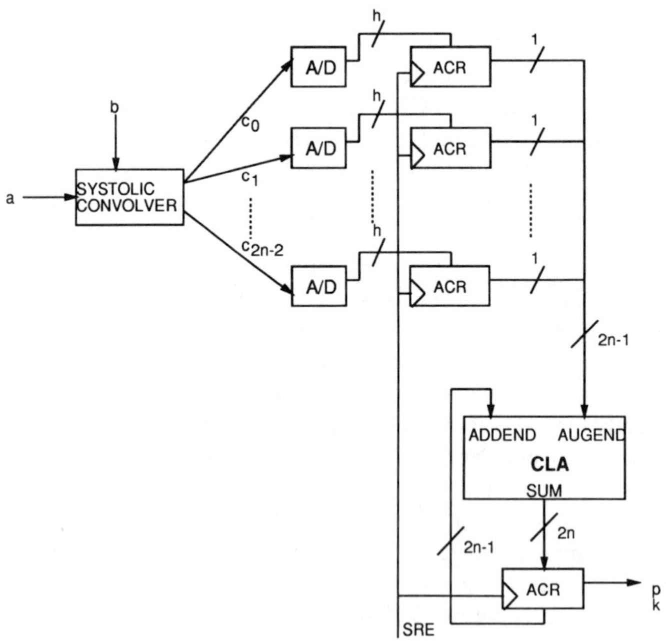

FIGURE 12.5 The smart quasi-serial postprocessor.

TABLE 12.1 Multiplication Delay (*D*) and CLA Adder Size (*S*) Versus *n*

		n						
Scheme	Measures	8	16	32	64	128	256	512
Add-Shift	$D = nL$	64	160	384	768	1536	3072	6144
	$S = n$	8	16	32	64	128	256	512
Booth's	$D = nL/2$	32	80	192	384	768	1536	3072
	$S = n$	8	16	32	64	128	256	512
Mort's	$D = 24 + (2n - 1)L$	144	272	528	1040	2064	4112	10254
$m = 1$	$S = h$	3	4	5	6	7	8	9
Mort's	$D = 24 + (2n - 1)L/2$	84	148	276	532	1299	2579	5139
$m = 2$	$S = q$	5	6	7	8	9	10	11
Mort's	$D = 24 + (2n - 1)L/3$	64	107	192	448	874	1728	3434
$m = 3$	$S = q$	6	7	8	9	10	11	12
SQS	$D = hL + 4$	72	84	96	108	120	132	144
	$S = 2n - 1$	17	31	63	127	255	511	1023

12.4 MATRIX OPERATIONS

A matrix can be interpreted as the two-dimensional representation of numbers or operators. Since optical multiplication can be performed simply by light passing through a modulator, expressing the matrix as a two-dimensional spatial light modulator or simply as a transparency, one can perform a multiplication of a matrix by a scalar constant. When a collimated beam of light is passed through the two-dimensional matrix, such a matrix multiplication is executed provided the optical intensity has been used to represent the magnitude of the scalar multiplier. As long as the multiplier is a single scalar number, representing its value by the intensity of the light does not pose any fundamental problem, for either coherent or incoherent LED sources. The numbers in the matrix could be encoded either by the transmittance of the transparency or, for a binary type of modulator, by the relative area of a pixel that is opaque compared to the transparent region. The ratio of the transparent area to the total available region can represent an analog number between 0 and 1. When it is practically difficult or when gray level modulators are unavailable, this latter approach is more desirable. To represent a negative or complex number, more than one pixel may be used.

The advantage of optics could be appreciated by noting that the two-dimensional input can be represented in a parallel format such that both the data access and arithmetic operation on the data are realized in parallel. If this simple multiplication of a matrix by a scalar constant is done in a digital computer, first the computer must store the matrix in some form, say, as floating point numbers, in the memory. Then the computer must bring one element at a time from the memory (this takes some time, called memory access time) to the central processing unit, where some time must be spent to execute the multiplication, and afterward the result has to be stored in some other location. After all these steps, we have only multiplied one element of the matrix by a constant. Now if the matrix dimension is 1000×1000, it will take 10^6 processing cycles to finish the whole process. In comparison, in an optical processor, after the matrix has been encoded on the transparency, the multiplication of all the matrix elements by the input is performed in an instant.

Now consider the case of a vector–matrix multiplication. Note that the elements of the input vector must be intensity modulated by changing the input power in the case of individually addressed LEDs or laser diodes. But for the case of a single laser source, the intensity modulation is achieved by a spatial light modulator via area modulation of the input light. In that case, the input will also be represented by modulation of the source by means of an array of one-dimensional spatial light modulators. In particular, when both the multiplicand and the multiplier are area modulated, usually one of the two operands is modulated using vertical area modulation and the other using horizontal area modulation. For a 4×4 matrix, one notes that the result of multiplication of matrix A by a vector $\mathbf{b}$ is given by

$$\begin{bmatrix} y_1 \\ y_2 \\ y_3 \\ y_4 \end{bmatrix} = \begin{bmatrix} a_{11} & a_{12} & a_{13} & a_{14} \\ a_{21} & a_{22} & a_{23} & a_{24} \\ a_{31} & a_{32} & a_{33} & a_{34} \\ a_{41} & a_{42} & a_{43} & a_{44} \end{bmatrix} \begin{bmatrix} b_1 \\ b_2 \\ b_3 \\ b_4 \end{bmatrix} \tag{12.12}$$

The multiplication is accomplished by the following formula:

$$\begin{bmatrix} y_1 \\ y_2 \\ y_3 \\ y_4 \end{bmatrix} = \begin{bmatrix} a_{11}b_1 + a_{12}b_2 + a_{13}b_3 + a_{14}b_4 \\ a_{21}b_1 + a_{22}b_2 + a_{23}b_3 + a_{24}b_4 \\ a_{31}b_1 + a_{32}b_2 + a_{33}b_3 + a_{34}b_4 \\ a_{41}b_1 + a_{42}b_2 + a_{43}b_3 + a_{44}b_4 \end{bmatrix} \tag{12.13}$$

For example, to generate the element $y_1 = a_{11}b_1 + a_{12}b_2 + a_{13}b_3 + a_{14}b_4$, one notes that the light originating from source b_1 should fall on element a_{11}; similarly, light originating from source b_2 should fall on element a_{12}; and so on. In general, light originating from b_i should fall on the ith column. Thus if the vector is represented by point sources, then by using a lens the collimated column of light may be forced to fall on the matrix columnwise. This accomplishes the "element-by-element" multiplication. Then one needs to perform one-dimensional summation of the partial products, as shown in Eq. (12.13), to generate the resulting vector elements. To accomplish this step, one may use one-dimensional

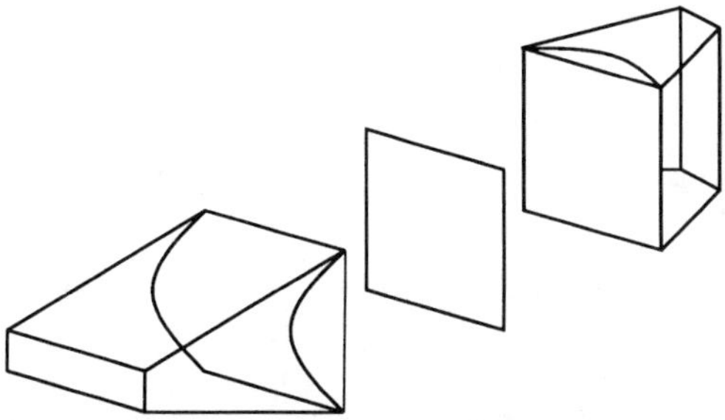

FIGURE 12.6 A vector–matrix multiplication setup.

spatial integrator such as a cylindrical lens. The overall system may appear as that shown in Fig. 12.6. Instead of using free-guided optics, a fiber optic version of the matrix–vector multiplier may also be built, where the optics is confined within the fibers. The advantage of such a system may be reduced space and arbitrary shape of the processor. The setup of this vector–matrix multiplier requires a two-dimensional spatial light modulator. If instead one has only a single cell spatial light modulator and only one LED source, then the whole matrix has to be generated time sequentially. The particular circuit of a sequential adder design may be used to this effect. As will be shown later, the matrix–vector architecture is useful in programmable logic array, content addressable logic array, and neural network applications.

Note that by using the vector–matrix processor many different inner product operations may be performed in parallel. By reducing the dimension of the matrix, one may obtain a single inner product. An outer product operation, on the other hand, produces a matrix from a pair of vectors. Mathematically, an outer product is defined as

$$\begin{aligned} y_{ij} &= \sum_{i=1}^{3} \sum_{j=1}^{3} a_i b_j \\ &= [a_1 a_2 a_3]^T [b_1 b_2 b_3] \\ &= \begin{bmatrix} a_1 b_1 & a_1 b_2 & a_1 b_3 \\ a_2 b_1 & a_2 b_2 & a_2 b_3 \\ a_3 b_1 & a_3 b_2 & a_3 b_3 \end{bmatrix} \end{aligned} \tag{12.14}$$

where a_i, b_j are elements of vector **a** and **b.** Note that each element of **a** multiplies with all elements of **b.** Therefore each element of **a** must broadcast light to all elements of **b,** such that all elements of vector **b** receive light from every element of vector **a.** For a binary type of vector, the outer product function is analogous to repeating the vector **b** by means of each element of vector **a.** For a three-element vector–vector outer product, only three scalar–vector multiplications are needed. Consequently, two one-dimensional SLMs may be placed at orthogonal orienta-

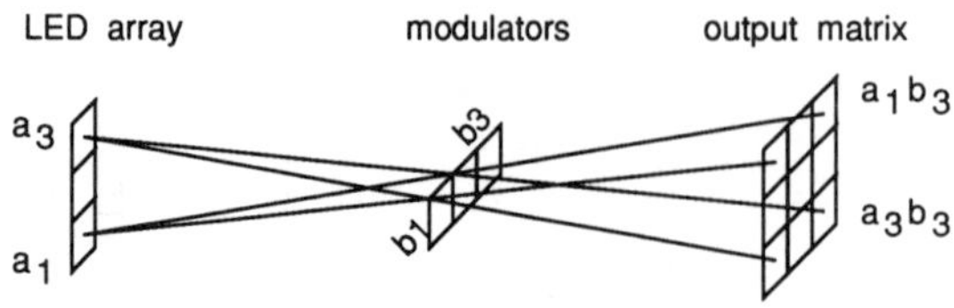

FIGURE 12.7 A vector–vector outer product multiplication setup.

tions such that the vertical vector **a** is imaged in the horizontal direction and focused in the vertical direction. The output plane should contain a two-dimensional array of detectors. The resulting optical setup for performing the vector–vector outer product is shown in Fig. 12.7.

A matrix–vector multiplier can be extended to perform matrix–matrix multiplication. The vector consisting of the first row elements of $m \times p$ matrix A multiplies with the $p \times n$ matrix B to generate the first row of the resultant $m \times n$ matrix Y. Thus m rows of matrix A generate the corresponding m rows of the resultant matrix Y. Accordingly, spatially displaced vectors in the output plane are generated by spatially displacing the vector in the input plane. The resulting matrix elements y_{ij} may be expressed as

$$y_{ij} = \sum_{k=1}^{m} a_{ik} b_{kj}, \qquad \text{where } i = 1, 2, 3, \ldots, p$$
$$j = 1, \ldots, n \qquad (12.15)$$

A spatially (angular) multiplexed matrix–matrix setup is shown in Fig. 12.8. Also, by decomposing the matrix–matrix multiplication as the summation of outer product operations, we note that it is possible to perform such operations by time multiplexing each outer product and then time integrating the result on detectors.

As an example of realizing Eq. (12.15), a systolic architecture may be employed. A systolic array is a flow through a computing architecture that computes harmonically (controlled by a clock) while the data are being passed through modular computing elements with uniform interconnections between them. As a result of the interconnections between processors, the computations can be pipelined and

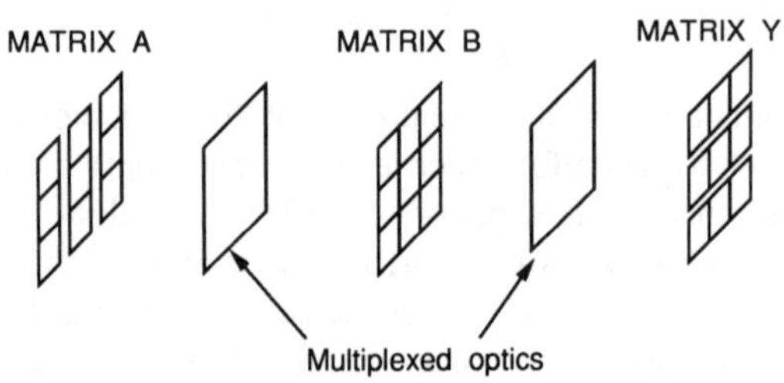

FIGURE 12.8 A matrix–matrix multiplication setup.

more efficiency may be achieved. For a 3 × 3 case of matrix–matrix multiplication, we get

$$\begin{bmatrix} a_{11} & a_{12} & a_{13} \\ a_{21} & a_{22} & a_{23} \\ a_{31} & a_{32} & a_{33} \end{bmatrix} \begin{bmatrix} b_{11} & b_{12} & b_{13} \\ b_{21} & b_{22} & b_{23} \\ b_{31} & b_{32} & b_{33} \end{bmatrix} = \begin{bmatrix} c_{11} & c_{12} & c_{13} \\ c_{21} & c_{22} & c_{23} \\ c_{31} & c_{32} & c_{33} \end{bmatrix} \tag{12.16}$$

which can also be written

$$A[b_1 \quad b_2 \quad b_3] = [c_1 \quad c_2 \quad c_3] \tag{12.17}$$

where A is the matrix containing a_{ij} and

$$b_i = \begin{bmatrix} b_{i1} \\ b_{i2} \\ b_{i3} \end{bmatrix} \tag{12.18}$$

The resulting matrix C is calculated one column at a time: c_1 then c_2 and so on. An acousto-optic setup to perform these operations is shown in Fig. 12.9. In this figure, vector $\mathbf{a}_i$ corresponds to the columns of matrix A. The AO cell is modulated by vectors $\mathbf{a}_i$, when it is illuminated by light vector elements $b_{11} b_{21} b_{31}$, thus producing the first column of matrix C. As the elements of vector $\mathbf{a}$ keep moving up, the next column of matrix C is generated.

12.4.1 Linear Algebraic Processors

Several applications of the optical matrix processors from linear algebra may be implemented. One such problem is determination of the largest *eigenvalue* of a matrix. An eigenvalue of a matrix A is defined as the quantity λ such that

$$A\mathbf{x} = \lambda\mathbf{x} \tag{12.19}$$

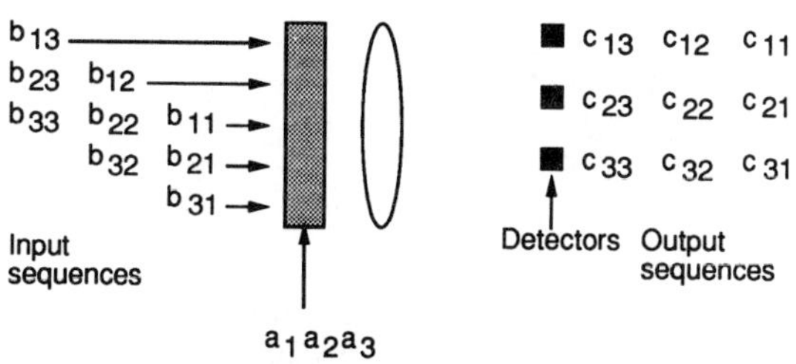

FIGURE 12.9 A matrix–matrix multiplication setup with AO cell.

where $\mathbf{x}$ is known as the eigenvector corresponding to λ. One of the iterative schemes known as the power method could be used to find the largest eigenvalue. In this method, the matrix is multiplied by any arbitrary starting vector, say, $\mathbf{q}$, then the output vector is normalized and remultiplied by the matrix. After repeated multiplication, the term $A^n\mathbf{q}$ converges to the required eigenvector. The multiplier yields the required eigenvalue as the scaling factor. Then after having removed the largest eigenvalue from the matrix, the second highest eigenvaluc may be obtained. The same process may then be repeated for determining other eigenvalues. Figure 12.10 shows a possible optical implementation of eigenvalue determination. An arbitrary input vector may be represented by LEDs, which are then used to multiply the matrix A, encoded in a spatial light modulator. The output is detected on photodetectors, which is then electronically fed back to the input LEDs.

12.4.2 Singular Value Decomposition

Singular value decomposition is useful for image compression in a least-square sense. From the singular value decomposition theorem, one notes that if A is a matrix of size $n \times m$ (with $n \geq m$) and has a size r, then there exists a matrix U of size $n \times n$, Σ of size $n \times m$, and V of size $m \times m$, such that U and V are orthogonal and the singular value decomposition of a matrix A is given by

$$A = U \Sigma V^T$$

$$= U \left[\begin{array}{cccccc|c} \sigma_1 & 0 & 0 & \ldots & & \ldots & 0 \\ 0 & \sigma_2 & 0 & \ldots & & \ldots & 0 \\ 0 & 0 & \sigma_3 & & & & \\ & & & & & & \\ 0 & & & & & \sigma_r & \\ \hline 0 & 0 & 0 & 0 & & 0 & 0 \end{array}\right] V^T \tag{12.20}$$

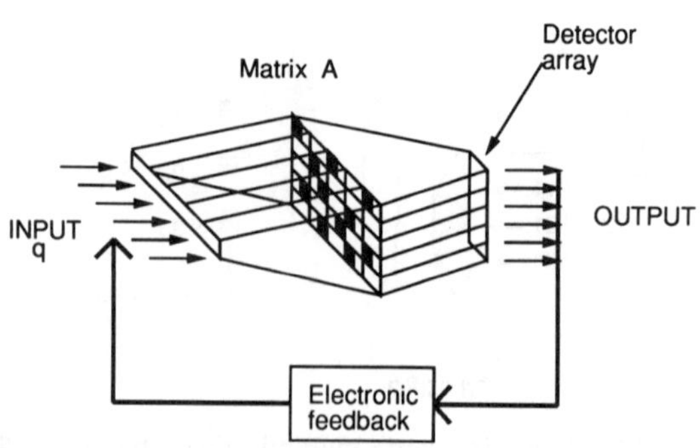

FIGURE 12.10 Optical determination of largest eigenvalue.

where $\sigma_1 \geq \sigma_2 \cdots \geq \sigma_r > 0$ and $V(U)$ contains the right (left) singular vectors $v_1, \ldots, v_m$ $(u_1, \ldots, u_n)$. By evaluating AA^T and A^TA, in terms of U, V, and Σ, it can be shown that $v_1, \ldots, v_m$ are eigenvectors of A^TA, $u_1, \ldots, v_n$ are eigenvectors of AA^T, and $\sigma_1^2, \ldots, \sigma_r^2$ are the nonzero eigenvalues of AA^T and A^TA. Mathematically,

$$AA^TU = U(\Sigma\Sigma^T) \tag{12.21}$$

and

$$A^TAV = V(\Sigma^T\Sigma) \tag{12.22}$$

Thus the singular values (i.e., σ_i's) of matrix A can be found by finding the eigenvalues of either A^TA or AA^T by the power method. Now applying the optical power method described in the previous section, one can find the eigenvalues of any of the above matrices. Taking the square root of the eigenvalues then provides the required singular values.

12.4.3 Parabolic Partial Differential Equation Solver

The parabolic partial differential equation can be decomposed into a discrete form, which can be solved numerically. Usually, in one of the numerical forms, a two-dimensional parabolic partial differential equation can be expressed in terms of a matrix–vector multiplication. Such equations are useful in determining, for example, the heat distribution in a two-dimensional form.

A parabolic equation in two dimensions is generally given by

$$\frac{\partial u}{\partial t} = \gamma\left[\frac{\partial^2 u}{\partial x^2} + \frac{\partial^2 u}{\partial y^2}\right] \tag{12.23}$$

The first derivative of u with respect to time may be expressed by a backward difference operator as

$$\begin{aligned}\frac{\partial u}{\partial t} &= \frac{u(t + \Delta t) - u(t)}{\Delta t} \\ &= \frac{u_{i,j}^{k+1} - u_{i,j}^{k}}{\Delta t}\end{aligned} \tag{12.24}$$

while the second derivative of u with respect to x is given by

$$\begin{aligned}\frac{\partial^2 u}{\partial x^2} &= \frac{u^k(x + \Delta x, y) - 2u^k(x, y) + u^k(x - \Delta x, y)}{(\Delta x)^2} \\ &= \frac{u_{i+1,j}^{k} - 2u_{i,j}^{k} + u_{i-1,j}^{k}}{(\Delta x)^2}\end{aligned} \tag{12.25}$$

Similarly, the y derivative may also be expanded. Note that here i and j are used to denote the x and y directions and superscript k is used for indicating time dependence. Assuming $\Delta x = \Delta y$ and substituting $r = \gamma \Delta t/(\Delta x)^2$, one obtains

$$u_{i,j}^{k+1} = r(u_{i+1,j}^{k} + u_{i-1,j}^{k} + u_{i,j+1}^{k} + u_{i,j-1}^{k}) + (1 - 4r)u_{i,j}^{k} \quad (12.26)$$

where the maximum value of r is restricted to 0.25 for stable solution. The matrix representation of Eq. (12.26) is given by

$$u_{i,j}^{k+1} = \begin{bmatrix} & 1 & \\ 1 & (1-4r) & 1 \\ & 1 & \end{bmatrix} U^k \quad (12.27)$$

where its central element corresponds to the coordinate $u_{i,j}$. Thus by supplying the initial distribution of u at any time, the recursive equation can be developed to estimate u at a later time. Since the partial differential equation is expressible in matrix form, it is implementable using optics, thus providing a fast solution to the problem under consideration.

12.5 CELLULAR LOGIC ARCHITECTURE

A cellular logic processor is particularly well suited for optics mainly because of its two-dimensional data arrangement and its parallel computation capability. A cellular processor, in general, is composed of an n-dimensional interconnected cell structure. The state of a cell as a function of time is basically governed by an equation, which is in turn a function of the state of the neighboring cells. A simple example suffices here to illustrate how a particular cellular logic may be defined. Both optical shadow-casting based operations and symbolic substitution can be shown to be special cases of cellular logic.

Figure 12.11a shows a cellular automata rule (or substitution rule) that can be used repetitively to perform an AND operation. Note that the cells labeled with an inner d represent data pixels while those with a c represent control cells. The rules described in Fig. 12.11a can be applied to the first image structure in Fig. 12.11b to realize an AND logic. When data inputs are introduced at the location of d cells in this structure and the rule described in Fig. 12.11a is applied, the second structure is obtained. Successive such substitutions will result in an image structure in the fourth step so that a least one datum appears at the output. Note that if only one of inputs were present, then no output would be produced. This is equivalent to an AND operation.

Application of cellular logic in image processing can also be achieved via rules such as erosion and dilation described in Fig. 12.12. A specific combination of these rules may be applied to perform removal of small unwanted noise from an image. Figure 12.12 shows four particular morphological operations: erosion, dilation, opening, and closing.

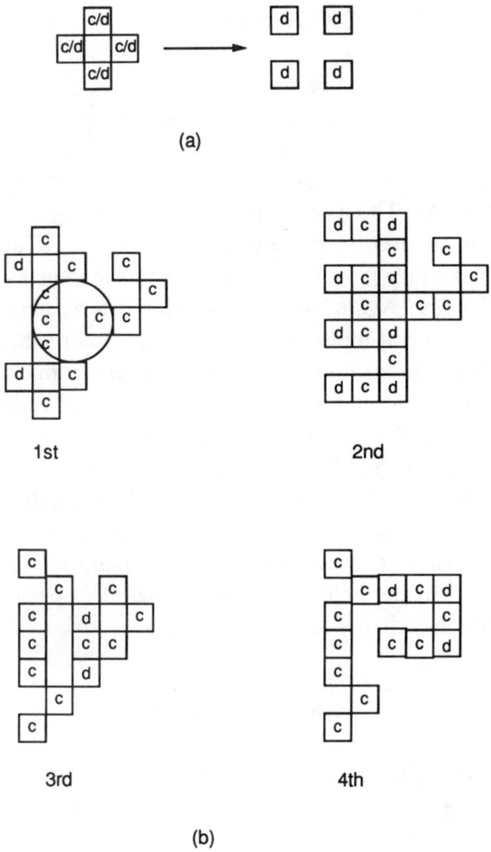

FIGURE 12.11 (a) A cellular automata rule. (b) Application in the construction of an AND gate.

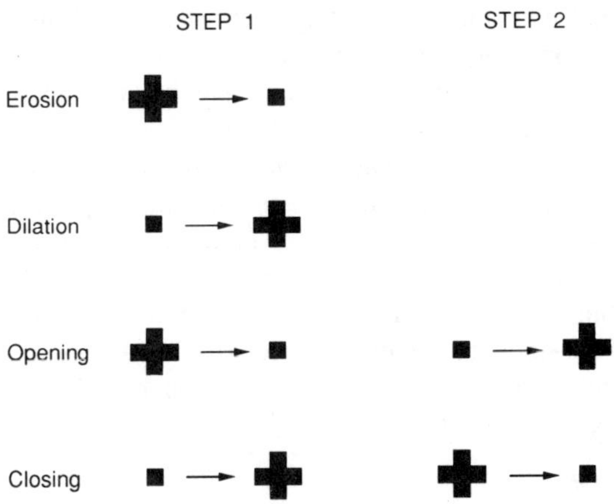

FIGURE 12.12 Cellular automata rule for morphological transformation.

12.6 PROGRAMMABLE LOGIC ARRAY

Programmable logic array (PLA) is an electronic device that consists of a set of AND gates followed by a set of OR gates, which can be organized in AND–OR format suitable for realizing any arbitrary general purpose logic operation. The optical analog of these electronic devices may be referred to as optical programmable logic array (OPLA). The motivation behind designing an OPLA is the ease with which optics can be used to perform any arbitrary logic operation. For example, AND logic is readily performed when the transparencies representing the inputs are just overlapped. Again, an OR operation can be performed by having transmitted light from several inputs converge to a common detector. The optical shadow-casting system, for example, uses a setup where the overlap of inputs results in an AND operation, and then the OR operation is realized by projecting multiple shadows to a common point. One problem with this approach is that since each input is represented by SLM, as the number of inputs increases the number of spatial light modulators increases proportionately. In a slightly modified version of OPLA, however, each input is represented by light sources instead, and then instead of performing AND via modulators, it is suitably realized using OR and NOR operations. We consider an example to illustrate this point.

Say a logical function F that is to be implemented is given by the sum of products form

$$F = AB + BC\overline{D} + A\overline{C}\,\overline{B} \tag{12.28}$$

where A, B, and C are logical inputs to the device. Using De Morgan's law, one notes that by performing an inversion operation twice on each of the product terms, Eq. (12.18) can be written

$$F = \overline{\overline{A} + \overline{B}} + \overline{\overline{B} + \overline{C} + D} + \overline{\overline{A} + C + B} \tag{12.29}$$

Thus the AND operation may be obtained by inverting the input first and then performing OR and then finally inverting each of the OR outputs. Finally, by performing the OR operation on the product term, the sum of product F is obtained. In other words, a sum of products term may be evaluated by an INVERT–OR–INVERT–OR sequence. The major advantage of the above formulation is that OR can be realized rather efficiently using optics. The inversion, on the other hand, is accomplished in dual logic by simply interchanging the pixels, or by selecting the pixel that contains the inverted positive logic input. Three architectures for implementing the above function may be considered. In one of the solutions, a free-space interconnection using a matrix–vector multiplier may be used; in the second case, a fiber-optic based interconnection may be used; and thirdly, an acousto-optic systolic array may be implemented rather easily.

12.6.1 Free-Space

The free-space interconnection may be used to create high fan-in OR gates. Mathematically, the OR operation can be done by performing an inner product of the terms of a binary input data vector and another binary to-be-determined vector mask. The vector containing the input literals usually contains all possible inputs, that is, all the inputs and their complements. Consider, for example, $(\overline{A} + \overline{B}) = [1 \quad 0 \quad 1 \quad 0][A \quad \overline{A} \quad B \quad \overline{B}]^T$. The function of the binary vector mask $[1 \quad 0 \quad 1 \quad 0]$ is to select which input literals have to be ORed. In a similar way, all the inverted AND terms of Eq. (12.29) may be generated by selecting a different binary mask corresponding to the different to-be-ORed terms.

The matrix–vector representation, Eq. (12.29), becomes

$$\begin{pmatrix} y_1 \\ y_2 \\ y_3 \end{pmatrix} = \begin{pmatrix} 0 & 1 & 0 & 1 & 0 & 0 & 0 & 0 \\ 0 & 0 & 0 & 1 & 0 & 1 & 1 & 0 \\ 0 & 1 & 0 & 0 & 1 & 0 & 1 & 0 \end{pmatrix} \begin{bmatrix} A \\ \overline{A} \\ B \\ \overline{B} \\ C \\ \overline{C} \\ D \\ \overline{D} \end{bmatrix} \tag{12.30}$$

Using the above binary mask, one can perform a selective AND operation between the input literals. The next complement and subsequent OR step could be realized by at least three ways. One of the ways involves performing an inversion using a LCLV, and then again an optical OR via a cylindrical lens.

An alternate scheme to accomplish the total processing involves performing first electronic detection of the matrix–vector product term corresponding to Eq. (12.30), then inverting vector y and feeding back the signal for a subsequent OR operation by the same setup. A third scheme may involve implementing the inner product optically and then conducting electronic inversion and an electronic OR operation. These schemes are shown in Fig. 12.13. It may be noted that to make the logic array programmable one has to have a controllable mask or spatial light modulator in place of the fixed mask. Since the fixed mask only provides a fixed logic operation between the input literals, a programmable mask can make the logic programmable.

12.6.2 Fiber-Optic

The example of a vector–matrix multiplier that was selected earlier could also be handled using a fiber-optic interconnection. In a vector–matrix multiplier, the interconnection was accomplished by a cylindrical or spherical lens. Accordingly, one of its disadvantages may be that it takes up too much space. It may be rather difficult to miniaturize the whole setup. However, if fibers are used, the input LEDs could be connected directly to the proper detectors without going through any spatial mask since the function of the spatial mask may be performed by the fixed connections. For example, to generate an OR operation of $\overline{A}$ and $\overline{B}$, the

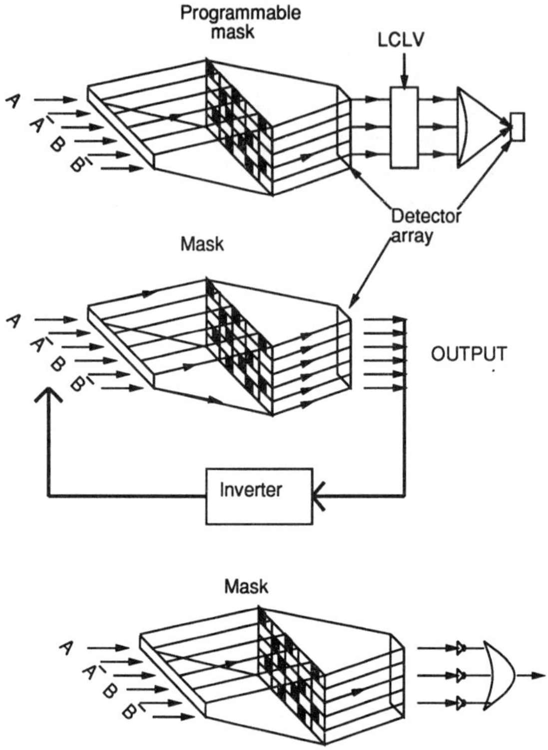

FIGURE 12.13 Free-space programmable logic arrays.

output from the LEDs representing $\overline{A}$ and $\overline{B}$ could be taken directly to the detector for generating y_1, and so on. Thus the complete setup could be completed by providing interconnections between LEDs and detectors by fibers. Suppose we want to design a logical unit capable of generating four logical functions AND, NAND, OR, and NOR. One may generate all the minterms (AB, $A\overline{B}$, $\overline{A}B$, and $\overline{AB}$) using fiber-optic interconnections as shown in Fig. 12.14. Then using a second level inverter followed by an OR gate, selective minterms can be ORed together to generate the necessary function. For example, to generate NAND logic, one only has to invert the AND and consequently obtain the NAND logic.

12.6.3 Acousto-optic

The product operation could be done by an acousto-optic modulator. An acousto-optic modulator can modulate the intensity of light by causing the cells to be driven by some electric driver circuit, such that the output intensity is proportional to the input to the cell driver. The second capability is that it deflects a beam of light in different directions. Both of these capabilities could be exploited in different OPLA matrix–vector multiplication architectures. A typical matrix–vector multiplication setup is shown in Fig. 12.15.

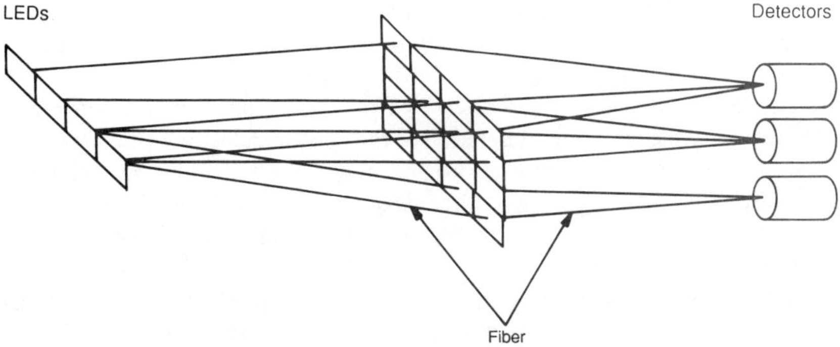

FIGURE 12.14 A fiber-optic programmable logic array.

Consider a 3 × 3 example of a vector–matrix multiplication as follows:

$$\begin{bmatrix} y_1 \\ y_2 \\ y_3 \end{bmatrix} = \begin{bmatrix} a_{11} & a_{12} & a_{13} \\ a_{21} & a_{22} & a_{23} \\ a_{31} & a_{32} & a_{33} \end{bmatrix} \begin{bmatrix} x_1 \\ x_2 \\ x_3 \end{bmatrix}$$

$$= \begin{bmatrix} a_{11}x_1 + a_{12}x_2 + a_{13}x_3 \\ a_{21}x_1 + a_{22}x_2 + a_{23}x_3 \\ a_{31}x_1 + a_{32}x_2 + a_{33}x_3 \end{bmatrix} \tag{12.31}$$

where vector **b** may represent input literals and matrix A may represent the binary mask. The implementation of Eq. (12.32) may utilize the beam modulation capability of an AO cell. The processor consists of an input laser diode array, collimation lens for individual sources, AO cell, imaging system, and a linear detector array. The element x_1 of the input vector is applied to the driver, which then produces a short diffraction grating that propagates across the modulator cell. When

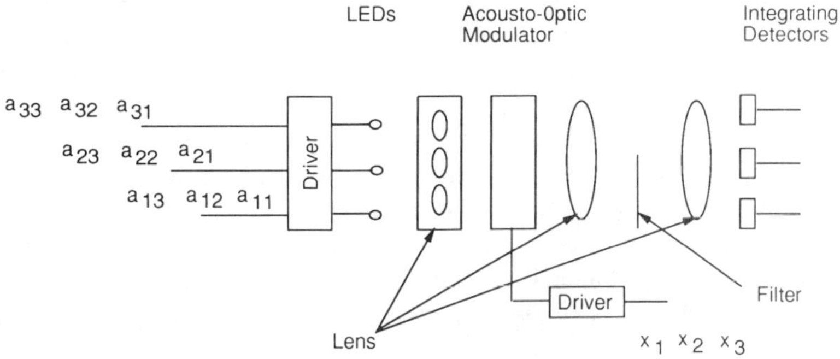

FIGURE 12.15 Systolic architecture PLA using a matrix–vector multiplier.

it is in front of laser diode 1, it is pulsed with an intensity proportional to the matrix element a_{11}, and then the product of $a_{11}x_1$ is formed, which is then collected by the detector. When x_1 travels further down, a grating produced due to x_2 arrives in front of diode 2, when it is addressed by a_{12}. At the same time a_{21} gets multiplied by x_1. Now the temporal integration of the data on detector 1 provides $a_{11}x_1 + a_{12}x_2$. Next, as x_1 continues to propagate, it gets multiplied by a_{31}, while a_{22} multiplies with x_2 and a_{31} gets multiplied by x_1. Now the detector y_1 has the sum of all three partial products, while the y_2 detector has the partial sum $a_{21} + a_{22}x_2$ and detector y_3 has just $a_{31}x_1$. In the next two clock cycles, y_2 and y_3 will be completely calculated. As shown earlier, any PLA is expressible in terms of matrix–vector multiplication, and therefore it is possible to perform PLA calculations on an AO processor.

12.7 PROBLEMS

1. Design a mask to implement an OPLA to obtain $C = AB + BC + CA$, and $S = A + B + C$.
2. Consider $A = (1101)_2$ and $B = (1011)_2$, and perform digital multiplication between them. Then by convolving the 2-bit streams show that the multiplication result may be obtained from it as stated by Eq. (12.3).

12.8 SUGGESTED READINGS

Arrathoon, R., and S. Kozaitis, Architectural and Performance Considerations for a 10^7-instructions/sec Opto-electronic Central Processing Unit, *Opt. Lett.*, **12,** 956 (1987).

Athale, R. A., Optical Matrix Algebraic Processors: A Survey, *Proceedings of the 10th International Optical Computing Conference*, pp. 24–31, IEEE No. 83CH1880-4, 1983.

Baranoski, E. J., and D. P. Casasent, High Accuracy Optical Processors: A New Performance Comparison, *Appl. Opt.*, **28,** 5351–5357 (1988).

Boker, R. P., Matrix Multiplications Using Incoherent Optical Techniques, *Appl. Opt.*, **13,** 1670 (1974).

Casasent, D., Acousto-Optic Linear Algebra Processors: Architectures, Algorithms, and Applications, *IEEE Proc.*, **72,** 831 (1984).

Eichmann, G., Y. Li, P. P. Ho, and R. R. Alfano, Digital Optical Isochronous Array Processing, *Appl. Opt.*, **26,** 2726 (1987).

Floyd, T. L., *Digital Fundamentals*, 2nd ed., Bell Howell, Columbus, OH, 1982.

Golshan, R., and J. S. Bedi, Implementation of Carry-Look Ahead Adder with Spatial Light Modulators, *Opt. Commun.*, **68,** 175 (1988).

Goodman, J. W., A. R. Dias, and L. M. Woody, Fully Parallel, High Speed Incoherent Optical Method for Performing Discrete Fourier Transforms, *Opt. Lett.*, **2,** 1 (1978).

Hill, F. J., and G. E. Peterson, *Digital Systems: Hardware Organization and Design*, Wiley, New York, 1978.

Hopper, D. G., The Control Operator Method, *SPIE*, **639,** 89 (1986).

Huang, H. X., L. R. Liu, Y. Z. Yin, and L. Y. Zhao, Fast Parallel Complex Discrete Fourier Transforms Using a Multichannel Optical Correlator, *Opt. Commun.*, **68,** 408 (1988).

Johnson, E. L., and M. A. Karim, *Digital Design: A Pragmatic Approach*, PWS–Kent Publishing, Boston, 1987.

Karim, M. A., A. Smart Quasi-Serial Post-Processor for Optical Systolic Systems, *Appl. Opt.*, **30,** 910–912 (1991).

Kung, S. Y., On Supercomputing with Systolic/Wavefront Array Processor, *IEEE Proc.*, **72,** 867 (1984).

Kung, S. Y., *VLSI Array Processors*, Prentice-Hall, Englewood Cliffs, NJ, 1988.

Li, Y., G. Eichmann, and R. R. Alfano, Fast Parallel Optical Digital Multiplication, *Opt. Commun.*, **64,** 99 (1987).

McAulay, A. D., Deformable Mirror Nearest Neighbor Optical Digital Computer, *Opt. Eng.*, **25,** 76 (1986).

Mort, M. S., Modified Quasi-Serial Multiplier, *Appl. Opt.*, **24,** 1396–1398 (1985).

Mowle, F. J., *A Systematic Approach to Digital Logic Design*, Addison-Wesley, Reading, MA, 1977.

Psaltis, D. Two-Dimensional Optical Processing Using One-Dimensional Input Devices, *IEEE Proc.*, **72,** 962 (1984).

Psaltis, D., and R. A. Athale, High Accuracy Computation with Linear Analog Optical Systems: A Critical Study, *Appl. Opt.*, **25,** 3071–3077 (1986).

Rhodes, W. T., and P. S. Guilefoyle, Acousto-optic Algebraic Processing Architectures, *IEEE Proc.*, **72,** 820 (1984).

Watkins, D. S., *Fundamentals of Matrix Computation*, Wiley, New York, 1991.

Yu, F. T. S., and M. F. Cao, Digital Optical Matrix Multiplication Based on a Systolic Outer-Product Method, *Opt. Eng.*, **26,** 1229 (1987).

CHAPTER THIRTEEN

Artificial Intelligent Computations

13.1 INTRODUCTION

While memoryless parallel computing is a specific way of achieving intelligent computations, artificial intelligent machines always involve a feedback of some sort. This feedback makes a machine seem like an intelligent one. In fact, by using feedback, such machines are able to correct their own past mistakes by incorporating the error signal. The concept of feedback is an ancient one. The formal study of such systems using engineering tools is found in control theory, which has now found use in adaptive signal processing, where instead of retrieving an error signal from the system it is capable of obtaining the error estimate from the surroundings. With the advancement of parallel computing capability using both optics and VLSI technology, adaptive signal processing has now given birth to the field of neural networks. The difference now, however, is that more researchers than ever before have plunged into this field. Interestingly, these newer efforts are interlinking areas as remote as chemistry, biology, linguistics, mathematics, physics, engineering, surgery, and computer science. When the initial glitter fades away, it is believed, important contributions in this area will surface.

There is more than one way to look at artificial intelligence (AI). It often refers to a system that can perform rule based inference. A simple correlator in that sense could be made to sound like an AI system that also makes some decision based on some rules. A system of linear correlators coupled with nonlinear elements can perform complex intelligent decision-making. It is not just a coincidence that the autocorrelation peak is nothing but an inner product between two objects. The majority of AI tasks are geared toward automated target recognition problems in which optics has proved to be quite applicable, which is another reason for the interdependence of optics and AI areas.

13.2 NEURAL NETWORKS

One of the important features of living organisms is their continual capability to learn. Learning has been hypothesized to take place while elementary brain cells known as neurons modify their interconnection weights, otherwise known as synapses. Even an infant brain shows its remarkable capability to recognize or classify patterns or objects from among noise and clutter. This capability cannot be repeated even by the most powerful computer. The key element for this capability of a human brain relies on the collective decision-making capability of densely connected neuron cells, which represents a totally novel computing architecture. Neural networks, if properly designed, are expected to achieve high computational speed by exploiting their massive parallelism and greater degree of fault tolerance. The potential application includes target recognition, classification, algorithm development to solve problems for which no algorithm exists, capturing expertise, and data summarizing. One of the main problems of neural network based computing, however, is its implementation of a huge number of required interconnections, while the other problem of programming this parallel computing machine is an inherent problem involving the bottlenecks of this otherwise serial or quasi-serial machine. This implementation problem can be solved perhaps by optics, since it possesses the capability to provide an immense number of interconnections. Several models of neural networks have already been proposed for optical implementation.

The smallest computational element, or neuron, forms a weighted sum of the inputs and then passes the result on to the next neuron through a nonlinear function. The said nonlinear function could be a hard limiter, such that when the weighted sum is greater than zero, the output is unity and zero otherwise. Such threshold elements have already been discussed in Chapter 7. A soft threshold, known as a sigmoid function, may also be used sometimes and it can be expressed as

$$f(z) = \frac{1}{1 + e^{-z}}. \tag{13.1}$$

The simplest form of neural networks are the perceptrons, which are just single-layer networks whose outputs are directly connected. It can be applied to classification on an input surface. The input surface could be an n-dimensional feature space. The output of such a system is given by

$$y = f\left(\sum_{i=0}^{N-1} w_i(t)x_i(t) - \tau\right) \tag{13.2}$$

where w_i is the weight from input i at time t and τ is the threshold at the output node. A two-input perceptron, as shown in Fig. 13.1, makes a decision boundary between the two categories of objects in the feature space. One disadvantage of

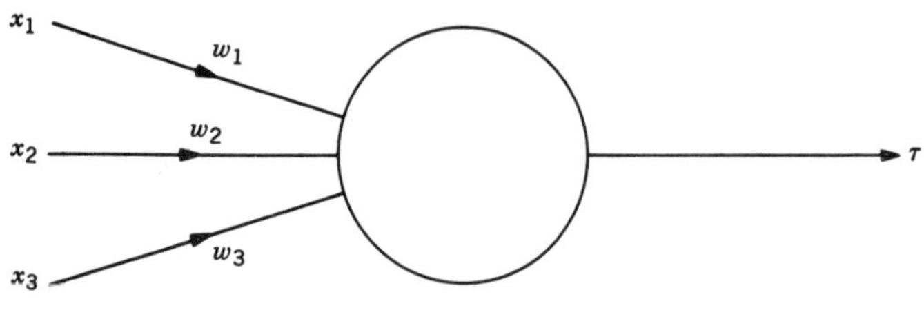

FIGURE 13.1 Example of single-layer perceptron.

the single-layer perceptron is that it cannot form a nonlinear decision boundary for classification. This can be alleviated, however, by using a multi-layer network or higher-order neural network.

Multi-layer perceptrons are feedforward nets with one or more layers of nodes between the input and output (Fig. 13.2). These layers are also called hidden layers. The extra strength of multi-layers results from the nonlinearity associated with the hidden layers. These nets are trained using a back propagation training algorithm, which is a generalized version of the Widrow–Hoff least mean square algorithm. Training is a method of adjusting the interconnection weights of a neural network, so that the error between the target output and the actual output continues

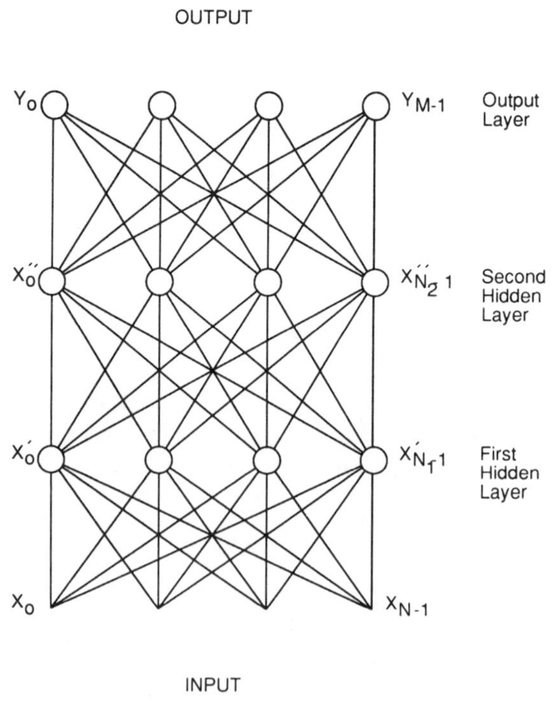

FIGURE 13.2 Example of multi-layer perceptron.

to decrease. It uses an iterative gradient search approach to minimize the error between the desired and the actual output. Usually a continuous sigmoid described by

$$f(z) = \frac{1}{1 + e^{-(z-\tau)}}. \tag{13.3}$$

provides the nonlinearity.

The back propagation algorithm runs as follows:

1. Initialize the weights and the offset τ to small random values.
2. A set of desired input vector $u_0, u_1, \ldots, u_{N-1}$ and the corresponding desired output $v_0, v_1, v_2, \ldots$ are specified.
3. The actual output is calculated by calculating the respective output for each layer:

$$y = f\left(\sum_{i=0}^{N-1} w_i(t)u_i(t)\right) \tag{13.4}$$

4. Training is done on the weights by starting from the output nodes and working back to the first hidden layer. Weights are altered by the general rule, also known as the delta rule, which is given by

$$w_{ij}(t+1) = w_{ij}(t) + \eta\delta_j x_i' \tag{13.5}$$

where $w_{ij}(t)$ is the weight from the hidden node i or from an input to node j at time t, x_i' is either the output of node i or is an input, η is a gain term, and δ_j is an error term for node j. If node j is an output node, then

$$\delta_j = y_j(1 - y_j)(v_j - y_j) \tag{13.6}$$

where v_j is the desired output while y_j is the actual output. For a hidden node j,

$$\delta_j = x_j'(1 - x_j')\sum_k \delta_k w_{jk} \tag{13.7}$$

where k is over all nodes in the layer above j. Convergence is accelerated by adding a momentum term to Eq. (13.5):

$$w_{ij}(t+1) = w_{ij}(t) + \eta\delta_j x_i' + \alpha(w_{ij}(t) - w_{ij}(t-1)) \tag{13.8}$$

The momentum term $(0 < \alpha < 1)$ determines the magnitude of weight change between each iteration and is weighted toward changing the weights in the next iteration.

13.2.1 Hopfield's Model

Hopfield's model uses the outer product of the vector with itself as a memory matrix for that vector and thereafter adds such memory matrices of each of the nearly orthogonal M stored vectors. The diagonals of these memory matrices are deliberately set to zero to indicate that the connection of a neuron with itself is nonexistent. Symbolically, the memory matrix element for M vectors can be expressed, as in the outer product model, by

$$T_{ij} = \begin{cases} \sum_{k=1}^{M} \mathbf{V}_i^{\mathrm{st},k} \mathbf{V}_j^{\mathrm{st},k}, & \text{if } i = j \\ 0, & \text{otherwise} \end{cases} \tag{13.9}$$

where $\mathbf{V}_p^{\mathrm{st},m}$ represents the pth element of the mth stored vector. For matching and retrieval, the memory matrix is multiplied by an input vector $\mathbf{V}^{\mathrm{in}}$, which may or may not be one of the stored vectors. The resultant ith element of the estimated vector $\mathbf{V}^{\mathrm{es}}$ is thus given by

$$\begin{aligned} \mathbf{V}_i^{\mathrm{es}} &= \mathrm{sgn}\left[\sum_{j-1}^{n} T_{ij} \mathbf{V}_j^{\mathrm{in}}\right] \\ &= \mathrm{sgn}\left[\sum_{k=1}^{M} \mathbf{V}_i^{\mathrm{st},k}\left[\sum_{j=1}^{n} \mathbf{V}_j^{\mathrm{st},k} \mathbf{V}_j^{\mathrm{in}}\right]\right] \end{aligned} \tag{13.10}$$

where n is the length of the vectors and the diagonal elements of the T_{ij} matrix are nonzero. The above representation of the associative memory is classified as the inner product type. The name arises from the fact that Eq. (13.10) involves an inner product as opposed to an outer product, such as that in Eq. (13.9).

The multiplication of the matrix T with the $\mathbf{V}^{\mathrm{in}}$ vector is equivalent to matching the said input vector with all the stored vectors. The sgn (i.e., hard limiter) operation is basically a threshold operation that removes noise and clutter and retrieves the closest estimated vector. The estimated vector is then fed back to the memory as the new input. Such an iterative process results in a newer estimate. After many repeated iterations, the estimated vector is expected to converge to the stored vector to which the input vector is the closet.

The characteristics of the estimated vector can be examined from another point of view. The term within the internal square bracket of Eq. (13.10) represents an inner product of the input vector with the kth stored vector. Consequently, the estimated vector can be expressed as a sum of weighted stored vectors where the weights correspond to the inner products. Through subsequent iterations, one of these weights is expected to increase, finally reaching a maximum indicating the occurrence of a convergence. The estimated vector is expected to converge to the particular vector whose inner product (or the weight) following the very first iteration was a maximum. Let a_i represent the inner product of the ith stored vector with the input. Then the estimated vector can be rewritten

$$\mathbf{V}^{\text{es}} = \text{sgn}\left[\sum_{k=1}^{M} a_k \mathbf{V}^{\text{st},k}\right] \tag{13.11}$$

Suppose, for simplicity, $a_k > a_i$ (when $i = 1, \ldots, n$, and $i \neq k$) represents the largest inner product. Then the estimated vector is given by

$$\mathbf{V}^{\text{es}} = \text{sgn}\left[a_k \mathbf{V}^{\text{st},k} + \sum_{i \neq k}^{M} a_i \mathbf{V}^{\text{st},i}\right] \tag{13.12}$$

When the stored vectors are nearly orthogonal, the inner product between the input and one of the stored vectors will become a maximum while those with the remaining stored vectors will be less than the maximum. The condition of orthogonality assures that, as the iterated vector moves closer to one of the stored vectors, it tends to move away from the remaining stored vectors. Evolution of the estimated vector through repeated iterations should correspond to increasing a_k while decreasing the a_i's. Appropriately, the first term of Eq. (13.12) can be referred to as the signal vector while the second term can be referred to as the noise vector. One very obvious method of retrieving the desired stored vector is to choose the stored vector corresponding to a_k. In other words, the process is equivalent to thresholding the scalars by a variable threshold that adapts itself to different inputs. One of the ways to accomplish this scheme may involve picking the maximum of the weights from the inner products by electronics and software. This approach is the same as correlating the input vector will all the stored vectors by forming inner products and then choosing the one having the highest correlation.

Carefully examining the noise term of Eq. (13.12), one notes that it has two effects. It may increase the iterated signal (by addition) at some of its element positions where the noise value has a sign identical to that of the signal. But sometimes it may also decrease the signal at some of its element positions by negation. The probabilistic approach to this problem comes from the fact that if the length of the vector is very large compared to the number of vectors so that $n > M/0.15$, the signal gain term a_k is significantly large compared to the noise gain terms (i.e., the remaining weights), provided each of the vectors is nearly orthogonal (i.e., have large Hamming distance between the vectors).

13.2.2 Higher-Order Neural Network

A first-order neural network is generally expressed as

$$y_i(x) = T\left[\sum_{j=1}^{N} W(i, j)x(j)\right] \tag{13.13}$$

where $\mathbf{x} = \{x(j)\}$ represents an input vector, $W(i, j)$ is a set of adaptive weights, N is the total number of elements in the vector $\mathbf{x}$, and T is a thresholding function such as a sigmoid function. Note that Eq. (13.13) represents a linear sum of the input vectors. Without the cross-product term present in Eq. (13.13) it is unable to inherit information about the higher-order correlation that may exist in the input data. Including higher-order terms in the weighted average, one obtains

$$y_i(x) = T\left[W_0(i) + \sum_j W_1(i, j)x(j) + \sum_j \sum_k W_2(i, j, k)x(j)x(k) + \cdots\right] \tag{13.14}$$

where $W_2(i, j, k)$ are higher-order weights corresponding to higher-order correlations. A threshold logic unit, which includes terms up to and including degree k, is the kth order unit. Thus an elementary higher-order neuron consists of a high-order threshold logic element. This can be expressed by a simpler relationship by replacing the weighted summation terms by defining

$$\begin{aligned} T_1(i) &= \sum_j W_1(i, j)x(j) \\ T_2(i) &= \sum_j \sum_k W_2(i, j, k)x(j)x(k) \end{aligned} \tag{13.15}$$

where $x(j)$ is the jth input of the ith higher-order neuron, and $W_n(i, j, k)$ is representative of the nth order product of inputs and the output of the threshold logic unit. One then obtains a simpler formula of a higher-order threshold logic network as

$$y_i = T[T_0(i) + T_1(i) + T_2(i) + \cdots + T_k(i)], \tag{13.16}$$

where y_i is the output of the ith higher-order neuron unit. $T_n(i)$ is the nth order term in Eq. (13.16) for the ith unit, and k is the order of the unit. The zero-order term represents the adaptive threshold. During learning the weights of the neural network are adjusted by a particular learning rule as it responds to a set of training data. A second-order weight, for example, may be updated by the rule

$$W_2'(i, j, k) = W_2(i, j, k) + [t(i) - y(i)]x(j)x(k) \tag{13.17}$$

where $t(i)$ is the target output and $y(i)$ is the actual output of the ith unit for an input vector **x**. By adjusting the weights using Eq. (13.17), it is possible to extract the correlation inherent in the input–output mapping. As an example, suppose that a neural network is to learn an X-NOR problem. Accordingly, the following table must have to be satisfied.

$x(1)$	$x(2)$	$t(\mathrm{x})$
−1	−1	1
−1	1	−1
1	−1	−1
1	1	1

Representing the two binary levels in a bipolar binary yields zero average value for input and output. Accordingly, we may define a second-order neural network as

$$y(x) = \mathrm{sgn}[W_1(1)x(1) + W_1(2)x(2) + W_2x(1)x(2)] \tag{13.18}$$

where sgn performs a hard thresholding, since $\mathrm{sgn}[x] = +1$, if $x \geq 0$, and $\mathrm{sgn}[x] = -1$, if $x < 0$. The corresponding learning rules are

$$W_1'(i) = W_1(i) + [t(x) - y(x)]x(i) \tag{13.19}$$

and

$$W_2'(i) = W_2(i) + [t(x) - y(x)]x(1)x(2) \tag{13.20}$$

The output does not depend on $x(i)$ alone. Since no correlation between x_1 or x_2 and the desired output $t(x)$ exists, the $W_1(i)$ terms average to 0. However, since $x(1)$ and $x(2)$ are perfectly correlated with the output, W_2 takes on a nonzero value. Thus the relationship $y(x) = \mathrm{sgn}[x(1)x(2)]$ could easily be verified algebraically. Using a similar approach, invariances such as position could be incorporated into the higher-order neural network.

13.3 ASSOCIATIVE MEMORY

13.3.1 Inner Product Model

One of the important applications of the neural network is associative memory. Associative memory is different from a regular location addressable memory both in terms of its storing mechanics and addressing scheme. The data are stored in such a way that the memory may be addressed by the content itself. By definition, an associative memory retrieves its information content by associating the address with the information content. The human brain is the best example of an associative memory.

One of the easiest ways to perform association or matching is overlap. In analytical geometry, for example, two triangles are considered congruent only when they overlap completely with each other. The same thing when performed for information strings is referred to as correlation. If two signals are correlated and a match occurs, a peak is observed at the origin of the correlation plane or axis. Note that we need to look at the origin only; the other information is simply redundant. The information at the origin of the correlation axis is nothing but the sum of the overlap of two signals when they are overlapped completely with each other. In other words, correlation corresponds to the inner product of the two signals. An example could make this clear. Suppose we have a vector $\mathbf{A} = [1, -1, 1, -1]$ with which two other vectors given by $\mathbf{B} = [1, -1, 1, -1]$ and $\mathbf{C} = [1, -1, 1, 1]$ are matched. Note that an inner product of two identical signals will result in the largest possible coefficient. This is why $\mathbf{A} \cdot \mathbf{B} = 4$ but $\mathbf{A} \cdot \mathbf{C} = 2$. Thus to associate a number of stored vectors we only need to take the inner product with all the stored vectors and then choose the one with the highest inner product as the desired memory output. One may see that Hopfield's memory in fact accomplishes the same objective. The repeated feedback is only a way to make the match stronger and the mismatch weaker.

13.3.2 Hopfield's Memory Modified to Inner Product Model: A Study

In order to fully utilize the parallelism of optics, the memory elements should be such that they can be accessed simultaneously and in parallel on the basis of data content rather than by specific address or location. Such an associative memory delivers the output even when it is stimulated by a partial input of the stored information. This kind of memory when built in a neural network fashion becomes highly fault-tolerant. The nonlinear decision-making operation and internal feedback mechanisms involved in the process make it possible for an associative memory to handle ill-posed problems such as signal reconstruction from incomplete information.

There are three processes involved in an associative memory for memorization and retrieval of information: (1) storing, (2) matching, and (3) decision-making. The storing mechanism generally affects the process of retrieval. The matching process depends entirely on how the information was stored and how stimuli are introduced to the memory. Accordingly, an a priori decision of which should match to which often determines how the input should be represented and stored. The decision-making process may be carried out in one step without feedback or in multiple steps with feedback. In a human brain exactly how the information is stored is not completely known. But models have been proposed to mimic a human brain. It is understood that the interconnections rather than the contents play a dominant role in human associative memory.

Hopfield's discrete model, in spite of its many problems such as false convergence, nonconvergence, and less storage capacity, involves a memory matrix that is very easy to implement. It is attractive because it requires comparatively less time for learning and retrieval. Hopfield's discrete model for associative memory provides a prescription for both storage and retrieval. Often bipolar binary numbers are used for storing the vectors and unipolar binary numbers are used to address the memory. When matching the input with the stored information, the Hamming distance is generally used for measuring the closeness of two vectors. However, in cases other than that of the bipolar binary input, the Hamming distance criterion does not directly represent the ongoing matching process.

Basically, the estimated vector is the sum of the products of three vectors. The elements of each vector can be either unipolar binary (consisting of 0 and 1) or bipolar binary (consisting of -1 and 1). Assuming that the input and output vectors are represented in like manner, there are four representations possible for associative storage and retrieval. Table 13.1 shows the combinations of these vectors that may appear in the inner product. Each of these four combination types

TABLE 13.1 Stored and Input Vector Representations[a]

	Vector Elements			Inner Products	
Type	Stored	Input	Weight	Exact	Expected
Unipolar–unipolar	0	0	0	S	$(n - H)/2$
	0	1	0		
	1	0	0		
	1	1	1		
Bipolar–unipolar	−1(0)	0	0	$S - d$	$(n/2) - H$
	−1(0)	1	−1		
	1(1)	0	0		
	1(1)	1	1		
Bipolar–bipolar	−1(0)	−1(0)	1	$n - 2H$	$n - 2H$
	−1(0)	1(1)	−1		
	1(1)	−1(0)	−1		
	1(1)	1(1)	1		
Unipolar–bipolar	0	−1(0)	0	$S - S'$	$(n/2) - H$
	0	1(1)	0		
	1	−1(0)	−1		
	1	1(1)	1		

[a] The vector element values within parentheses are the actual values; the values outside parentheses correspond to the bipolar values.

generates a particular type of weight, which in turn is associated with a maximum of two kinds of vectors (either unipolar binary or bipolar binary) for yielding the estimated vector as a linear combination of stored vectors. This amounts to four different kinds of associative memories. Furthermore, if one considers the flexibility of having nonzero diagonal elements for the T matrix, there would be up to eight different associative memories. Also, consider that the T_{ij} matrix of the outer product model can be either clipped or unclipped; thus a total of 16 different kinds of associative memories are possible.

When both the stored and input vectors are represented by unipolar binary numbers, the resultant weight is equal to the number of 1's, S, that are present at identical positions of the two vectors. It is thus obvious that the weight is not related to the Hamming distance, H, but to the number of elements having matching 1's (i.e., having 1 at the identical locations of both the stored and input vectors). Note that the Hamming distance H is given by [n − (number of matching 1's + number of matching 0's)]. The degree of similarity thus ranges from a minimum of 0 through a maximum of $n - H$, with the expected value of the weight being in the neighborhood of $(n - H)/2$. Note that the matching of a 0 with a 0 is ignored in this weight estimation, since the product of 0 with 0 is 0. A 1(0) in a location of the input vector and a 0(1) at the corresponding position of the stored vector does not contribute to the weight. But if the stored vector is represented in bipolar binary form while the input vector is represented in unipolar binary form, an appearance of 1 and -1, respectively, in the identical locations of the input and the stored vectors reduces the weight by unity. If d represents the number of elements having a 1 in the input vector and a -1 (representing 0) in the corresponding location of the stored vector location, the exact sum of weights is given by $S - d$. If 50% of $n - H$ vector element combinations correspond to matching 1's and if 50% of H relates to the appearance of a 1 in the input while the corresponding stored value is -1, the expected sum of weights is given by $[\{(n - H)/2\} - (H/2)] = \{(n/2) - H\}$.

Instead of the Hamming distance, the number of matching 1's represents the degree of similarity for measuring the closeness of vectors. In general, the exact sum of weights is not directly related to the Hamming distance. This is why the input vector is sometimes seen to converge to a stored vector having a higher Hamming distance instead of converging to one having a smaller Hamming distance. Such an apparent convergence anomaly can be understood by studying the weights that determine the correlation between the input and stored vectors. The underlying mathematics is actually driving each input vector toward a certain destiny, which is not determined by the Hamming distance at all. In fact, weight (i.e., inner product) plays a more important role in determining vector convergence.

The Hamming distance directly affects the weights of the vectors only when the input vector representation is bipolar binary. This may also be seen as a process of increasing the weights of the vectors having small Hamming distances. With both the input and stored vectors in bipolar binary, the exact sum of weights on each vector is $n - 2H$ when the absolute value of the sum of weights is less than or equal to n. The maximum sum of weights occurs when the input vector is identical to one of the stored vectors, while the minimum occurs when the input vector is totally complementary to one of the stored vectors. Interestingly, matching 1's just as well as matching 0's in bipolar representation contributes toward increasing the sum of weights, whereas mismatching of elements decreases the overall sum of weights.

For the case in which the stored vector is unipolar binary while the input vector is bipolar binary, a -1 in the input vector location and a 1 in the corresponding stored vector location decrease the sum of weights. Defining S' as the number of locations having a -1 at the input vector and a 1 at the corresponding location of the stored vector, the sum of weights is $S - S'$. If 50% of H relates to the appearance of a -1 (representing 0) at the input while the corresponding stored value is 1, and 50% of $n - H$ is matching 1's then the sum of weights is $\{(n/2) - H\}$.

Comparing all these different cases in Table 13.1, one may note that any one representation is not enough to handle all cases of associative recall. Each of these representations has its own use. In case 1 of Table 13.1, the presence of 1's at the correct location (i.e., in those locations where the stored values are also 1) of the input increases the sum of weights. The remaining mismatch can be ignored. When the input is a partial portrait of the stored vector with some of its elements having correct 1's, the stored vector with which the maximum number of 1's match should be chosen as the output. Note that the presence of a 1 at the input while there is a 0 at the stored vector in the corresponding location does not alter the sum of weights. However, at the same time, such a 1 at the input can increase the sum of weights of other stored vectors provided there is a 1 at the corresponding positions. To make sure that the presence of a false 1 in the partial input vector reduces the sum of weights for the corresponding stored vector, the stored vector should be represented in bipolar binary form. This is accomplished in case 2.

Consider, for example, that a portion of the stored vector is available as a partial input, in which case both matching 1's and matching 0's should be allowed to contribute to the sum of weights since 0's also carry information just as 1's. In such a scenario, it may be desirable to represent both input and stored vectors in bipolar binary form as in case 3. Now consider the case when we have a complete portrait of the stored vectors in addition to one or more false 1's. It may then be desirable not to decrease the sum of weights owing to the presence of a false 1. Rather, the presence of a false 0 in the input should reduce the sum of weights. Case 4 achieves this particular objective.

For measuring the degree of similarity, the Hamming distance can be used only in the bipolar–bipolar case, while in all other cases the weights directly represent the degree of similarity. When the input is random, the associative memory acts as a pattern classifier, wherein the search is for the nearest-neighbor stored vector. In this case, the Hamming distance serves well as a measure of the closeness of vectors. Therefore the bipolar–bipolar representation is best for pattern classification.

Mathematically, for each case, the output should be the stored vector for which the sum of weights is a maximum. Without going through further iterations, one could choose V_k corresponding to a_k. The process thus becomes software dependent rather than hardware dependent, which thus lowers the computation speed. The maxima can be picked iteratively by Hopfield's threshold and remultiplying scheme.

The iterative method begins to pose a problem when the sums of the weights are relatively close and no single sum of the weights surpasses the others. The most disappointing problem, however, in the case of the unipolar binary input vector is that very often the highest sum of weights may not correspond to the lowest Hamming distance. Consequently, the sum of weights should be used in these cases to identify the closeness of vectors. During the retrieval process, in these cases, the noise vector when compared to the signal vector becomes relatively large.

When the input and stored vectors are represented in bipolar binary form, the sum of weights directly corresponds to the Hamming distances. Consider the case when the stored vectors are either nearly orthogonal or have large Hamming distances. Now if the memory is addressed with an input vector that is closer to a stored vector, then that input vector is usually further away from the other stored vectors. Moreover, when there is a large Hamming distance between the input and one of the stored vectors (i.e., when $2H > n$), the sum of weights may become negative. In the extreme case, a negative sum of weights may even lead to a convergence to a vector that is the complement of the stored vector having the largest Hamming distance. A similar result is also obtained with unipolar binary inputs.

As suggested earlier, one would like to use a variable threshold in the inner product domain so as to choose the maximum sum of weights. For Hamming distances $H_1, H_2, \ldots, H_n$, of the input vector from n stored vectors, the sum of weights are, respectively, $(n - 2H_1), \ldots, (n - 2H_n)$. When $H_i > n/2$, the sum of the weights becomes negative. With slight modification, one could use a zero-crossing thresholding in the inner product domain. This may be referred to as IP thresholding. If, for example, an input vector is more than $n/2$ Hamming distance away from any one of the stored vectors, then by scalar thresholding in the inner product domain one could prevent its convergence to the complement of the stored vector. Thus zero-crossing thresholding is analogous to forgetting the vectors(s) of very high Hamming distance, which increases the chance of a correct convergence to the true stored vector. This scheme is analogous to the function of a human brain in which an input stimulus does not compete directly with all the stored information for locating its closest neighbor, but competes only with a select class of information totally disregarding the rest of the classes. Consequently, the ith element of the estimated vector $\mathbf{V}^{\text{es}}$ can be modified to give

$$\mathbf{V}_i^{\text{es}} = \operatorname{sgn}\left[\sum_{k=1}^{M} \mathbf{V}_i^{\text{st},k} \operatorname{sgn}\left[\sum_{j=1}^{n} \mathbf{V}_j^{\text{st},k} \mathbf{V}_j^{\text{in}}\right]\right] \tag{13.21}$$

This is equivalent to introducing a nonlinear operation in the correlation domain.

As stored vectors, consider the following four vectors:

$$a_1 = (1, -1, 1, 1, -1, 1, 1, 1, -1, 1, 1, 1, -1, 1, -1, -1, 1, -1, -1, 1)$$

$$a_2 = (-1, 1, -1, 1, 1, -1, -1, 1, -1, -1, -1, -1, 1, 1, 1, -1,$$
$$-1, -1, 1, 1,)$$

$$a_3 = (1, 1, -1, -1, -1, 1, -1, 1, 1, 1, -1, 1, 1, -1, 1, 1, 1, 1, -1, -1)$$

$$a_4 = (-1, 1, 1, 1, -1, 1, 1, -1, 1, -1, 1, -1, -1, -1, 1, -1, 1, 1, -1)$$

When an input vector is applied to the memory, it is actually related to all the stored vectors rather than to just one stored vector. For example, an input vector that is at a Hamming distance x from the stored vector p may actually be closer to another stored vector q (at a Hamming distance of y) when $x > y$. Accordingly, the input vectors are identified by a set of four Hamming distances for the case under consideration.

Consider further that $a_i(m:n)$ denotes a vector whose m through n elements are identical to those of the stored vector a_i while the remaining elements are zeros. Table 13.2 includes simulation results pertaining to several random inputs. For each of the input vectors, the corresponding Hamming distances and the sums of weights obtained immediately after the first iteration, for both bipolar and unipolar representations, are listed, respectively, under the second and third columns. The entries in the remaining columns indicate to which stored vector the input vector converges and after how many iterations, while the FC entries represent false convergences. The entry $a_i(j)$ indicates that the input vector converges to the stored vector a_i after j iterations. Consider the input vector $a_1(1:7)$ and the bipolar–bipolar format for the stored and input vectors. From the Hamming coordinate one can conclude that it is closest to a_1 and after three iterations it actually converges to a_1. If IP thresholding had been used instead in the inner product domain, the vectors a_2 and a_3 would have been eliminated from the retrieval process after the very first iteration, leading to a faster convergence. With bipolar binary and uni-

TABLE 13.2 Simulation Results

				Converges When			
				Stored: Bipolar Input: Bipolar		Stored: Bipolar Input: Unipolar	
		Starting Weight with Input		and When IP Threshold Is		and When IP Threshold Is	
Input Vector	Hamming Distance	Bipolar	Unipolar	Absent	Present	Absent	Present
$a_1(1:7)$	(7,12,13,8)	(6,−4,−6,4)	(5,−3,−1,3)	$a_1(3)$	$a_1(2)$	FC[a]	$a_1(2)$
$a_2(1:7)$	(13,6,13,10)	(−6,8,−6,0)	(−1,3,−1,1)	$a_2(3)$	$a_2(2)$	$a_2(2)$	$a_2(2)$
$a_3(1:7)$	(11,10,9,10)	(−2,0,2,0)	(1,−1,3,1)	$a_3(3)$	$a_3(2)$	FC	$a_3(2)$
$a_4(1:7)$	(9,10,13,6)	(2,0,−6,8)	(3,−1,−1,5)	$a_4(3)$	$a_4(2)$	$a_4(2)$	$a_4(2)$
$a_1(1:4)$	(9,10,13,10)	(2,0,−6,0)	(3,−1,−1,1)	$-a_3(2)$	$a_1(2)$	$a_1(2)$	$a_1(2)$
$a_2(1:4)$	(12,7,12,9)	(−4,6,−4,2)	(0,2,0,2)	$a_2(3)$	$a_2(2)$	$a_4(3)$	FC
$a_3(1:4)$	(12,9,10,11)	(−4,2,0,−2)	(0,0,2,0)	FC	$a_2(2)$	$a_3(2)$	$a_3(2)$
$a_4(1:4)$	(11,8,13,8)	(−2,4,−6,4)	(1,1,−1,3)	FC	$a_2(2)$	$a_4(3)$	$a_4(2)$
$a_1(16:20)$	(10,9,12,11)	(0,2,−4,−2)	(2,0,0,0)	$-a_3(3)$	$a_2(2)$	$a_1(2)$	$a_1(2)$
$a_2(16:20)$	(12,7,14,11)	(−4,6,−8,2)	(0,2,−2,0)	$a_2(3)$	$a_2(2)$	FC	$a_2(2)$
$a_3(16:20)$	(13,12,9,10)	(−6,−4,2,0)	(−1,−3,3,1)	$a_3(3)$	$a_3(2)$	FC	$a_3(2)$
$a_4(16:20)$	(13,10,11,8)	(−6,0,−2,4)	(−1,−1,1,3)	FC	$a_4(2)$	$a_4(3)$	$a_4(2)$
$a_1(13:20)$	(9,8,13,12)	(2,4,−6,−4)	(3,1,−1,−1)	$-a_3(4)$	$a_2(2)$	$a_1(3)$	$a_1(2)$
$a_2(13:20)$	(13,4,13,12)	(−6,12,−6,−4)	(−1,5,−1,−1)	$a_2(2)$	$a_2(2)$	$a_4(2)$	$a_2(2)$
$a_3(13:20)$	(15,10,7,10)	(−10,0,6,0)	(−3,−1,5,1)	$-a_1(2)$	$a_3(2)$	$a_3(2)$	$a_3(2)$
$a_4(13:20)$	(14,9,10,7)	(−8,2,0,6)	(−2,0,2,4)	$-a_1(3)$	$a_4(2)$	$a_4(3)$	$a_4(2)$
$a_1(1,2,18:20)$	(10,9,12,13)	(0,2,−4,−6)	(2,0,0,−2)	$-a_4(2)$	$a_2(2)$	$a_1(3)$	$a_1(2)$
$a_2(1,2,18:20)$	(13,6,13,10)	(−6,8,−6,0)	(−1,3,−1,1)	$a_2(3)$	$a_2(2)$	$a_2(2)$	$a_2(2)$
$a_3(1,2,18:20)$	(13,10,9,10)	(−6,0,2,0)	(−1,−1,3,1)	$-a_1(2)$	$a_3(2)$	$a_3(2)$	$a_3(2)$
$a_4(1,2,18:20)$	(15,8,11,8)	(−10,4,−2,4)	(−3,1,1,3)	$-a_1(3)$	$a_2(3)$	FC	$a_4(2)$
Random	(9,12,13,10)	(2,−4,−6,0)	(3,−3,−1,1)	$a_1(3)$	$a_1(2)$	FC	$a_1(2)$
Random	(9,10,13,10)	(2,0,−6,0)	(3,−1,−1,1)	$a_3(3)$	$a_1(2)$	$a_1(2)$	$a_1(2)$
Random	(9,8,15,10)	(2,4,−10,0)	(3,1,−3,1)	$-a_3(3)$	$a_2(2)$	$a_1(2)$	$a_1(2)$
Random	(8,9,14,11)	(4,2,−8,−2)	(4,0,−2,0)	$-a_3(3)$	$a_1(2)$	$a_1(2)$	$a_1(2)$
Random	(8,11,14,9)	(4,−2,−8,2)	(4,−2,−2,2)	$-a_3(3)$	$a_1(2)$	$a_1(3)$	$a_1(2)$
Random	(11,8,13,12)	(−2,4,−6,−4)	(1,1,−1,−1)	$a_2(3)$	$a_2(2)$	$a_1(4)$	FC
Random	(12,9,12,11)	(−4,2,−4,−2)	(0,0,0,0)	$a_2(3)$	$a_2(2)$	FC	FC
Random	(11,10,11,12)	(−2,0,−2,−4)	(1,−1,1,−1)	$-a_4(3)$	FC	FC	FC
Random	(10,7,14,11)	(0,6,−8,−2)	(2,2,−2,0)	$-a_3(3)$	$a_2(2)$	FC	FC
Random	(9,8,15,10)	(2,4,−10,0)	(3,1,−3,1)	$-a_3(3)$	$a_2(2)$	$a_1(5)$	$a_1(2)$
Random	(11,8,13,8)	(−2,4,−6,4)	(1,1,−1,3)	FC	$a_4(3)$	$a_4(3)$	$a_4(2)$
Random	(9,10,13,6)	(2,0,−6,8)	(3,−1,−1,5)	$a_4(3)$	$a_4(2)$	$a_4(2)$	$a_4(2)$

[a] FC represents false convergence.

polar binary representations, respectively, for the stored and input vectors, no convergence is achieved; instead, the output stabilizes to a false vector $6a_1 - 10a_2 + 6a_3 + 2a_4$ (denoted by FC). The initial sum of weights is (5, −3, −1, 3), but since 5 is less than the absolute sum of the other three sums of weights, these sum of weights drive $a_1(1:7)$ toward a false vector. Obviously, IP thresholding would have favored a_1 since a_2 and a_3 would have been out of consideration and, clearly, $5a_1 + 3a_4$ would have converged to a_1. Sometimes, large negative sums of weights, such as the one in row 5, drive the vector to the complement of the vector (a_3 in this case). Such problems can be avoided simply by IP thresholding.

Table 13.2 supports the assumption that, in the case of unipolar binary representation for the input, the Hamming distance criterion does not determine the vector convergence, since the input vector usually converges to a stored vector that corresponds to the highest sum of weights. For example, the input vector $a_2(1:4)$ when represented in unipolar binary form converges after three iterations to the stored vector a_4 even through the Hamming distance criterion suggests that it should converge to a_2.

The input vectors using bipolar binary representation usually converge to a stored vector for which the sum of weights is the largest in magnitude. Whenever the largest sum of weights is negative the convergence leads to a vector that is the complement of the stored vector for which the negative sum of weights is maximum. For example, in the absence of thresholding the input vector $a_1(1:4)$ converges to $-a_3$ instead of converging to a_1. This can be predicted by observing the sum of weights value −6 for a_3 while the sum of weights value is only 2 corresponding to a_1. This problem of convergence to a complementary vector can be resolved by thresholding the negative sum of weights. By means of such thresholding, convergence of the input vector to the stored vector that has the lowest Hamming distance is achieved.

When a partial stored vector is introduced as an input vector, it could actually be close to a different stored vector. For example, input vector $a_3(1:4)$ is closer to the stored vector a_2 in terms of Hamming distance. With bipolar binary representation, this input vector converges (with IP thresholding) to a_2. However, with unipolar binary representation the input converges to a_3. The latter neuron representation only leads to the desired stored vector.

With bipolar binary input and thresholding, the convergence outcome of the associative memory always agrees with the minimum Hamming distance criterion. At the same time, the bipolar binary representation prevents the input from converging to a negative vector. On the other hand, the use of a binary unipolar representation seems to recognize a partial input irrespective of its Hamming distance. Examining the convergence mechanism and the sum of weights, we note the following:

1. When the partial input vector is represented in unipolar binary, the largest sum of weights is usually contributed by those partial input vector elements (known) that are in common with the corresponding elements of a stored vector.

2. The sum of weights corresponding to the unipolar binary input comes only from the matching 1's of the known elements.
3. When partial inputs are represented in bipolar binary, the largest sum of weights is contributed both by the known as well as unknown elements of the vector since both the unknown elements and known 0's are being represented by -1. Consequently, known 0's contribute toward increasing the sum of weights while the unknown elements contribute toward increasing/decreasing the sum of weights when the number of matching 0's is greater/less than the number of mismatched 1's.
4. The advantage of having 0's in the unknown portion of the partial inputs (represented in unipolar binary) is that the 0's do not contribute toward either increasing or decreasing the sum of weights.

To combine the merits of both unipolar and bipolar neuron representations therefore, a different neuron representation can be established. To make the sum of weights more significant, one can make use of matching 0's; but in order to do that only the elements of the partial input that are in common with those of the stored vector should be represented in bipolar binary and the remaining elements should be set to zero. The sum of weights corresponding to the partial input vector will then be equal to the number of known elements. Thereafter, one can apply IP thresholding to accelerate convergence and to avoid convergence to the negative vector. This mode of representation, where the known elements are in bipolar binary and the remaining elements are in unipolar binary, can be referred to as trinary representation.

To measure closeness, the Hamming distance may be used only for the partial elements and not for the whole vector. Since it is understood that one does not know of the remaining elements of the stored vector, there is no use comparing the rest of the stored vector with the rest of the input vector. Thus, in the first iteration, the unknown portion of the input may not be considered. Only after the first iteration is the input represented in bipolar binary for accelerating the convergence process.

In binary representation, for example, the input vector $a_3(1:4)$ would have been represented as (1, 1, 0, 0, 0, 0, 0, 0, 0, 0, 0, 0, 0, 0, 0, 0, 0, 0, 0, 0) in unipolar binary and as (1, 1, -1, -1, -1, -1, -1, -1, -1, -1, -1, -1, -1, -1, -1, -1, -1, -1, -1, -1) in bipolar binary, whereas in trinary representation, it will be given by (1, 1, -1, -1, 0, 0, 0, 0, 0, 0, 0, 0, 0, 0, 0, 0, 0, 0, 0, 0), where the nonzero numbers are known and the zero are unknown. The sum of the weights after the first trial is (-2, 0, 4, -2), while the sum of the weights with IP threshold is (0, 0, 4, 0). The first estimate prior to vector thresholding is $-2a_1 + 0a_2 + 4a_3 - 2a_4$, which gives (4, 4, -8, -8, 0, 0, -8, 4, 4, 4, -8, 4, 8, -4, 4, 8, 0, 4, -4, -4) without IP thresholding and (4, 4, -4, -4, -4, 4, -4, 4, 4, 4, -4, 4, 4, -4, 4, 4, 4, 4, -4, -4) with IP thresholding. Thus the first estimate after vector thresholding gives (1, 1, -1, -1, -1, -1, -1, 1, 1, 1, -1, 1, 1, -1, 1, 1, -1, 1, -1, -1) with IP thresholding. Accordingly, the sum of weights after the second trial is (-6, -4, 18, -4). Since the sum of weights after IP thresholding is (0, 0, 18, 0), the second estimator prior to vector thresholding given by $-6a_1 - 6a_2 + 18a_3 - 4a_4$ gives (20, 16, −24, −32, −12, 12, −24, 12, 24, 20, −24, 20, 24, −24, 16, 32, 12, 24, −20, −24) without IP thresholding and (18, 18, −18, −18, −18, 18, −18, 18, 18, 18, −18, 18, 18, −18, 18, 18, 18, 18, −18, −18) with IP thresholding. Therefore the second estimate after vector thresholding gives (1, 1, −1, −1, −1, 1, −1, 1, 1, 1, −1, 1, 1, −1, 1, 1, 1, 1, −1, −1) for both cases. Finally, the sum of weights after third trial is (−4, −6, 20, −2), which after IP thresholding gives (0, 0, 20, 0).

The weight on the third vector now exceeds the sum of absolute sums of weights of the remaining vectors, thus indicating that the input has finally converged to a_3. Table 13.3 lists the results of a few other inputs. The trinary neuron representation is dependable for the partial inputs provided the positions of the partial input elements are known. However, when the input is totally random and it is not known which part of the input contains the correct information, bipolar binary representation for the inputs along with scalar thresholding is preferable for achieving convergence to the expected stored vector having minimum Hamming distance.

Most visual information, however, is two-dimensional. Accordingly, a trinary associative memory may be used to retrieve two-dimensional stored images. The trinary associative memory can store a $m \times n$ two-dimensional image in the form of an mn-element vector. Figure 13.3 shows two-dimensional stored images of four machine objects, S1, S2, S3, and S4, each having 11×20 pixels. The set of Hamming distances for the chosen stored images are (0, 67, 56, 55), (67, 0, 77, 94), (56, 77, 0, 67), and (55, 94, 67, 0), respectively, for S1, S2, S3, and S4. Therefore the four stored images are reasonably similar as far as their relative Hamming distances are concerned. Figure 13.4 shows nine such partial inputs, P1 through P9.

Note that P2 and P6 are the partial inputs of S1; P1, P3, and P7 are the partial inputs of S2; P4 and P8 are the partial inputs of S3; and P5 and P9 are the partial inputs of S4. To compare the performances of different neuron representations, the partial inputs may be encoded using unipolar binary, bipolar binary, the trinary

TABLE 13.3 Simulation Results with the Proposed Modification

Input Vector	Hamming Distance	Starting Weight with Input		Converges When: Stored: Bipolar, Input: Ternary, and When IP Threshold Is		Converges When: Stored: Bipolar, Input: Unipolar, and When IP Threshold Is	
		Ternary	Unipolar	Absent	Present	Absent	Present
$a_3(1:4)$	(3,2,0,3)	(−2,0,4,−2)	(0,0,2,0)	$a_3(3)$	$a_3(2)$	$a_3(2)$	$a_3(2)$
$a_4(1:4)$	(2,1,3,0)	(0,2,−2,4)	(1,1,−1,3)	$a_4(3)$	$a_4(2)$	$a_4(3)$	$a_4(2)$
$a_1(16:20)$	(0,2,3,3)	(5,1,−1,−1)	(2,0,0,0)	$a_1(2)$	$a_1(2)$	$a_1(2)$	$a_1(2)$
$a_1(13:20)$	(0,4,6,5)	(8,0,−4,−2)	(3,1,−1,−1)	$a_1(2)$	$a_1(2)$	$a_1(3)$	$a_1(2)$
$a_1(1,2,18:20)$	(0,3,3,5)	(5,−1,−1,−5)	(2,0,0,−2)	$a_1(3)$	$a_1(3)$	$a_1(3)$	$a_1(2)$
$a_4(1,2,18:20)$	(5,2,2,0)	(−5,1,1,5)	(−3,1,1,3)	$-a_1(3)$	$a_4(2)$	FC[a]	$a_4(2)$

[a] FC represents false convergence.

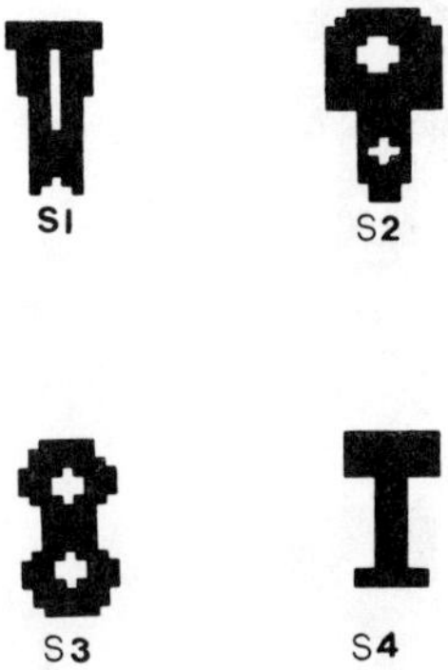

FIGURE 13.3 Two-dimensional stored images.

representation. Figure 13.5 shows nine retrieved images corresponding to the nine partial inputs when they are represented using unipolar binary numbers. Similarly, Figs. 13.6 and 13.7 show the retrieved images corresponding to the partial inputs when they are represented, respectively, using bipolar and trinary numbers. Note that adaptive thresholding has been used in the trinary case. The inner product is left unchanged when it is greater than or equal to the average of inner products; otherwise it is set equal to zero.

It can be seen that each of the trinary partial inputs readily converges to a stored vector to which it corresponds. From among the unipolar binary partial inputs only P1 and P3 converge to the stored vector S2 while the remaining partial inputs result in false convergences. These false images are different from the stored images since they are formed from linear combinations of the stored images. Surprisingly,

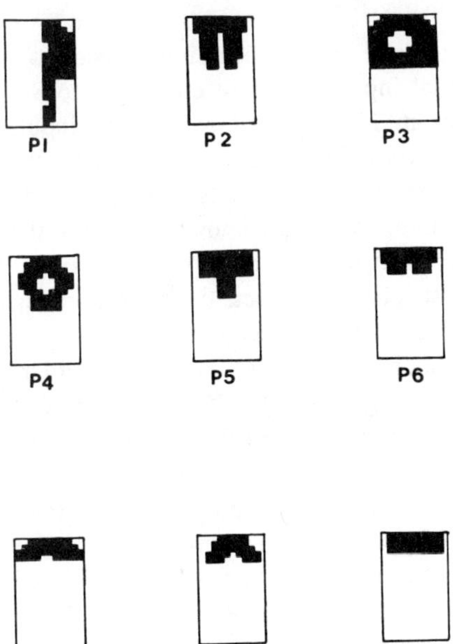

FIGURE 13.4 Partial input images.

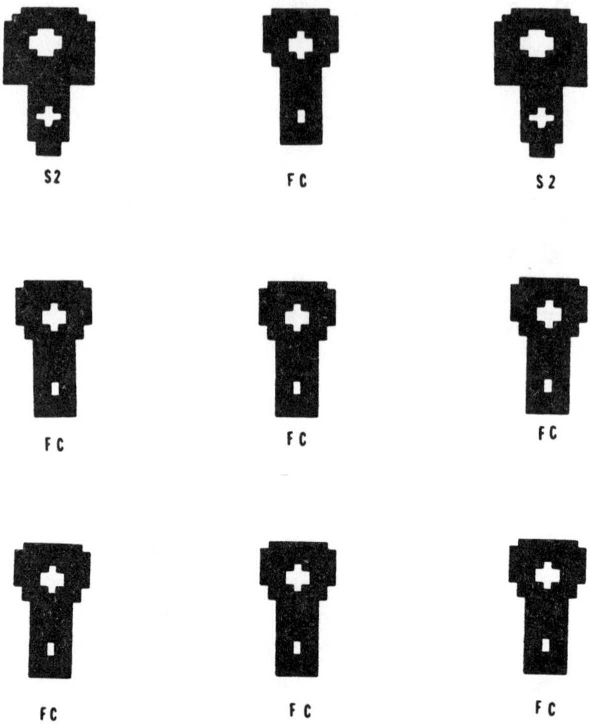

FIGURE 13.5 Retrieved two-dimensional images when partial inputs are in unipolar binary with no IP thresholding.

from among the bipolar binary partial inputs none of the partial inputs leads to a correct convergence, while the bipolar partial input P3 almost converges to S2. Two interesting cases deserve special attention. The bipolar binary partial inputs P5 and P9, both of which correspond to S4, do indeed converge to S4 after the first iteration but with subsequent iterations they converge to false images.

In summary, the trinary neuron representation, where the known portion of the partial input is represented in bipolar binary and the unknown portion is represented by a string of 0's, is perhaps the most reliable. In fact, the corresponding associative memory is well suited for error-free character recognition.

The optical implementation of trinary associative memory could use a polarization-encoded optical shadow-casting (POSC) system. The POSC systems can be designed to accommodate mixed mode numbers for optical computing applications. The input vector can be represented by an array of LED source patterns, while the stored vectors can be coded on transparencies or in a liquid crystal television or spatial light modulator. A possible optical implementation of the inner product model is shown in Fig. 13.8. The matrix is used to store all the individual vectors and the input LED may be used to address the memory as a multiplying vector. Two LEDs may be used to represent each trinary input. Two pixel subcells

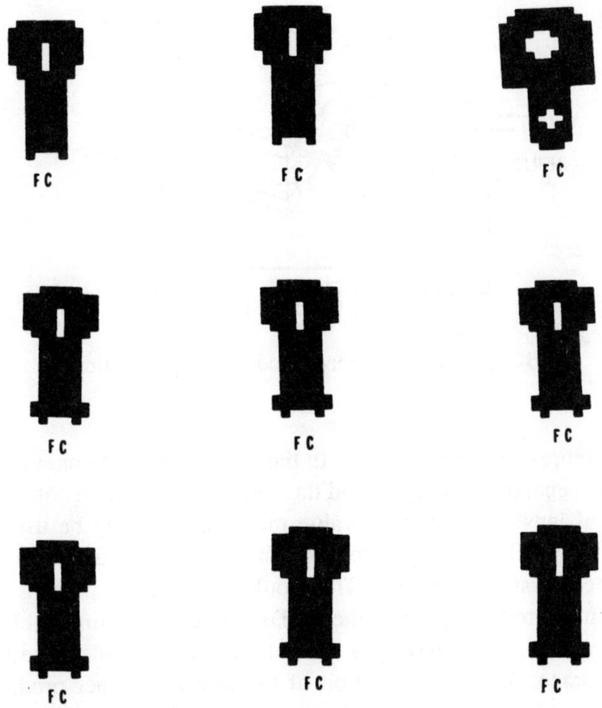

FIGURE 13.6 Retrieved two-dimensional images when partial inputs are in bipolar binary without IP thresholding.

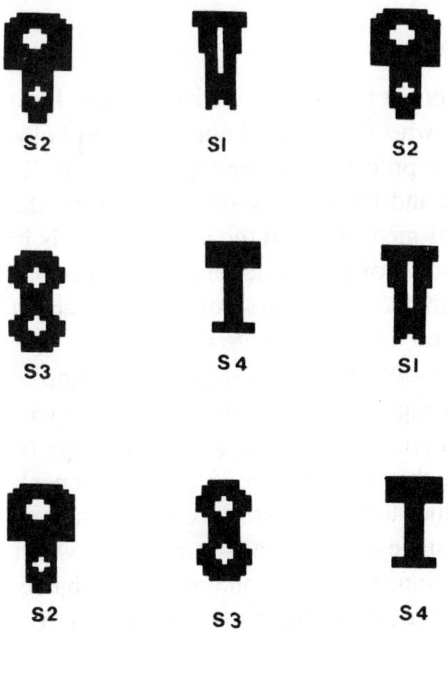

FIGURE 13.7 Converged two-dimensional images when partial inputs are represented in trinary with IP thresholding.

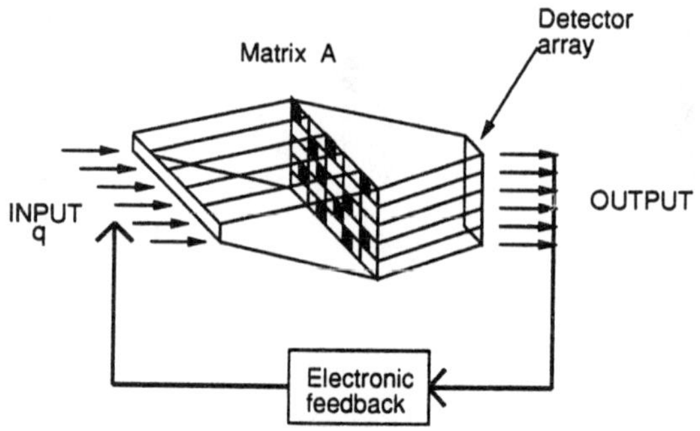

FIGURE 13.8 An inner product implementation.

are needed to represent bipolar values. In the inner product domain a beam splitter may be used to separate the outputs and then an average may be obtained by means of a cylindrical lens. This average value may appropriately be used to perform average thresholding by electronic means. These inner products may then be used to reilluminate the stored vectors. This could be done by passing the resultant thresholded inner product again to the LEDs, or just reflecting back through the same storage matrix, or by having a second matrix identical to the first one just after the first stage. The multiplication of the threshold inner products with the stored vectors results in the estimated vector after a final zero-crossing thresholding. This is then reused until convergence.

13.4 OPTICAL IMPLEMENTATIONS

13.4.1 Holographic Associative Memory

Holographic implementations of the inner product type of associative memory have also been demonstrated. Gabor was the one who first pointed out the use of holograms as associative memory. The recording process of holograms automatically performs the association between the object and reference beam used to store the hologram. The only difference in this type of memory from the previous one is in the way the information is being stored. Instead of a discrete location for each of the memories, this memory has distributed memory structure—one overlapped on top of the other. Angular multiplexing may be used to separate them.

The holographic associative memory may operate in two steps. For example, in Fourier hologram memories, as shown in Fig. 13.9, the partial input addresses the whole memory as in a matched filter. In fact, the purpose of the first stage of such a memory is to store the filters for each object. Then when a correlation peak is obtained corresponding to the location of the stored objects, the correlation peak is masked and used to illuminate the second stage of the correlator memory, which stores the objects. As a result of the illumination, the corresponding stored objects are obtained, more appropriately, as a weighted combination of those objects. By

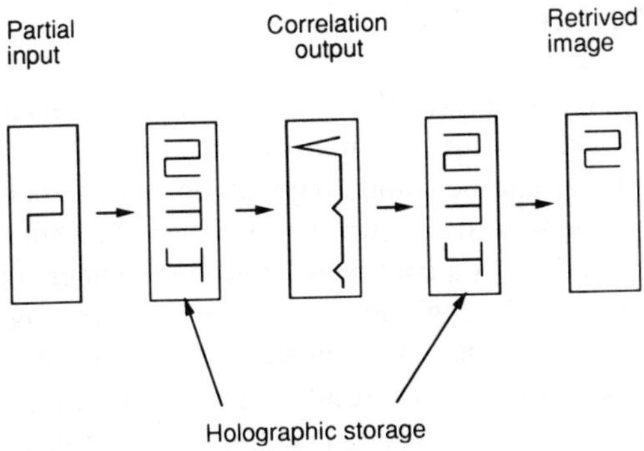

FIGURE 13.9 A holographic associative memory.

applying a threshold in the output plane to eliminate all the intensity below some noise level, a cleaner output image may be obtained. To increase the degree of association and obtain a better convergence, one may feed back this result and recalculate an improved output image. It may also be noted that this operation is similar to the symbolic substitution operation. Holographic associative memory may also be interpreted as an inner product based memory, since the inner product result is obtained when the output in a complex matched filter setup is sampled at the correlation location.

13.4.2 Optical Implementation of Hopfield's Memory

As stated earlier, Hopfield's memory is constructed by performing the summation of the outer product of individual to-be-stored vectors, with zero diagonal terms of the resulting memory matrix. The memory matrix contains all the vectors. The memory retrieval process described in Section 13.2 involves only multiplication of the stored memory matrix by an input vector. Then the vector–matrix multiplication yields an output vector that after thresholding serves as the estimated vector. This estimated vector is later fed back as the new input. This can be implemented very simply in optics by having a matrix–vector multiplication setup. The memory matrix consisting of positive and negative values are thresholded to $+1$, 0, and -1 values. The values -1 and $+1$ are encoded using dual-rail code and 0 is encoded using two opaque pixel subcells. The binary input vector is represented simply by LEDs. The output is connected through detector arrays. When the upper cell of the dual-rail code is transparent with the input vector, the value is $+1$; when the lower cell is transparent, it represents -1. A pair of detectors connected in reverse direction is used behind the two pixel subcells to perform the necessary multiplication by -1. Afterward, an operational amplifier may be used to perform the thresholding operation. Following thresholding, the estimated vector is fed back through electronic means to the input LEDs. After such repeated feedbacks when stability is obtained, the circuit is said to converge to a state corresponding to the retrieved vector.

13.5 INTERCONNECTIONS

One of the most significant drawbacks of today's digital computer is the interconnection bottleneck, which is just one motivation for pursuing optical computing. A proposal for digital electronic computing with optical interconnections has been put forward to reap the benefits of two well developed technologies. In order to reduce the gate delay, there is a continual effort in the integrated circuits industry to scale down the size of the logic gates; however, the propagation delay of the connecting wire does not reduce as a consequence because the *RC* time constant of the wires remains constant. As a result, the interprocessor and intraprocessor communication still remains a big bottleneck for digital computing. In electronics the speed of the electronic signal is not constant but rather governed by the capacitance. The density of connections is often limited due to the interference problem that may occur. Interferenceless photons can provide a large number of channels through a small medium. Furthermore, electronic interconnections are two-dimensional, whereas optical interconnections are naturally three-dimensional.

Several possibilities exist where optics can play a role in solving the interconnection problem of electronics. Consider the problem of clock distribution in a chip. A fiber-optic faceplate can be used that can take a laser input and then distribute it to specified detectors, where it may be converted back to electronic pulses. A holographic element that is capable of generating multiple point sources from a single source may be used to broadcast light pulses to detector arrays situated at the back plane of a chip. A reflective element may also be used to reflect light when both the source and detector are located in the same plane.

13.5.1 Holographic Interconnections

Optically generated or synthetic computer generated holograms (CGHs) may be designed to modify the direction of ray propagation in any arbitrary direction. For the regular type of interconnections, elements such as a Damman grating and hololens can be used to achieve this effect. For example, by recording the interference pattern of two beams, one can use it to create dual beams from a single beam. Or by using a halftone contact screen method, one can record the summation of several frequencies of the cosine transmittance function on a glass plate. When illuminated by a plane wave and then using a Fourier transforming lens, the Fourier transform of the cosine function will give rise to a pair of delta functions. Thus the summation of n frequencies of cosine will produce $2n$ equal intensity points in one dimension. Using a two-dimensional grating one can thus generate $2n \times 2n$ interconnections with a single light beam. Basically, by recording the interference pattern of the desired input beam and the output beam, one can generate any arbitrary interconnection patterns. Using a computer and a laser printer, one can also fabricate such CGHs. For example, by drawing Fresnel-type holograms repetitively, one can fabricate a lenslet array, which can also be used for interconnections.

13.5.2 Waveguide Interconnections

Optical fibers are more often used for longer distance guided wave interconnections but they are also attractive for short distance interconnections inside integrated optical circuits. One of the major bottlenecks of today's electronic components is the limited amount of interconnections between different processors due to the minimum width that must be provided and the interaction between interconnection lines that must be avoided. Free-space optical interconnections by means of holograms inside the integrated circuit could make efficient hybrid opto-electronic processors.

13.5.3 Reconfigurable Interconnections

In signal processing, optical computing, communications, and holographic memory lookup, the pattern recognition capability of reconfigurable interconnections is an extremely desirable feature. For real-time pattern recognition, it is often necessary to store a large number of patterns of various sizes and orientations and be able to rapidly compare these with an incoming signal. An alternate way to improve the correlation speed is to project multiple copies of the target images shown in Fig. 13.10 onto all matched filters and then find the best match therefrom. In optical computing operations, such as in optical shadow-casting and symbolic substitution, it is often necessary to perform shift and overlap of input images for executing logical operations. All these operations can be realized also by an optical interconnection device. A crossbar switch is the most flexible switching device capable of connecting any one of n inputs with n outputs. The setup for the parallel matrix–vector multiplier could be used for a crossbar switch. By changing the matrix, one can perform reconfigurable interconnections. The reconfiguration time is then a function of the setting time for an electrically addressable spatial light modulator representing the matrix.

The other way to produce a dynamic reconfiguration is by using a real-time device to store the interference fringes. The interference patterns are generated by

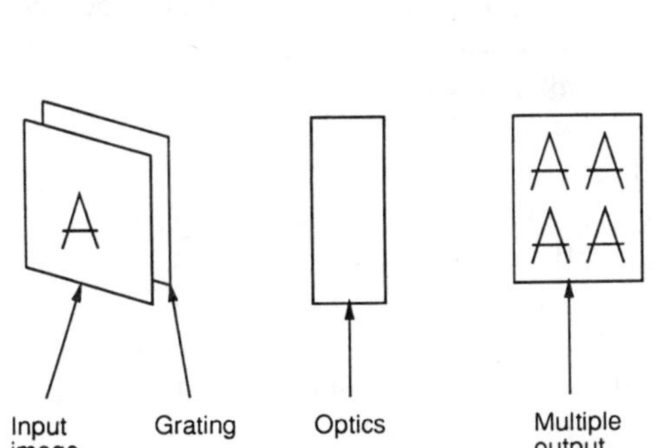

FIGURE 13.10 Multiple copies of a single image via an interconnection network.

multiple sources of light. A collimating beam can be split and then reflected by piezoelectric controlled mirrors. The reflected beams are then allowed to interfere with one another and then the interference pattern can be recorded on a recording device, say, a LCLV. The interference pattern, when illuminated by a plane wave, will provide an interconnection between the input and different locations of the output. The angular orientation of the mirrors can be controlled by changing the actuation input to the piezoelectric mirrors or electronic shutters may be used to prohibit certain beams to interfere. Thus by changing the interference pattern, one can reconfigure the interconnection. The limiting factor in such an approach may be the minimum resolution of the LCLV.

The diffraction grating necessary to generate multiple point sources from a single point source may be simulated in a digital computer. The gratings may be optimized by changing different parameters, such that all the points in the output plane receive equal amounts of energy. Finally, the corresponding diffraction grating can be fabricated. The binary modulation required for the grating can be derived from the computer simulation and then used to construct a real-time interconnection network by supplying the gray level modulation information to SLM. The other medium used to store such interference gratings is the nonlinear crystal used for four-wave mixing or suspension of colloids in a liquid, which forms a real-time grating structure when an interference pattern is formed over it.

13.6 ARTIFICIAL INTELLIGENCE

Artificial intelligence is concerned with the manipulation of stored information to make intelligent decisions. It maintains a huge database on a specific knowledge base and applies specific rules to infer intelligent conclusions from it. Optics can make a significant impact in artificial intelligence by allowing the parallel search of huge databases and associative capabilities. Optical associative memories such as holographic memories, linear algebra processors, correlators, and spatial light modulators have been used for this purpose. An inference machine based on learning may be proposed for optics which has to make a decision based on earlier training. For example, for a set of physical symptoms, if the result is a specific diagnosis, then this machine may be used later as an expert medical advisor. Similarly, based on the location, velocity, and distance of an incoming missile, an expert system based on a neural network may be devised to advise the pilot of a jet whether to launch an attack or eject out of the plane. An expert system could simply be of the form "If this is (*A*) then that is (*B*)."

The expert system rules can be seen as simple symbolic substitution operations. Here the rules needs to be identified and then the corresponding inference needs to be spelled out. Or the rule can be stored in a neural network associative memory. Even a multi-channel correlator could be used to identify a set of rules connected by an AND logic and then perform a substitution rule by means of another correlator memory.

For an expert system that uses a set of rules to infer new rules, a neural network training may be executed to learn the association between an input vector containing some conditions and another output vector containing the inferences. Then if a partial input condition is presented to the expert system, it can interpolate by associating the input with its previous knowledge base to provide a plausible answer. One of the problems associated with such a system is knowledge representation. Knowledge can be converted into vectors or numbers and then processed in an optical numeric processor. This field is rather new and has enormous potential.

13.7 PROBLEMS

1. Use a Hopfield's memory (with nonzero diagonal values) to store $A = (1\ \ 1\ \ 0\ \ 0)$, $B = (1\ \ 0\ \ 1\ \ 1)$, and $C = (1\ \ 1\ \ 0\ \ 1)$. Calculate the Hamming distance from each other. Now consider an input $C' = (0\ \ 1\ \ 0\ \ 1)$, a partial input derived from C. Use unipolar binary, bipolar binary, and trinary (assume that the last three elements are known) representations to retrieve the image of C. What happens if an IP thresholding is used in the inner product domain?
2. Simulate a holographic memory of four stored characters, A, B, C, and D, using a 12×10 matrix. Using the approach of Peak et al. (1987), show the performance of the memory. You may use the program supplied in Chapter 1 for performing the two-dimensional FFT.

13.8 SUGGESTED READINGS

Anderson, D. Z., and M. C. Erie, Resonator Memories and Optical Novelty Filter, *Opt. Eng.*, **26,** 434–444 (1987).

Athale, R. A., H. H. Szu, and C. B. Friedlander, Optical Implementation of Associative Memory with Controlled Nonlinearity in the Correlation Domain, *Opt. Lett.*, **11,** 482–484 (1986).

Awwal, A. A. S., and M. A. Karim, Associative Polarization-Encoded Optical Shadow-Casting: Gray-Level Image Encoding for Serial and Parallel Operations, *Appl. Opt.*, **28,** 284–290 (1989).

Awwal, A. A. S., M. A. Karim, and A. K. Cherri, Polarization-Encoded Optical Shadow-Casting Scheme: Design of Trinary Combinational Logic Units, *Appl. Opt.*, **26,** 4814–4818 (1987).

Awwal, A. A. S., M. A. Karim, and H. K. Liu, Recognition of Machine Objects Using Trinary Associative Memory, *Opt. Eng.*, **28,** 537 (May 1989).

Bayley, J. S., and M. A. Fiddy, On the Use of the Hopfield Model for Optical Pattern Recognition, *Opt. Commun.*, **64,** 105–110 (1987).

Casasent, D., and B. Telfer, Key and Recollection Vector Effects on Heteroassociative Memory Performance, *Appl. Opt.*, **28,** 272–283 (1989).

Cheung, K. F., L. E. Atlas, and R. J. Marks II, Synchronous vs Asynchronous Behavior of Hopfield's CAM Neural Net, *Appl. Opt.*, **26,** 4808–4813 (1987).

Choudhury, A. H. F., Wood-Jezercak, B. J. Ackerson, M. A. Karim, and A. A. S. Awwal, All Optical Logic Gates Using Crossed Laser Beams in Colloids, *Microwave Opt. Technol. Lett.*, **1**(5), 175 (1988).

Domany, E., and H. Orland, A Maximum Overlap Neural Network for Pattern Recognition, *Phys. Lett. A*, **125,** 32–34 (1987).

Eichmann, G., and H. J. Caulfield, Optical Learning (Inference) Machines, *Appl. Opt.*, **24,** 2051–2054 (1985).

Farhat, N. H., and D. Psaltis, Optical Implementation of Associative Memory Based on Models of Neural Network, in *Optical Signal Processing*, J. P. Horner (ed.), pp. 129–162, Academic Press, Orlando, 1987.

Farhat, N. H., D. Psaltis, A. Prata, and E. Paek, Optical Implementation of the Hopfield Model, *Appl. Opt.*, **24,** 1469–1475 (1985).

Fisher, A. D., W. L. Lippincott, and J. N. Lee, Optical Implementations of Associative Networks with Versatile Adaptive Learning Capabilities, *Appl. Opt.*, **26,** 5039–5054 (1987).

Giles, C. L., and T. Maxwell, Learning, Invariance, and Generalization in High-Order Neural Networks, *Appl. Opt.*, **26,** 4972–4978 (1987).

Gindi, G. R., A. F. Gmitro, and K. Parthasarathy, Hopfield Model Associative Memory with Nonzero-Diagonal Terms in Memory Matrix, *Appl. Opt.*, **27,** 129–134 (1988).

Golub, G. H., and C. F. Van Loan, *Matrix Computations*, The Johns Hopkins University Press, Baltimore, MD, 1987.

Goodman, J. W., F. J. Leonberger, S. Y. Kung, and R. A. Athale, Optical Interconnections for VLSI, *Proc. IEEE*, **72,** 850–866 (1984).

Hong, J., and D. Psaltis, Storage Capacity of Holographic Associative Memories, *Opt. Lett.*, **11,** 812–814 (1986).

Hopfield, J. J., Neural Networks and Physical Systems with Emergent Collective Computational Abilities, *Proc. Natl. Acad. Sci. USA*, **79,** 2554–2558 (1982).

Jau, J. Y., Y. Fainman, and S. H. Lee, Comparison of Artificial Neural Networks with Pattern Recognition and Image Processing, *Appl. Opt.*, **28,** 302–305 (1989).

Karim, M. A., A. A. S. Awwal, and A. K. Cherri, Polarization-Encoded Optical Shadow-Casting Logic Units: Design, *Appl. Opt.*, **26,** 2720–2725 (1987).

Kohonen, T., *Content-Addressable Memories*, 2nd ed., Springer-Verlag, Berlin, 1987.

Kosko, B., Adaptive Bidirectional Associative Memory, *Appl. Opt.*, **26,** 4947–4960 (1987).

Kostuk, R. K., J. W. Goodman, and L. Hesselink, Optical Imaging Applied to Microelectronic Chip-to-Chip Interconnections, *Appl. Opt.*, **24,** 2851–2858 (1985).

Kung, S. Y., and H. K. Liu, An Optical Inner-Product Array Processor for Associative Retrieval, *SPIE, Nonlinear Opt. Applications*, **613,** 214–219 (1986).

Lalane, P., P. Chavel, and J. Taboury, Optical Inner-Product Implementation of Neural Network Models, *Appl. Opt.*, **28,** 377–385 (1989).

Liu, H. K., T. H. Chao, J. Barhen, and G. Bittner, Inner-Product Optical Neural Processing and Supervised Learning, *Abstracts of the 1st International Neural Network Society Meeting*, Boston, Vol. 1, p. 391 (Sept. 1988).

Liu, H. K., J. A. Davis, and R. A. Lilly, Optical-Data-Processing Properties of a Liquid-Crystal Television Spatial Light Modulator, *Opt. Lett.*, **10,** 635–637 (1985).

Liu, H. K., S. Y. Kung, and J. A. Davis, Real-Time Optical Associative Retrieval Technique, *Opt. Eng.*, **25,** 853 (1986).

Lohmann, W. W., Optical Bus Network, *Optik*, **74,** 30–35 (1986).

Macukow, B., and H. H. Arsenault, Modification of the Threshold Condition for a Content-Addressable Memory Based on the Hopfield Model, *Appl. Opt.*, **26,** 34–36 (1987).

Macukow, B., and H. H. Arsenault, Optical Associative Memory Model Based on Neural Networks Having Variable Interconnection Weights, *Appl. Opt.*, **26,** 924–928 (1987).

Marks, R. J. II, L. E. Atlas, J. J. Choi, S. Oh, K. F. Cheung, and D. C. Park, Performance Analysis of Associative Memories with Nonlinearity in the Correlation Domain, *Appl. Opt.*, **27,** 2900–2904 (1988).

McAulay, A. D., Real-Time Optical Expert Systems, *Appl. Opt.*, **26,** 1927–1934 (1987).

McEliece, R. J., E. C. Posner, E. R. Rodemich, and S. S. Venkatesh, The Capacity of the Hopfield Associative Memory, *IEEE Trans. Info. Theory*, **33,** 461–482 (1987).

Montgomery, B. L., and B. V. K. V. Kumar, Evaluation of the Use of the Hopfield Neural Network Model as a Nearest-Neighbor Algorithm, *Appl. Opt.*, **25,** 3759–3766 (1986).

Murakami K., and T. Aibara, An Improvement on the Moore–Penrose Generalized Inverse Associative Memory, *IEEE Trans. Syst. Man Cybern.*, **17,** 699–706 (1987).

Murdocca, M. J., A. Huang, J. Jahns, and N. Steibl, Optical Design of Programmable Logic Arrays, *Appl. Opt.*, **27,** 1651–1660 (1988).

Nielsen, R. H., Counterpropagation Networks, *Appl. Opt.*, **26,** 4979–4984 (1987).

Oh, S. H., T. H. Yoon, and J. C. Kim, Associative-Memory Model Based on Neural Networks: Modification of Hopfield Model, *Opt. Lett.*, **13,** 74–76 (1988).

Owechko, Y., Optoelectric Resonator Neural Networks, *Appl. Opt.*, **26,** 5104–5411 (1987).

Owechko, Y., G. J. Dunning, E. Marom, and B. H. Soffer, Holographic Associative Memory with Nonlinearities in the Correlation Domain, *Appl. Opt.*, **26,** 1900–1910 (1987).

Peak, E. G., and D. Psaltis, Optical Associative Memory Using Fourier Transform Holograms, *Opt. Eng.*, **26,** 428–433 (1987).

Psaltis, D., D. Brady, and K. Wagner, Adaptive Optical Networks Using Photorefractive Crystals, *Appl. Opt.*, **27,** 1752–1759 (1988b).

Psaltis, D., and N. Farhat, Optical Information Processing Based on an Associative-Memory Model of Neural Nets with Thresholding and Feedback, *Opt. Lett.*, **10,** 98–100 (1985).

Psaltis, D., C. H. Park, and J. Hong, Higher-Order Associative Memories and Their Optical Implementations, *Neural Network*, **1,** 149–163 (1988a).

Rumelhart, D. E., and J. L. McClelland, *Parallel Distributed Processing*, The MIT Press, Cambridge, MA, 1988.

Sawchuk, A. A., and B. K. Keith, Dynamic Optical Interconnection for Parallel Processors, *SPIE Proc.*, **625,** 243–253 (1986).

Sawchuk, A. A., and T. C. Strand, Optical Computing, *Proc. IEEE*, **72,** 758–779 (1984).

Schmidt, R. A., and W. T. Cathey, Optical Implementations of Mathematical Resolution, *Appl. Opt.*, **26,** 1852–1858 (1987).

Szu, H. H., and H. J. Caulfield, Optical Expert System, *Appl. Opt.*, **26,** 1943–1947 (1987).

Takeda, M., and J. W. Goodman, Neural Networks for Computation: Number Representations and Programming Complexity, *Appl. Opt.*, **25,** 3033–3045 (1986).

Ticknor, J. T., and H. H. Barrett, Optical Implementations in Boltzmann Machines, *Opt. Eng.*, **26,** 16–21 (1987).

Vidal, J. J., Implementing Neural Nets with Programmable Logic, *IEEE Trans. Acoust. Speech Signal Process.*, **36,** 1180–1190 (1988).

Warde, C., and J. Kottas, Hybrid Optical Inference Machines: Architectural Considerations, *Appl. Opt.*, **25,** 940–947 (1986).

Widrow, B., R. G. Winter, and R. A. Baxter, Layered Neural Nets for Pattern Recognition, *IEEE Trans. Acoust. Speech Signal Process.*, **36,** 1109–1118 (1988).

Widrow, B., and R. Winter, *Adaptive Signal Processing*, Prentice Hall, Englewood Cliffs, NJ 1985.

Zhou, Y. T., R. Chellappa, A. Vaid, and B. K. Jenkins, Image Restoration Using a Neural Network, *IEEE Trans. Acoust. Speech Signal Process.*, **36,** 1141–1151 (1988).

Index